I0034538

SCIENCE AND MATHEMATICS EDUCATION IN MULTICULTURAL CONTEXTS:

New Directions in Teaching and Learning

Edited by Sufian A. Forawi

SCIENCE AND MATHEMATICS EDUCATION IN MULTICULTURAL CONTEXTS:

New Directions in Teaching and Learning

Edited by Sufian A. Forawi

COMMON GROUND RESEARCH NETWORKS 2020

First published in 2020
as part of The Learner Book Imprint
http://doi.org/10.18848/978-1-86335-225-3/CGP (Full Book)

Common Ground Research Networks
2001 South First Street, Suite 202
University of Illinois Research Park
Champaign, IL
61820

Library of Congress Cataloging-in-Publication Data

Names: Forawi, Sufian A., editor.
Title: Science and mathematics education in multicultural contexts : new directions in teaching and learning / [edited by] Sufian A. Forawi.
Description: Champaign, IL : Common Ground Research Networks, 2020. | Includes bibliographical references. | Summary: "The proposed book is an addition to the science and mathematics education research line, internationally and regionally. Researchers and graduate students would benefit from the proposed book as a reference with diverse topics and perspectives of teaching science and mathematics in the classroom. College instructors and policy makers can benefit from the proposed book as a core or recommended text in different modules and courses in areas of science education, mathematics education, STEM education, curriculum and instruction and comparative education"-- Provided by publisher.
Identifiers: LCCN 2020036414 (print) | LCCN 2020036415 (ebook) | ISBN 9780949313010 (hardback) | ISBN 9781863352246 (paperback) | ISBN 9781863352253 (adobe pdf)
Subjects: LCSH: Science--Study and teaching (Higher). | Mathematics--Study and teaching (Higher). | Multicultural education.
Classification: LCC Q181 .S3584 2020 (print) | LCC Q181 (ebook) | DDC 507.1/1--dc23
LC record available at https://lccn.loc.gov/2020036414
LC ebook record available at https://lccn.loc.gov/2020036415

Cover Photo Credit: Phillip Kalantzis-Cope

Table of Contents

Part 6

Part 7

Part 1

CHAPTER 1

Directions of Science and Mathematics Education

Sufian Forawi

ABSTRACT

The free, publicly provided education has been a central tenet of the social contract in every country in the Middle East and North Africa since independence. Several constituents can benefit from reading and using this book, such as graduate students and researchers in science education. This chapter aims to provide a preface on the book, highlighting the main aspects of each of the subsequent twenty chapters and the plethora of issues related to science and mathematics in multicultural contexts. The chapter examines the main directions of science and mathematics education, especially in the main context of the MENA, GCC, and UAE contexts. Also, it states briefly the main challenges faced in the development of research and practice in these fields.

INTRODUCTION

The free, publicly provided education has been a central tenet of the social contract in every country in the Middle East and North Africa since independence. Post-independence governments have significantly expanded their education systems, driven by rapidly expanding youth populations, the need to build nationhood and to establish political legitimacy and popular support for new regimes through making education a fundamental right of citizenship. It is well known that population growth in Arab countries is among the highest in the world, which makes providing basic education a major challenge (Forawi, 2015). However, education systems in the region, with a few exceptions, now provide basic education to most children.

One of the most pressing needs in the Gulf Cooperation Council (GCC) countries is to develop citizen competencies to lead the long-awaited development of their nations. While oil development is seen as a key factor for financial and social development, a lasting effect of such development can only be achieved if citizens are adequately prepared in all different walks of life and especially in education, sciences, and mathematics. The United Arab Emirates (UAE) is an active member of the GCC and may become a strong partner for the EU in a number of domains. In its recent country report debriefing, the European Union (EU, 2017) introduced the report by stating that the UAE has the most open and diversified economy in the region.

The book provides critical review and analysis of related research topics of science and mathematics education of their paramount importance to diverse readers and researchers in fields of science education, policy making, and academia. In 21 chapters, the main scope of the book highlights of theory and practice of these fields, contributes descriptively and empirically to the research, and provides recommendations and implications to similar contexts.

The book highlights the science and mathematics advancements that accommodate ever changing nature of the increasing development of research and practice in these fields. It further provides discourse on important topics for their currency and urgent needs, particularly for developing counties, such as the MENA. The book also has a major focus science education research and reflects on trends and issues as they relate to UAE experience. Therefore, the book devoted considerable attention to the ways science is planned, taught, assessed, and integrated in different contexts. The book provides knowledge, skills and dispositions on science education and related issues as they pertain to international research lines and local needs. In particular, 21 chapters provide diverse topics in science and mathematics education, such as reforms and framework, STEM, cooperative learning, inquiry-based learning, TIMSS, NGSS, cognition and metacognition, concept mapping, critical and creative thinking, bilingualism, nursing and health, probeware technology, etc.

Several constitutes can benefit from reading and using this book, such as graduate students and researchers in science education, as book styles combines both descriptive and empirical research findings and interpretations. It, also, provides policy makers, particularly in the GCC, MENA and developing counties framework and experience worth looking into when developing local science and mathematics education reforms and policy in their particular countries. The book, additionally, is a good textbook for science, STEM, IBL, cognitive science, bilingual science education, etc., which can provide appropriate discourse on these topics as they relate to designated graduate programs and studies. Authors of the book and chapters have written the book for rigorous research protocol and personal experience as they work in similar context of the target book audience who will relate to nicely.

This book is one of very few books provided in science and mathematics education with special focus on multicultural and developing counties. Governments in those countries may consider the book an important integral document for any discussion on science and mathematics reform and policy development. Several government officials, in UAE, for example, have indicated, anecdotally, their interest is such book. This book, through plethora of perspectives, discourse and recommendations, is suitable for developing counties, focusing on the UAE, and the GCC but which may have broader resemblance and implications for other counties in region as well as for international aspects of multicultural, diverse developed ones.

PREFACE

Through description and discussion in different chapters, the book sheds light on the major directions and topics in the reform agenda, describing challenges and paving road for future advancement in science and mathematics multicultural contexts. Below is the preface of the 21 chapters of the book which are organized in seven parts.

Part 1 is about Science Education Reform and Inquiry and includes two chapters.

Chapter 1, titled Directions in Science and Mathematics Education, provides an introduction to the book by shedding light on the main new directions in science and mathematics education, their status in the Middle East, MENA and the UAE, their policies, reforms and implementation, along brief description to preface subsequent chapters of the book.

Chapter 2, titles Reforming GCC Mathematics and Science Curricula, elucidates information with discussion on the reformation of education curriculum followed in most of the educational institutions of the GCC countries. Particularly, focuses on the newly reformed education curriculum which mainly concentrates on science and mathematics education with an aim to enable students to attain high scores in international assessment tests such as; TIMSS and PISA and attain knowledge and practices in new areas such as STEM and STEAM.

Part 2 focuses on STEM/STEAM Education in four chapters.

Chapter 3, Contextualization of Interdisciplinary Teaching, aims to investigate the influence of contextual teaching and learning approach in an interdisciplinary course on recognizing the purpose and values of learning. The philosophy of contextualization is based on the relevance of the content of learning to help students to recognize the purpose and value of basic skills development. This chapter points out the various possibilities and perception in order to continue researching the combined impact of contextualized teaching and learning method in an interdisciplinary course. Additionally, the setup of the interdisciplinary learning environments shows the groundwork for guiding pedagogical strategies to deliver NGSS and related frameworks.

Chapter 4, STEM Education Policy Development, provides a comprehensive discussion regarding education and economic advantages of STEM programs. It also suggests a universal policy of STEM education by benchmarking the already established and successful k-12 STEM programs. Hence, the chapter explains the role of policymaker in promoting STEM education. It also focuses on the reforms introduced by federal governments in an attempt to create a more unified vision of STEM education to promote the awareness of the importance of incorporating the four main disciplines in the national education system. Additionally, the study discusses how the STEM education offers the students an opportunity to become more innovative and flexible to survive in the highly competitive business environment in the contemporary world.

Chapter 5, titledAssessing Creativity in STEM Education, is about the STEAM education which reflects the fusion of the disciplines, science, technology, engineering, art and mathematics as an essential paradigm for creative approach of teaching and learning. In STEM classes students develop their cognitive and metacognitive thinking in addition that they are intrinsically motivated. Furthermore, adding “A” to STEM can enhance students’ creativity and has a positive impact on students’ attitudes and interests. The chapter answers two main questions, what are the teachers’ perceptions about the factors that affect teaching and assessing creativity? and, to what extent STEAM education foster students’ creativity?

Chapter 6, the purpose of this chapter is to present a brief overview of the implementation of STEAM education in schools in developed countries such as China, Australia, United Kingdom and United States of America and to provide a roadmap of its implementation in the context of the United Arab Emirates. Also, this chapter adopts a qualitative approach whereby purpose sampling of secondary data is collected, compiled and analyzed. Themes are generated after coding the content: implementation of STEAM, challenges related to STEAM application and implementation and requirements for success implementation. For the purpose of ensuring proper integration of STEAM in UAE educational system, a roadmap is proposed with policy drafting recommendations, such as curriculum reform, technology integration, teacher professional development and financial funds.

Part 3 is on Science and Math Cognition and Thinking and has four chapters.

Chapter 7, Mathematics Metacognition, presents discussion of current state of research as it relates and reflects on metacognitive strategy and awareness and self-efficacy in mathematics and physics education. The comprehensive literature review presented the following major topics: the theoretical review of metacognition, the recent studies on metacognitive strategies and inventories to assess metacognition, and reflection and self-regulation in mathematics and physics. This is followed by how these research studies were translated into educational practice.

Chapter 8, Critical and Creative Thinking, provides a critical analysis creative and critical thinking research and how students can acquire it. It is widely believed by the psychologists and the educators internationally, and as presented through many of the research works, that the critical and creative thinking can be taught like other learning stuff. Critical thinking is a thought process that involves gathering and evaluating information to make decisions and solve problems that you encounter. While creative thinking is probably most likely where there is an optimal amount of information more is not always better, and when the individual knows to treat information as contextual rather than absolute. Therefore, the chapter presents the divergent thinking it nourished creativity which is fostered in convergent thinking, critical thinking, associational and analogical thinking. So, the chapter discusses the critical and creative thinking through its status and the possibility of imparting students with the creative -not the ordinary- critical thinking skills.

Chapter 9, Nursing Curriculum and Critical Thinking, aimed to critically study the critical thinking and decision-making skills in BSc nursing curriculum among nursing students, through a review of related literature. The nurse's clinical decision-making skills can have a great impact on the nursing profession. Wise clinical decisions can influence the patients' health either positively or negatively. For that, the literature had stressed on the appropriate educational strategies to prepare a nurse critical thinker. This was discussed in the context of the UAE to provide active non-threatening environment and updating the teaching strategies with advanced simulations and planed clinical training programs found to improve student critical thinking skills positively. Thus, nursing educational systems have to play a major role in empowering the health system with professional nurses.

Chapter 10, Concept Mapping and Chemistry Achievement aims to provide discourse and results on use of concept mapping with middle school chemistry. The chapter's relevant contributions include, understanding concept mapping as an effective instruction and chemistry curriculum design in the UAE. Chemistry educators must acquaint with concept mapping in classrooms in the absence of well-equipped laboratories and lack of appropriate instructional aids for hands- on learning as it can be a best alternative to laboratory experience which provides relationships between concepts that will enhance students' performance in the subject. Textbooks and other learning resources should include concept maps at the end of each lesson as a summary of the learning material. Curriculum planners, science educators and researchers must take cognizance of this innovative method as an effective pedagogical strategy in chemistry classes.

Part 4, Science Instruction, includes four chapters.

Chapter 11, Science Education Innovative Trends, aimed to answer the main question, how effective is the science-inquiry teaching approach on the improvement of students' achievement in the UAE public schools following a standard-based curriculum? Effective instructional strategies and effective teaching both are derived from a psychological perspective on thinking about teaching. The UAE government requires teachers to commit to excellence in teaching. Thus, provided them with pedagogical support and training opportunities to improve teaching skills and to help them develop the student maximum potentials. Therefore, the chapter emphasis is placed on identifying innovative, observable science education trends and behavior in the classroom, such as inquiry teaching approaches that can be linked to observable outcomes that make a difference in the student live.

Chapter 12, Formative Assessment of Science Inquiry, demonstrates through analysis of descriptive and empirical data the use of inquiry-based learning (IBL) in science education. The IBL is considered one of the effective strategies to develop students' critical thinking skills and train them to become problem solvers for future problem. The purpose of this chapter is to investigate students' and teachers' perceptions towards formative assessment (FA) and inquiry-based learning in science classrooms, in addition to investigating how the formative assessment of inquiry-based learning is currently implemented and finding out the best practices in the educational field. The major results indicated that formative assessment strategies were implemented effectively in science lessons. While inquiry-based activities were implemented with less efficiency in all science classrooms. A positive relationship between formative assessment and inquiry-based practices was found from teachers' perspective.

Chapter 13, Inquiry learning and TIMSS, presents a comprehensive review of the literature from several angles related to IBL instruction. IBL is one of the most substantial key elements of many reform efforts in science education because of its impact on accelerating students' acquisition of scientific skills within the spectrum of 21st century skills as well as enhancing students' achievements through overcoming misconceptions, and acquiring the essential content knowledge which can be measured by standardized tests. Particular attention is paid to the historical and theoretical research backgrounds that shed light on IBL, followed by the nature of IBL and its effectiveness on education. Constructivism is then compared to IBL as an active learning instruction. The relationship between IBL and student acquisition of scientific skills is discussed as well. Factors affecting its implementation and IBL between theory and practice will be briefly mentioned. Finally, the relationship between TIMSS and strategies of inquiry model will be interpreted.

Chapter 14, Science Project-based Learning, presents an in-depth descriptive review of selective empirical research studies that are related to implementing the project-based learning (PBL) approach in science. It also provides main findings and discussion of the significant issues related to implementing project- based learning in science instruction. The review covers the 0following topics: the definition of project-based learning, the effective implementation of the science PBL in science, the common features of PBL compared to the inquiry-based learning, the models of PBL and their instructors' and students' roles and perceptions.

Part 5, Science and Health Education consists of three chapters.

Chapter 15 is about the Student Conceptions of Science. The nature of science has evolved in the last century due to the complex and dynamic nature of the scientific knowledge. The UAE is an active member of the GCC and a strong partner for the EU and USA in a number of domains, including sciences and education. This chapter provides extensive literature review of the nature of science (NOS), its connection to science education, student understanding, pedagogy and inquiry instruction, and professional development. It also presents major results of a study on UAE students' conceptions of NOS. Finally, it concludes by critical discussion and implication of multicultural contexts.

Chapter 16, Science Cooperative Learning Strategies, presents a comprehensive review of the literature related to cooperative learning strategies and the context of science. Cooperative learning is one of the most significant teaching strategies to have a substantial positive effect on developing student learning skills. The chapter has a closer look at the use of science cooperative learning strategies that play a vital role in promoting students' academic knowledge and social skills. Particular attention is paid to the historical and theoretical research backgrounds that shed light on the meaning of cooperative learning, followed by the effects of cooperative learning strategies in science education. The pedagogy of cooperative learning and its influence on effective analysis of learning habits is discussed as well. Finally, the benefits of using cooperative learning strategies in science classes are interpreted to help teachers in their learning process.

Chapter 17, Learning Factors of Health Sciences, purpose is to explore the student perceptions of factors influencing their learning of health sciences. The chapter provides future scholars with tools to promote students' learning of health science education, and to help the preparation of health work force being equipped with the needed knowledge and skills in this dynamic and rapid changing field. The chapter presents answer to the main research question: What factors are influencing students' learning of health sciences at the CHS? Broad review of literature has been carried out to explore learning approaches used in the field of health sciences, particularly, critically review general issues of health sciences, and career opportunities. The sections of the chapter describe and discuss the approaches to learning, case-based learning (CBL), and critical thinking in health sciences education. The last section sheds light on the barriers to interactive strategies in health sciences education.

Part 6, Science and Mathematics Technology Instruction has two chapters.

Chapter 18 is titled Web 2.0 Applications in Math Instruction, Technology provides an extensive range of possibilities for teachers in establishing a collaborative, interactive mathematics class. The worldwide practice of reforming mathematics instruction has emanated from findings regarding college students and others entering the job market, who, according to recent studies, lack the competency to analyze situations despite their adequate knowledge of the subject matter. This chapter aims to provide comprehensive description and discourse on teacher implementation of contemporary techniques that integrate mathematics and technology as essential to scientific and economic development. Web 2.0 technology can be applied in the classroom in various ways and forms, three of which are addressed for the purposes of this study.

Chapter 19 is about Science Handheld Technology Instruction.

The rapid movement toward the use of palm and handheld devices for communication makes it likely that these particular technologies will play an important role in advancing teacher and student science experiences in the near future. The major purpose of the study is to investigate the effectiveness of the use of sensor probeware technology and guided inquiry in teaching and learning of sciences. To achieve this goal, six in-service science teachers and their 38 students participated in this study during a summer camp in Al Ain, United Arab Emirates. Science teachers were trained as part of this study to use the sensor probeware technology and guided-inquiry instruction to investigate water quality. Then each group of students investigated water quality at different locations. Qualitative as well as quantitative data were collected from teachers and students. Results indicated that students and teachers had benefited from the use of sensor probeware technology (t- test 13.784 $p<.000$). Participants showed great interest in conducting science guided-inquiry activities and using the sensor probes.

Part 7 is about English Medium in Science Instruction. It has two chapters.

Chapter 20 is about the Language of Tertiary-Level Science Instruction. As a result of the supremacy of English that globally sweeps across the higher education landscape, English is seen to have played a crucial role in scientific literacy in its fundamental sense which refers to coaching students on how to synthesize scientific texts, and its derived sense or students' being well-founded and knowledgeable in science. It is widely believed that in order to attain scientific literacy, learners should be proficient in the language through which science is delivered. Science is distinguished from other epistemologies because of the consistent need for experiential standards, rational justifications, sound arguments, and plausible analysis and reasoning.

Chapter 21, Language of School-Level Science Instruction, presents a critical analysis of a textbook that is part of the UAE Ministry of Education, Madares Al Ghad (MAG) system, that is aimed to teach science curriculum in English medium. The research was conducted on a textbook sample from a collection of textbooks that are used in grades one to nine in the MAG systems. These textbooks were prepared in terms of how the second language is integrated with teaching another subject that was normally taught in Arabic in most of the government schools. A qualitative research method was considered as a methodology of the study through considering the science textbook analysis as an artifact. Three themes were identified which are content, structure and objectives, English skills, and the strategies. The study outcomes demonstrated the importance of language proficiency in understanding scientific concepts. It also suggested some pedagogical implications for teachers. It recommended further consideration to the use of IBL in teaching science using the target language. It also recommended focusing on issues facing teachers when teaching another subject using the second language as in the UAE.

Education in the UAE

The UAE is currently undergoing a rapid expansion, with Emirati government fueling growth by investing the proceeds of high oil prices and international investments into a huge array of public and public-private enterprises. Recognizing that this increase in available funds is not in itself a long-term growth strategy, the UAE government has made a number of public commitments to strengthen the country's macroeconomic foundations including emphasis on education reform, resulting in the country leading the chart in the Global Competiveness Index in the region as in the figure below. This was argued in depth as a result of the World Economic Forum report that has not fully been implemented:

> "The United Arab Emirates has consolidated its leadership in the region as the main tech and innovation hub. In 2017, its entrepreneurship and start-up ecosystem attracted global attention with the acquisition of Souq.com, a local e-commerce platform, by Amazon, for an estimated value of $650-700 million. Education and innovation: One of the UAE's key challenges is to ensure that its education system provides nationals with the skills demanded by its growing private sector, thereby helping to diversify the country's industries and redressing the demographic imbalance in its workforce." (World Economic Forum, 2020).

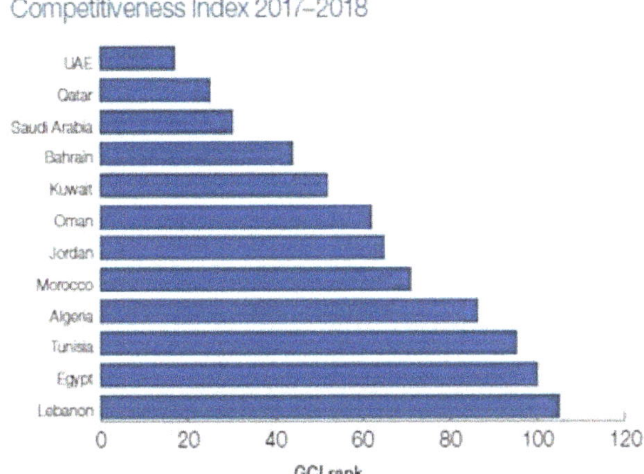

Figure 1: Arab world country rankings, Global Competitiveness Index 2017-2018

Science education must be considered as a main focus of educational reforms in the UAE. This can only be true if the UAE strategically plans to be successful as a nation in developing its citizens and advance as a nation, then science education should have a major emphasis in reform agendas. Science literacy, standard-based education, guided-inquiry instruction, authentic assessment and other important new directions have been the focus all over the world for several years now (US, UK, Canada, Australia, and Singapore). These directions provide substantive frameworks for curriculum development and policy making to better educate children and citizen. In order to meet the challenge of instituting and implementing such new science frameworks, teachers are expected to keep up to date with the advancement in science and technology to continue transforming themselves from technicians to professionals.

They will move from merely delivering textbook knowledge to students, to intellectual professionals who are capable of facilitating student active learning and making decisions on curricular, instructional, and assessment issues to effectively impact their students. To help teachers grow professionally in this new context, there is a pressing need for developing and instituting a new model for professional development that can foster the growth of teachers from technicians to professionals (Fwu & Wang, 2012).

Teacher shortage has become a global concern. Similarly, there has been an acute shortage of Emirati male teachers in schools in general, but especially in the public sector as the Knowledge and Human Development Authority (KHDA, 2011) revealed that although 90% of female teachers in Dubai's public schools are Emirati, only 4% of Emirati teachers are males. This is a major concern as Dubai is the second largest Emirate. There is such a heavy reliance on expatriate teachers which may create, on one hand, a major concern with regard to educational reform and 'emiratization' of jobs. On other hand, this situation of the lack of Emirati male teachers, especially for boys' schools, begs the need for professional development of teachers to help familiarize them with the UAE educational system and further assist them to implement the new curricula. Science, mathematics, and English are regarded to be of paramount interest in the reform agenda of the UAE. While there has been a growing debate on striking a balance among school subjects, especially Islamic and Arabic studies, emphasis is still on English, science, and mathematics. This is unique to the country and differs from global reform which still has major emphasis only on reading, writing and math (e.g. No Child Left Behind, US). In particular, several ambitious initiatives were launched to improve students' science attainment. Yet, teachers are still the cornerstone for any improvement that may be foreseen for students.

Education is a profession that is based on systems, philosophies, and standards for instruction which are all reflected within teacher professionalism. Initial teacher qualifications can potentially have significant effect on the quality and impact of the educational system. More importantly teachers should continuously be developed to be presented by the up-to-date strategies and the new pedagogical content knowledge to positively impact students' learning experiences.

Science Education in the UAE

Education is considered as a key for the sustainable development of the UAE. The ministry of education has emphasised the goal of the strategic plan of 2014 to build a knowledge-based society. The UAE has one of the lowest student-to-teacher ratios in the world (15:1). Education is compulsory to ninth grade, although, according to the U.S. Department of Education, this requirement is not enforced. As some small percentage of children pass school age without being admitted in schools or starting school at late age, while others attend after evening schools due to parental neglect or financial or other reasons.

The UAE children are required to attend gender-segregated schools in all the three public school levels, elementary, middle, and secondary. However, students are segregated after grade four in the private sector. The education budget has increased steadily over the past 20 years. In 2012, 47 % of the UAE's federal budget was channelled onto social development (public education, higher education, health, social affairs, Islamic and cultural affairs, youth and community development) (EU, 2012). In order to improve education in the UAE, the government has implemented policies of free public schools, colleges and universities for UAE citizens. Therefore, the government is continuously attempting to improve public education, which is considered lagging behind private education, by building new schools, institutes, colleges, and universities in all the seven emirates in the country, teacher preparation and qualification, teaching quality, staffing salary, and technology upgrade.

The private education system is also well developed and attended by 40 % of the students from K-16 with diversified curricula that are modelled on different systems, e.g. US, UK, Australia, India. The structure of the educational system in the UAE mainly follows a model of primary education that starts from the age of five which is obligatory from kindergarten to grade six. The secondary education from 12-18 years is composed of two cycles of three years each (preparatory school from grade 7 to grade 9 and secondary education programme from grade 10 to grade 12. Tertiary education 31 is also well developed. There are also many foreign universities, including representation from almost all the European, American, and Australian institutions, such as Sorbonne and Harvard, in the UAE. However, science teaching is faced with challenging problems such as rigorous teacher training, suitable and sustainable curricula, implementation, and the lack of English preparation, which turns Emirati school children off science and mathematics.

Mathematics Education in UAE

The development of mathematical knowledge covers not just understanding of concepts, facts and procedural skills, but also ability to determine appropriate problem-solving strategy. Some tutoring systems allow students to write all solutions step by step as if they were solving on paper. Knowledge about different problem-solving strategies and ability to decide when and how to use a particular strategy is termed as meta-cognition (Jaafar, Wan, and Ahmad, 2010).

Most non-mathematics major students fail to appreciate the true nature of mathematics when they encounter endless drill exercises. The tediousness of problem solving can be removed with the help of appropriate use of technology and tools. There always have been arguments in favour and against the use of technology in education. In the latest review, it is found that iPads are mainly used for communication, for accessing learning resources and for completing online assessments.

Derar and Almeqdadi (2020) investigated UAE students' perceptions of the advantages and disadvantages of using two web-based homework management systems: MyMathLab and WebAssign. Enhancing students' conceptual understanding and encouraging them to be involved in the classroom discussions are important for instructors of mathematics. The use of technology was an essential part of the teaching and learning in mathematics courses. Web-based homework management systems are possible alternatives to the traditional pen-and-paper based approach. Homework systems provide a flexible instructional tool that offers students immediate feedback which helps students learn and understand mathematical concepts. In addition, these systems track student performance, and facilitate a student-centered environment where students are able to rework the questions multiple times thus enhancing the learning process. Somewhat similar situation is faced by teachers in the federal universities in the UAE. The teachers in these universities adopted iPads in their teaching more rigorously after the iPads were incorporated in the assessments (Gitsaki & Robby, 2015).

Directions of Science and Math Education

Tairab (2010) has indicated that increasing teachers' content knowledge of their discipline and their expertise in pedagogical content knowledge has been both challenging and overwhelming to many educators in the UAE. In elementary education, for example, the lack of strong content knowledge in science is particularly evident. One of the new reform directions that is aimed to strengthen science education and makes content relevant to multiple areas, is the interdisciplinary Science, Technology, Engineering and Mathematics (STEM) approach. In the UAE, there is no official STEM education embedded in school curricula. However, the students' interest in science and its related subjects has been sparked in their participation of school projects and at the college levels. A rare example is a10-day STEM program that was run at Khalifa University, Abu-Dhabi, that hosted 181 national students from private and government schools who participated in the Advanced Technology Investment Company (ATIC's 2nd TECH QUEST, 2012). While such program is not widely used, it is aimed to enhance participants' interdisciplinary research skills, introduce K-12 students to undergraduate research, and shape their career choices.

Investing heavily in science and mathematics education could help to elevate the UAE to the top ranks in global index. Dubai, in particular, is seen as the center of country's economic development, with Abu Dhabi is now rising as well. The country developed frameworks for all the school subjects including science and mathematics for K-12 in an attempt to strategically plan to develop a systematic science education reform. However, this movement is facing many obstacles and has not been seen as influential. The frameworks of science and mathematics seemed sufficient though with lack of real contextual and cultural connections as they are mostly drawn from western documents with little or no relevance to the country's culture and heritage. Also, with lack of follow-up and evaluation of the many educational initiatives, these reforms remain questionable and open to doubt.

History, grammar, literature and the Arabic language are taught in the native language (Arabic), whereas Mathematics and sciences are generally taught in English language. In the UAE, (i.e., Dubai & Abu Dhabi emirates), an in most Gulf countries, there is a dual language focus mainly in core subjects such as English, Math and Science. Recognizing a constant need for progress, the UAE has sought to implement and monitor high quality education standards by undertaking new policies, programs and initiatives (Eltanahy, et, al, 2020). Throughout the Middle East, educational advancement is often impeded by insufficient focus on the English language, inadequate provision of technology as well as modern techniques of instruction and methodology. Stressing the importance of "modern curricula with assorted and non-monotonous means of training and evaluation", the Emirates launched ambitious campaigns to develop each of these areas. The UAE Cabinet approved in October 2019 a Dh61.35 billion federal budget for 2020 and established an Education Support Fund to encourage partnerships with the private sector (Abbas, 2020).

The fund will ensure sustainability of educational development programs which will be managed by federal authorities and ministries. Through its Teachers of the 21st Century and a two hundred-million-dirham share of this budget, the UAE hopes to train 10,000 public school teachers within the next five years. In addition, the UAE government believes that a poor grasp of English is one of the main employment barriers for UAE nationals; as a first remedial step, the Abu Dhabi Education and Knowledge (ADEK) is developing an elementary school pilot program with Zayed University, which it hopes to extend to all schools in the emirate, to enhance student English language skills. This has also seen through our visits, supervision, and consultancy to schools and teachers in different emirates. Therefore, the prime minister has demanded that the ministers of education and higher education work to find innovative and comprehensive solutions.

In light of this, students' level of English in government schools is also cause of problem. Students in private schools do not seem to have any major issues with English proficiency while English is problematic for students in government schools. In private schools, English language is used as the medium for delivery of all the subjects, except the Arabic and Islamic studies with ample opportunities for intervention. Therefore, most students in the private schools are proficient in the English language, while the majority of students in public school lack adequacy in English language skills. In my observation of students and their use of English language, students at private schools, make about 70% of population of students in the UAE, extend their use of the English language beyond the classroom to become the language of communication in the school with teachers and their peers, since they come from diverse backgrounds and the English language is considered the common language. This in turn makes it all but impossible to take a degree in science or technology areas for local students who complete their education in governmental schools, because such areas require more knowledge of English than others as liberal arts and business. At many institutions in the UAE such as the United Arab Emirates University and the Petroleum Institute in Abu Dhabi, the students will stay in a foundation (remedial English) program for the maximum of two years and still not make it to undergraduate stage, which may make them feel depressed and give them a sense of failure.

The Abu Dhabi Education Council (ADEC) and the Knowledge and Human Development Authority (KHDA) have long acknowledged the problem. A KHDA initiative of providing Post Graduate Diploma in Education (PGDE) degrees to interested mid-career professionals in science, mathematics and English was run for three years starting 2008 in conjunction with the British University in Dubai but it was postponed due to shifting of authority from KHDA to the Ministry of Education and lack of clear strategic policy. To reverse the trend, in 2010, ADEC launched its New School Model based on bilingual teaching to better prepare students for university. By 2015, it is expected that all students in the capital's government schools will be taught science and mathematics in English.

Similarly, Malaysia started to teach science and math in the English language in 2003, instead of the Bahasa Malaysian (MB)mainly to improve the English language as a part of the students' future literacy in technology and science. In order to implement this kind of project, public school science and mathematics teachers were trained on English language courses yet all of them used the Malay language to explain scientific concepts and the English language for science terminologies and vocabulary. These practices enhance the learning of academically and linguistically strong students but negatively impact the content and language development of weaker students which created a challenge for developing counties that attempt to do the same. For example, in one of this book chapters, the discussion of MAG in the UAE, new schools catered to teach science and math in English, did not success and were stopped after some years of piloting. Nonetheless, the Malaysian government is recently (Bernama, 2019) conducted an in-depth study of a new system of teaching science and mathematics in English, with the lessons recorded by the best teachers replicated in every school through information technology. Additionally, Prytula and Weiman (2012) indicated that this issue was different in the Singaporean context. They explained that the vision of the Singaporean government to implement this kind of project was to help students express themselves in English once they attain higher grade levels and ease into their university years. It is known that to succeed in this kind of implementation, the content knowledge of the teacher is a first factor, as well as his/her English proficiency level, and finally the curriculum adaptation according to the new instructional language. Additionally, it should be mentioned that the urban and non-urban environment could play a crucial role in the outcome of the students as well as the proficiency level of the teacher. Such experiences are worth looking into as the UAE, GCC and the MENA countries strive to benchmark its educational system with higher standards and indeed reviewing these two examples, Malaysian and Singaporean, in its educational policy developing.

Finally, this book highlighted several challenges regarding science and mathematics education and their experiences in the UAE. While generally schools seem progressing, it many teachers seem to lack motivation. This was with regard to lack of effective professional development. Teachers also complained about work overload, job dissatisfaction, unwelcoming environment and low salaries. In some chapters, studies indicated that both male and female teachers shared these views. Yet, male teachers expressed greater job dissatisfaction. This may be due to the presence of a high percentage of expatriate male teachers than female teachers, especially in public schools. They also complained about the changes to the curriculum that were happening almost every year, which is very hard for them to manage.

REFERENCES

Abbas, W. (2020). *Budget: Education gets front bench in UAE.*

Al Khaleej Times. https://www.khaleejtimes.com/uae/abu-dhabi/budget-education-gets-front-bench-in-uae-

Bernama, B. (2019). New system under study to teach science, mathematics in English Mahathir. https://www.theedgemarkets.com/article/new-system-under-study-teach-science-mathematics-english-mahathir

Cronbach, L. J. (1951). Coefficient alpha and the internal structure of tests, *Psychometrika, 16*, 297-334

Derar, S. & Almeqdadi, F. (2020). Students' Perceptions of Using MyMathLab and WebAssign in Mathematics Classroom. *International Journal of Technology in Education and Science, 4*(1), 12-17.

Eltanahy, M., Forawi, S., Mansour, N. (2020). STEM leaders' and teachers' views of integrating entrepreneurial practices into STEM education in high school in the United Arab Emirates. Entrepreneurship Education, https://doi.org/10.1007/s41959-020-00027-3

European Union (2017). European Commission Report. https://ec.europa.eu/trade/policy/countries-and-regions/regions/gulf-region/

Forawi, S. (2017). STEM career aspirations of Emirati youth. *Humanities and Management, 9*, 336-339.

Forawi, S. A. (2015). *Science Teacher Professional Development Needs in the United Arab Emirates*. In N., Mansour & S. Al-Shamrani (Eds). Science Education in the Arab Gulf States: Visions, Sociocultural Contexts and Challenges. Sense Publishers. Rotterdam, the Netherlands.

Fwu, H & Wang, H. (2012). Bridging the gap between and beyond school science through collaboration: Promoting science teachers' professional development through diversity and equal partnership. Asia-Pacific Education Researcher ,http://archive-ph.com/ph/e/ejournals.ph/2012-12-13_955481_3/American_Studies_Asia_Scholarly_works_and_journals_of_Filipinos/

Gitsaki, C., & Robby, M. A. (2015). *Evaluating mobile technology in math education: A case study*. In H. Crompton & J. Traxler (Eds.). Mobile learning and STEM: Case studies in practice (pp. 29-42). Routledge.

Guarda, M., & Helm, F. (2017). "I have discovered new teaching pathways": The link between language shift and teaching practice. *International Journal of Bilingual Education and Bilingualism, 20*(7), 897–913.

Jaafar, Wan M., and Ahmad F. (2010). "Mathematics self- efficacy and meta-cognition among university students. *Procedia-Social and Behavioral Sciences*, 8, 519-524.

Posnanski, T. (2010). Developing understanding of the nature of science within a professional development program for inservice elementary teachers: Project nature of elementary science teaching. *Journal of Science Teacher Education, 21*(5), 589-621.

Prytula, M., & Weiman, K. (2012). An examination of changes in identity, and development: An examination of identity through the professional learning community model. *Journal of Case Studies in Education.* 1-18.

Swan, M. & Ahmed, A. (2012). Poor teaching turning Emirati students off science. *The National.* 9.22. http://www.thenational.ae/news/uae- news/education/poor-teaching-turning-emirati-students-off-science.

Tairab, H. (2010). Assessing Science Teachers' Content Knowledge and Confidence in Teaching Science: How Confident Are UAE Prospective Elementary Science Teachers? *International Journal of Applied Educational Studies*, 7(1), 59-71.

Thacker, S. (2019). Collaboration Road: Dubai's Journey towards Improved School Quality. A World Bank Review. https://www.khda.gov.ae/Areas/Administration/Content/FileUploads/Publication/Documents/English/20190324160314_CollaborationRoad.pdf

World Economic Forum (2020). 4 Innovation Hotspots in the Arab World. https://www.weforum.org/agenda/2018/08/4-innovation-hotspots-in-the-arab-world/

CHAPTER 2

Reforming GCC Mathematics and Science Curricula

Rehaf Madani

ABSTRACT

The present chapter sheds light on the reformation of education curriculum followed in most of the educational institutions of the GCC countries. The newly formed education curriculum is mainly focused on science and mathematics education with an aim to enable students to attain high scores in international assessment tests such as TIMSS and PISA. It further emphasizes on the importance of implementing STEM education in the education curriculum of GCC countries while intervening significant changes in the existing curricula. Several barriers were discussed along with way forward recommendations such as the need to conduct more research on effective implementation of mathematics and science towards greater STEM, standardized curriculum reformation.

INTRODUCTION

Until the middle of the 20th century, the education system in the Gulf Cooperation Council (GCC) countries was mainly based on religious scholarship covering teaching reading, writing, and religious scriptures. However, the education system in GCC countries began to take on a more modern and organized custom (Aziz, 2016). Today, Education in the GCC has become an issue of national concern and discussion and is viewed as a key factor for financial and social development (Almazroa, 2013). Globalization and the regional high economic growth rates have placed a lot of strain on educational organizations and its citizen's intellectual performance (Abouammoh, 2009). There are claims that the K-12 education systems in GCC countries are not preparing students adequately for post-secondary education or for 21st century work force. Officials in the education sectors blame outdated and rigid curriculum along with educational systems lack of vision and goals (Azizi, 2016).

Comprehensive reform strategies began to take place where, science and mathematics education has garnished interest like never before. Each GCC country has organized guides for K-12 instruction, including the development of mathematics and science curriculum standards that focus on preparing students as productive citizens (Aziz, 2016; Al-Mutawah & Fateel, 2018). A careful review of the well-developed standards in GCC countries reveal a strong emphasis on the fundamental elements of enquiry, critical thinking, reasoning and problem-solving techniques to ensure that students develop the ability to work creatively, think analytically and be able to solve problems (Alfadala, 2015).

Arab countries have encountered extensive educational reform in the previous three decades, yet the degree to which these changes have influenced the quality of science education has not been deliberately investigated (Dagher & BouJaoude, 2011). Curricular reformation in the GCC countries have helped students to improve in mathematics and science subjects, yet their performance levels in international assessment tests such as TIMMS and PISA remained below the average level, when compared to the international average (Abouammoh, 2009). Students' achievement in international assessment tests is influenced by many different variables such as schools, teachers, parents, socioeconomic status, culture, curricula etc. (Greenwald, Hedges & Laine, 1996).

Educational researchers have indicated the need to look at curricular reform systemically to comprehend pathways and barriers to achieve effectiveness (Anderson & Helms, 2001). In order to produce a human capital of scientists and innovators, the Gulf Cooperation Council (GCC) countries are now focusing more on the implementation of STEM education in schools and on to higher education (Zaher & Damaj, 2018).

Math and Science Education Reforms in GCC Countries

Educational system reform and development is a chief priority to GCC states economic and social development. Numerous actions have been taken to achieve goals and visions set by each country (Abouammoh, 2009).

In the UAE, the Ministry of Education has crafted a plan to achieve positive results in the TIMSS international assessments, ensuring alignment with the national agenda and the UAE 2021 vision. It has been further emphasized to place students among the top 15 countries in the next TIMSS (Thenational, 2016). The main goal is to reform education to become one of the best educational systems in the world, which will be measured by the country's performance in literacy, numeracy and scientific skills through TIMSS & PISA, as set out in the National Agenda Targets (Malaeb, 2016). According to Shaikh Mohammed bin Rashid, the national strategy of innovation focuses on knowledge integration in science, technology, engineering, mathematics and other fields, all of which contribute towards strengthening the knowledge economy. STEM education has been introduced through Abu Dhabi Educational Council (ADEC), where changes have been applied to school curriculum to apply STEM approach of teaching and learning. The UAE government is closely monitoring the progress of STEM implementation, encouraging schools through accreditation from the UAE's Ministry of Education (MOE), Dubai UAE and Abu

Dhabi Educational Council and the Knowledge and Human Development Authority (KHDA) (Soomro, 2019).

Several programs have been designed to encourage the implementation of STEM and create social awareness. As for instance; In 2013, UAE Abu Dhabi launched the TECHQUEST leadership program in shaping different attitudes towards STEM subjects, accompanied by school programs for children and STEM teachers' professional program that aim to equip teachers with new and effective teaching strategies (Wiseman et al., 2014).

Qatar values scientific knowledge and pedagogical practices as it shares a history of continuous educational policies development and educational reform, especially in the fields of science and mathematics. In 1995, the Qatar Foundation for Education, Science and Community Development was established. In 2001, RAND Corporation was assigned to initiate educational reform by proposing world-class educational framework that would meet nation's rapid changes and development. In 2004, with the help of international educational organizations, new curriculum standards were developed, covering mathematics, sciences, Arabic and English subjects. In 2008, the Qatar National Vision 2030 was introduced, including one of its key goals which is to focus on educational reform and improve citizens' scientific knowledge and invention to achieve economic goals leading towards diversity in Qatar's economy from being dependent on gas and oil (Said, 2016).

In translation to Bahrain's 2030 Economical Vision, with a focus to improve educational resources, the Ministry of Education (MOE) created numerous projects to advance students teaching and learning in STEM subjects, especially in mathematics (Al-Mutawah & Fateel, 2018). Further, to achieve competitiveness and empower its individuals, the MOE launched a new professional development policy to promote teachers' development and progress (Economic Development Board, 2008). It also stressed the essentiality and effectiveness of education and training to cope with the world's economical rapid change (Directorate of Training and Professional Development, 2009).

In 2008, MOE in Saudi Arabia introduced a reformed curriculum, the new mathematics and science curricula was an amended version of the curricula published by McGraw Hill (Obeikan for Research and Development, 2010). A constructivist theory was adapted, the aim was to make meaningful connections between students' lives and their educational experiences through the implementation of new instructional practices including, student centered, investigation strategies and problem-based learning. To achieve the Kingdom of Saudi Arabia's 2030 Vision goals, the "Ninth Development Plan (2010-2014)" aimed to continue improving the content and structure of school curriculums, teachers and students' content knowledge and development. In-order to sustain a knowledge-based economy and develop qualified students' capable of contending internationally in science and technology fields, more focus was directed towards the application of STEM education through the development of science, technology, engineering, and mathematics subjects (Mitchell & Alfuraih, 2018).

In an effort to redirect mathematics and science education towards STEM way of teaching and learning, Oman represented STEM Oman in 2018, which refers to a global education system designed by Rolls-Royce UK and adopted by the Oman Authority for Partnership for Development (OAPD). It was implemented in 18 public schools with an aim to reach 30 schools by 2019/2020 (Elayyan & Al-Shizawi, 2019). This educational reform in Oman took place through changes in the curriculum content and textbooks. Two issues were given specific consideration: first, the content of the curricula, and second, instructional techniques. The MOE embraced the new curriculum and encouraged the utilization of the new student cantered teaching techniques through development of teacher's professional programs and workshops to insure proper implementation. In order to achieve relevance of the curriculum, part of the new reform is the presentation of national environment life skills, to connect school learning with students' environmental characteristics and needs, making STEM subjects more relevant, realistic and motivational for students (Issan & Gomaa, 2010).

In Kuwait, several local programs have been introduced in support of STEM education implantation into the educational system. These local programs where comparable to many US project-based integrated STEM programs, which adopted the five stages of STEM implementation described by Dierdorp et al. (2014) which includes: reflection, research, discovery, application, and communication (Zaher & Damaj, 2018). The Programs included the Kuwait Institute for Scientific Research (KISR), the Ministry of Education, Ministry of Social Affairs and the Kuwait Foundation for the Advancement of Sciences (KFAS), which launched the Mathematics and Science Program to promote STEM education by supporting interactive teaching methods. The program intends to integrate all the aspects of the educational process into three main classifications, the messenger (educator), the message (instructional provisions), and the receiver (students). In-order to promote STEM education, the program further aims to enhance teachers' capacity, elevate Arabic educational content, design platforms to support mathematics and science students, design teachers' professional programs and support the use of interactive teaching materials (Zaher & Damaj, 2018).

The Concept of STEM Education

The acronym STEM (science, technology, engineering, and mathematics) is one of the most growing areas in educational reform of the twenty-first century and has become a major focus of the educational system for achieving global competitiveness (English, 2016; Slavit et al., 2016; Weis et al., 2015). Many programs have adopted STEM education to improve and reform their educational structures to positively influence student's achievement levels (Zeidler, 2016).

On the contrary, the conception of STEM varies among educators, educational researchers, curriculum developers and policy makers (Williams et al., 2015; English, 2016). Besides, educators have mixed views on STEM as, some view it as a new way of teaching STEM subjects while others view it as a new curricula or program (Herschbach, 2011). The STEM Task Force Report (2014), viewed STEM education as being significantly more than an "advantageous combination" of its four disciplines. Rather, it depends on real-world issues and problem-based learning that connects STEM four disciplines through a consistent active way of learning and teaching. The Report contended that STEM disciplines should not be instructed in isolation to one another, as they do not exist in separation in the real world or the workforce (Task Force Report, 2014).

STEM education is emerged as an integrated curricular design since 1920s, as a part of school programming era (Kilebard, 1987). At that time, educators and curricular developers thought that subject integration would take away intellectual characteristics from the integrated subjects. Therefore, more focus on the development of separate subject's direction was given. Additionally, emphasis was given on making school relevant to student's life experiences. Today, the educational agenda focuses mainly on the separate subject orientation methods, where students are limited by course content restrictions and boundaries. Moreover, academic achievement is evaluated through test results of each independent subject. The focus on making learning relevant to students' lives is fading away, as a result, students are losing interest in STEM subjects (Herschbach, 2011).

Importance of STEM Education

The significance of STEM education is presently notable around the world (Zaher & Damaj, 2018). In order to accomplish the abilities for economic growth and sustainable development, GCC policymakers, curriculum developers, economics and educators are approaching educational reform from a conventional and technical perspective. As a result, education in the GCC is moving towards STEM based education (Soomro, 2019; Wiseman & Alromi, 2007). Educators are now pressured to follow new creative teaching methods and move away from abstract problems and traditional instructions that have caused lack of motivation, anxiety and the overall low performance of GCC students' in STEM subjects and international assessment tests including TIMMS and PISA (Al-Mutawah and Fateel, 2018).

There are several benefits of using integrated i.e. STEM education in attaining high level of students' educational achievements. Research has revealed that integrated curricula provide students with an overall encouraging educational experience through relevant and connected disciplines that are less disintegrated and more relevant with daily life scenarios. Moreover, it improves the overall level of students' critical thinking and problem-solving techniques and therefore, their level of retention. Effective implementation of STEM education assist students in enhancing their competencies during problem solving, making them self-reliant, logical thinkers, and innovators (Hartzler, 2000). Moreover, assistance in terms of technical knowledge is also improved by proper STEM education.

Numerous researches have been conducted to point out the importance of students' involvement in the classrooms to overcome the international decline of enrolment rates within STEM fields (Lund & Stains, 2015; Asghar, 2012). However, it has been revealed that student's lack of motivation towards STEM subjects is mainly caused by the implementation of traditional way of instruction, where textbooks are overemphasized, while new instructional strategies as inquiry-based learning and problem-based methods are being under emphasized or neglected (Said, 2016). The national science education standards are utilizing more inquiry-based learning to enhance students' conceptual thinking that is relevant to real life issues. STEM way of learning is an ability that promotes students' critical thinking on how STEM concepts, ideas, standards, and practices are associated with daily life experiences (Roberta, 2015).

Benefits of STEM way of instruction are not limited to classroom only, and creates a bond between teachers and students, which aid in understanding relevance of STEM subjects with their environment, its impact on their communication, learning skills and future profession. Pursuing STEM is now considered as an economic imperative providing unique opportunities for improvement, future jobs and national welfare (Soomro, 2019; Issan & Gomaa, 2010).

STEM as a Curricular Concept

In spite of the global concern in STEM education, little considerations have been made regarding STEM's curricular concept and the manner in which it ought to be actualized in schools (Herschbach, 2011). The political, social and innovative history of every nation around the globe is unique, which results in different educational systems and application of STEM (Williams, 2011).

Early in 1998, Drake described the curricular integration theory through three directions. First, the multidisciplinary approach, where students are expected to make an association with respect to an issue that has been instructed among various branches of knowledge in various classes. Secondly, the interdisciplinary approach, where students are expected to apply their knowledge from different subject areas, as it is connected to a greater extent than a theme or an issue. The focus was on interdisciplinary content and skills. Thirdly, the trans-disciplinary approach, in which students are required to focus on real-life issues and have the ability to connect these issues to social, political, economic, international, and environmental concerns (Drake, 1998). In 2013, Vasquez et al. (2013) mentioned STEM subject's integration in forms similar to what was described earlier by Drake's (1998) three integration directions. However, the only difference was that Vasquez et al., (2013) added one more direction to the integration process i.e. the addition of disciplinary form of integration, where knowledge and skills gained from at least two strains in the interdisciplinary way of integration are connected to real-world issues and projects.

Further, Jacobs (1989) described several ways to integrate curricula, in this method only the appearance of the integrated subjects is changed, and no alteration is done to the actual content of the subjects. Teachers are supposed to reorganize the course's topics in a way to meet similar fields within the disciplines. Jacob's perspective of subject integration is also similar to Drake's multidisciplinary model. Both models agree on the same theme, the rearrangement of topics sequence in the integrated curricular to make connections among different disciplines. Comparably, in an article "STEM Education: Proceed with caution" the author describes the integration of all four STEM disciplines into one curriculum as confusing and favored the interaction between STEM subjects rather than their integration. A more reasonable approach was proposed, where interactions between STEM subjects are made by fostering cross-curricular links in a context without altering the integrity of each subject (Williams, 2011).

Implementation of STEM Education

STEM education has been envisioned in the United States since the 1990s as a way of strengthening mathematics and science curricula. Today, several decades later, its exact implication remains unclear (LaForce et al., 2016). Integrating subjects and connecting disciplines, as in STEM education, have been emphasized throughout past studies, especially with respect to the integration of mathematics and science curricula. Due to the lack of it solidifying perception and implementation framework, several ways have been identified to formulate a STEM program, while no specific curriculum model to follow and apply.

Globally, the structure of educational framework developed in internationally competitive nations becomes a model script for other nations to adapt into their educational system (Ramirez et al., 2006). Yet each country shares a different political, social and technological history; and therefore, possess different views on its educational system, specifically for STEM education and the use of technology (Cutucache et al., 2016). A research revealed that different conceptions of STEM implementation held by educators from different background and environment are reflected through the ways by which integrated STEM curriculum is developed and implemented within the region (Wiseman et al, 2014). Three considerations must be taken into account when designing, developing or reforming curricula, which includes the society, its students, and the subject of matter (Tyler, 1950).

Maroun et al. (2008) has described an efficient framework for increasing the chances of successful reforms in the GCC, particularly in Saudi Arabia. The framework combined three major dimensions central to educational reform. Firstly, it included socioeconomic environment, where social and economic priorities were translated into feasible educational strategies and related goals. Second is the educational sector-operating model, in which upright governance, operating entities, and funding to support educational goals has been emphasized. Third is the educational infrastructure that aided in achieving educational goals such as; the overall quality of teachers and curricula, reliable assessment measures and healthy learning environments.

Henderson et al. (2012) suggested a few noteworthy issues that ought to be contemplated while advancing instructional change. The emphasis of these factors is based on the changes of instructional system made in instructional practice that consequently influence instructional decisions. The first factor incorporates change like a process not an event; therefore, it requires time and patience. Next include the awareness of required adaptation plan based on the type of change needed, and appropriate strategies that should be selected. In order to facilitate the process of adaptation, different procedures maybe required for altering instructional beliefs. Lastly, viable change systems must consider different components of the instructional structure, for example, instructor feelings, assets and institutional settings.

Zeidler et al. (2005) described ten best teaching practices teachers should follow to achieve a successful mathematics and science integration process, hence, STEM education. The authors described that teachers should act more as facilitators rather than the main source of information. Also, to encourage students' hands on learning, concentrating on the use of cooperative learning, critical thinking and problem-solving techniques. Moreover, during class, teachers should depend heavily on questioning, assumptions and discussions. Further, it has been added that mathematics and science classes should be based on students' centred inquiry, where assessments are included with in the class. Additionally, more attention should be focused on the integration of technology (Zeidler et al., 2005). Dierdorp et al. (2014) discussed five aspects to enhance students' meaningful scientific concepts and contexts. These aspects included students' acknowledgment of the purpose to what they are learning. Second is motivation which refers to the students' level of engagement with the context understudy. Third is the application i.e. to achieve meaningful education, students must be able to apply the concepts learned in their daily life issues. Fourth includes authenticity where lessons become more meaningful to students when dealing with relevant and authentic contexts. Last includes connection, where knowledge from different subjects is integrated.

Barriers to STEM

The process of educational reform is much more difficult than it has been anticipated. Approaching the improvement of science and mathematics education by changing textbooks, buying new computers, or adding a new course simply is not enough (Alfadala, 2015). Educational developers often focus mainly on developing new instructional materials, while not taking into consideration the factors that may influence the adaptation process and its overall success (Stanford, et al., 2016; National Research Council, 2012).

The Concept of STEM and Framework of Application

The inconclusive meaning of STEM and the numerous potential methods for actualizing it, along with the absence of a unified framework of application and quality assessment of implementation are considered barriers to its application (Williams, 2011).

Relevance of Reformed Curriculum

Educational reforms, innovations and large-scale changes are considered as difficult tasks, due to the fact that on a larger scale, reform is about shared meaning. It does not only include individual change, but also social change, which is intrinsically difficult to accomplish, as the case in GCC where most reformed curriculum are adapted from different countries who share different religion, customs, heritage and environment. Many factors including political, social, cultural, economic and religious issues make school curricular structure rigid. This further makes it difficult for students to relate and acknowledge. According to Hopkins (1998), formal requirements, structures and event-based activities have been a major focus of reform strategies paying least attention to present cultures that requires specific values and practices.

Teachers' Perceptions and Content Knowledge of STEM

The shortage to successful STEM implementation does not rely solely on the number of students and teachers. But it is important to concentrate on the quality of teachers, as most STEM teachers are not 100% confident in what they are teaching due to lack of knowledge on STEM subject or its precise teaching methods (Hughes, 2009). Alabdulkareem (2004) argued that most researches conducted in science education has concentrated mainly on reform efforts, overlooking the complete image of interactions among teachers and curricular developers in an environment that encourages not only collaboration, but also discussion of scientific thinking. A successful integration of mathematics and science subjects is dependent heavily on teachers' understandings, content knowledge, and perceptions on the integration process. Students' content accomplishment, convictions, feasibility toward oneself, and inspiration can be negatively influenced if teachers inadequately implement STEM teaching strategies (Han et al., 2015).

Instructional Practices

Research indicated that student's lack of motivation towards STEM subjects is caused by the implementation of traditional way of instruction, where textbooks are overemphasized while new teaching strategies as inquiry-based learning and problem-based methods are being under emphasized or neglected (Said, 2016). Rigid structures of school curricular, standalone subjects that are taught in complete isolation to one another along with educational department's agendas and requirements are all barriers to STEM implementation (Kelley & Knowles, 2016).

Students

Most students believe that STEM subjects are too demanding, tedious or exclusionary causing lack of performance and interest in pursuing in STEM fields (Hall et al., 2011). Old distinct teaching methods and curriculum designs have neglected students from gaining a solid foundation in STEM subjects and related disciplines. This led to the lack of confidence and practice leaving students unprepared to cope with new instructional practices required for the implementation of STEM, as relying on their inquiry skills to come up with creative problem-solving ideas.

Recommendations

Several recommendations may be offered regarding educational reform and improvement in the GCC.

First, more research is required for educational reform and the implementation of STEM within the region through focusing on strengthening the areas such as; curriculum design and instruction especially in the field of STEM subjects and related disciplines, taking into consideration social, cultural and religious factors.

Secondly, GCC countries should work together to create a standardized curriculum for teaching STEM subjects and to develop a unified framework of its application that focuses on the areas of curriculum design and instruction. Thirdly, raise public awareness of STEM, its meaning, importance and means of application within the region. Developing and implementing a public awareness campaign to raise the prominence of STEM and the promising future it holds to individuals and society as a whole can achieve this. Fourthly, Ministries of Education and Higher Education should work together to create quality assurance centers that organize, assess, and evaluate educational outcome and assist in the reform and improvement of STEM education.

It is further suggested to reform educational policies, standards, instructional methods and proposals to provide unique opportunities of improvement to be implemented, in order to meet GCC countries' visions of success. Fifthly, further research and investment in the development of proper teachers' professional programs and workshops to attain proper implementation of STEM instructional methods and assure the application of constructive way of teaching and learning focusing on students' centered approach, critical thinking and problem solving techniques in order to fulfill the 21st century skills and requirements, while overcoming the distinct traditional way of instruction that causes students lack of motivation and outcome in STEM subjects. Redesigning current preparation programs to provide more instruction and concentration in science and mathematics; developing science preparatory measures and support for research-based STEM preparation can also achieve this.

Finally, emphasizing the need to address students' lack of motivation, commitment, and self-agency towards STEM subjects. STEM institutions need to introduce the concept of STEM innovation to students by providing after-school STEM-oriented programs and activities. This will aid in the introduction of an integrated approach to STEM learning and help the students consider the real-world applications of the knowledge acquired. Further, aim to improve the acceptability of student-faculty interaction, activities, and interaction with practitioners and employers within the field.

Conclusion

GCC countries are developing in economical, industrial, intellectual and social sectors. Globalization and technological changes have made human capital development increasingly important for a nation's economic progress. Science, technology, engineering, and mathematics (STEM) are closely tied to achieve economic success (Sellami et al., 2017). In order for GCC countries to achieve and maintain high international standards of economic success for its citizens, the percentages of students enrolled in STEM programs must increase as educational reform is considered a core element.

One of the main reasons of students' low achievement levels in STEM subjects and low enrollment in its fields is the distinct way STEM subjects are taught in the region. Implementing STEM education help in eliminating the traditional barriers between four educational subjects, aiding a teaching and learning experience that helps students to making sense of their education by linking it to real-world applications.

Despite the positive effect of STEM on students' achievement and content knowledge, incorporation of STEM should be approached with caution through a gradual long-term plan. Educational reform in the GCC towards the implementation of STEM should be more of updating school curriculum, which should be reorganized with coherence and relevance to each other especially between different STEM subjects making it more open and international with preservations on ethnic, heritage, social and religious inputs. Modernizing curriculum making it more relevant to students' lives and compatible with global reformation is not an easy task, yet alone is not enough. To address the problem, there is a two-fold approach leading to the development of economy knowledge in GCC countries; First is the functional part, including educational reform efforts to enhance the delivery of required knowledge and advanced skills for building human capital. Second includes the cultural factor, where region's culture should be infused in order for Gulf nationals to embrace capacity building (Wiseman, 2014)

STEM instructional teaching methods are critical parts of educational reform which usually consists of a three-part equation: the curriculum, the teachers and then the students at the receiving end of the equation. In order to create a generation capable of effectively participating in sustainable development and knowledge-based economy, proper reform should be applied. In-regards to curriculum, all GCC countries have invested heavily and allocated well experienced leading educational experts to improve their educational programs and curricular textbooks. The middle part which consist of teachers and applied instructional practices followed by the implementation process is the most vital to reach reform success. If the implementation is executed properly then students at the end of the equation should reflect their effectiveness in application.

In order to improve math and science education and inspire students to study and pursue their careers in STEM fields, it is necessary to have the government's commitment and support. There needs to be a radical change in curriculum implementation. Improving STEM education within Arab society is associated with providing opportunities for excellence to the next generation and the future of society as a whole.

REFERENCES

Abouammoh, A. M. (2009, June). The role of education: Trends of reforms and EU-GCC understanding. In *workshop entitled: The EU and the GCC: Challenges and Prospects under the Swedish Presidency of the EU*, Lund, Sweden.

Alabdulkareem, S. (2004). *Investigating science teachers' beliefs about science and science teaching: struggles in implementing science education in Saudi'* [Doctoral Dissertation. West Virginia University, West Virginia].

Alfadala, A. (2015). K-12 Reform in the Gulf Cooperation Council (GCC) Countries: Challenges and Policy Recommendations. The World Innovation Summit for Education (WISE).

Almazroa, H. (2013). *Professional development: A vision for Saudi science teachers*. Esera. org.

Al-Mutawah, M. & Fateel, M. (2018). Students' Achievement in Math and Science: How Grit and Attitudes Influence? *International Education Studies, 11*(2), 97-105.

Anderson, R. & Helms, J. (2001). The ideal of standards and the reality of schools: Needed research. *Journal of Research in Science Teaching, 38*(1), 3–16.

Ashgar, A., Ellington, R., Rice, E., Johnson, F. & Prime, G. (2012). Supporting STEM education in secondary science contexts. *Interdisciplinary Journal of Problem-based Learning*, 6(2), 85-125.

Aziz, H. (2016). Science and Mathematics Education in the GCC Countries. *Teacher Education and Curriculum Studies, 1*(2), 39-42.

Cutucache, C., Luhr, J., Nelson, K., Grandgenett, N. & Tapprich, W. (2016). NE STEM 4U: an out-of-school time academic program to improve achievement of socioeconomically disadvantaged youth in STEM areas. *International Journal of STEM Education, 3*(1), 1-7.

Dagher, Z., & BouJaoude, S. (2011). Science education in Arab states: Bright future or status quo? *Studies in Science Education*, 47, 73-101.

Dierdorp, A., Bakker, A., van Maanen, J. & Eijkelhof, H. (2014). Meaningful statistics in professional practices as a bridge between mathematics and science: an evaluation of a design research project. *International Journal of STEM Education, 1*(1), 1-15.

Directorate of Training and Professional Development. (2009). *The Guide to Training and Professional Development Programs.* Kingdom of Bahrain: Ministry of Education.

Directorate of Training and Professional Development. (2011). *The Guide to Training and Professional Development Programs.* Kingdom of Bahrain: Ministry of Education.

Drake, L. D. (1998). *Design and development of a computer-based simulation authoring system for problem-solving instruction.* Utah State University. 1-116.

Economic Development Board (EDB). (2008). *The Bahrain Economic Development Board.* In Regional Pioneer to Global Contender: The Economic Vision 2030 for Bahrain. Available at: https://www.evisa.gov.bh/Vision2030Englishlowresolution.pdf

Elayyan, S. R., & Al-Shizawi, F. I. (2019). Teachers' Perceptions of Integrating STEM in Omani Schools. *Journal of Education, 8*(1), 16-21.

English, L. (2016). STEM education K-12: perspectives on integration. *International Journal of STEM Education, 3*(3), 1-8.

Greenwald, R., Hedges, L. & Laine, D. (1996). The effect of school resources on student achievement. *Review of Educational Research, 66*(3), 361–396.

Hall, C., Dickerson, J., Batts, D., Kauffmann, P. & Bosse, M. (2011). Are we missing opportunities to encourage interest in STEM Fields? *Journal of Technology Education, 23*(1), 32- 46.

Han, S., Yalvac, B., Capraro, M. & Capraro, R. (2015). In-service teachers' implementation and understanding of STEM project-based learning. *Eurasia Journal of Mathematics, Science & Technology Education, 11*(1), 63-76.

Hartzler, D. (2000). *A meta-analysis of studies conducted on integrated curriculum programs and their effects on student achievement.* [Doctoral Dissertation, Oregon State University, Oregon].

Henderson, C., Cole, R., Froyd, J. & Khatri, R. (2012). Five claims about effective propagation. *A White Paper prepared for January* 30-31.

Herschbach, D. (2011). The STEM initiative: Constraints and challenges. *Journal of STEM Teacher Education, 48*(1), 96-122.

Hopkins, K. (1998). *Educational and psychological measurement and evaluation*. Boston: Allyn & Bacon.

Issan, S., & Gomaa, N. (2010). Post basic education reforms in Oman: A case study. *Literacy Information and Computer Education Journal, 1*(1), 19-27.

Jacobs, H. H. (1989). *Interdisciplinary curriculum: Design and implementation.* Association for Supervision and Curriculum Development, 1250 N. Pitt Street, Alexandria, VA 22314.

Kelley, T. & Knowles, J., (2016). A conceptual framework for integrated STEM education. *International Journal of STEM Education, 3*(1), 1-11.

Kliebard, H.M. (1986). *The struggle for the American curriculum*. Routledge.

LaForce, M., Noble, E., King, H., Century, J., Blackwell, C., Holt, S., Ibrahim, A. & Loo, S. (2016). The eight essential elements of inclusive STEM high schools. *International Journal of STEM Education, 3*(1), 1-11

Lund, T. & Stains, M. (2015). The importance of context: an exploration of factors influencing the adoption of student-centered teaching among chemistry, biology, and physics faculty. *International Journal of STEM Education, 2*(1), 1-21.

Malaeb, I (2016). Mathematics Education Helps Creative, Critical Thinking, Khaliji Times. Available at: www.khaleejtimes.com/nation/education/science.

Maroun, N., Samman, H., Moujaes, C.N., Abouchakra, R. & Insight, I. (2008). How to succeed at education reform: The case for Saudi Arabia and the broader GCC region. *Abu Dhabi, Ideation Center, Booz and Company*, 109-13.

Ministry of Economy and Planning, Kingdom of Saudi Arabia. (2010). *Ninth Development Plan (2010 - 2014).*

Ministry of Education, (2008). *National report on education development in the Kingdom of Saudi Arabia.* Riyadh: Ministry of Education.

Mitchell, B., & Alfuraih, A. (2018). The Kingdom of Saudi Arabia: Achieving the Aspirations of the National Transformation Program 2020 and Saudi Vision 2030 Through Education. *Journal of Education and Development, 2*(3), 36.

Obeikan Education (2012). General information. Available at: http://www.obeikaneducation.com/en/content/k-12-education.

Obeikan for Research and Development. (2010). *Project of mathematics and natural sciences.* Available at: http://msd-ord.com/project.htm

Roberta, N. (2015). Promoting innovative thinking. *American Journal of Public Health, 105*(1), 114-118.

Said, Z. (2016). Science Education Reform in Qatar: Progress and Challenges. *Eurasia Journal of Mathematics, Science & Technology Education, 12*(8).

Sellami, A., El-Kassem, R. C., Al-Qassass, H. B., & Al-Rakeb, N. A. (2017). A path analysis of student interest in STEM, with specific reference to Qatari students.

Slavit, D., Nelson, T. & Lesseig, K. (2016). The teachers' role in developing, opening, and nurturing an inclusive STEM-focused school. *International Journal of STEM Education, 3*(1),1-17.

Soomro, T. (2019). STEM Education: United Arab Emirates Perspective. In *Proceedings of the 2019 8th International Conference on Educational and Information Technology.*

Stanford, C., Cole, R., Froyd, J., Friedrichsen, D., Khatri, R. & Henderson, C. (2016). Supporting sustained adoption of education innovations: The designing for sustained adoption assessment instrument. *International Journal of STEM Education, 3*(1), 1-13.

Tyler, R. (2013). *Basic principles of curriculum and instruction.* University of Chicago Press.

Vasquez, J., Sneider, C., & Comer, M. (2013). *STEM lesson essentials, grades 3–8: integrating science, technology, engineering, and mathematics*. Heinemann.

Weis, L., Eisenhart, M., Cipollone, K., Stich, A., Nikischer, A., Hanson, J., Ohle Leibrandt, S., Allen, C. & Dominguez, R. (2015). In the guise of STEM education reform: Opportunity structures and outcomes in inclusive STEM-focused high schools. *American Educational Research Journal, 52*(6), 1024-1059.

Williams, C., Walter, E., Henderson, C. and Beach, A. (2015). Describing undergraduate STEM teaching practices: a comparison of instructor self-report instruments. *International Journal of STEM Education, 2*(1), 1-14.

Williams, J. (2011). STEM education: Proceed with caution. *Design and Technology Education: An International Journal, 16*(1).

Wiseman, A., Alromi, N. & Alshumrani, S. (2014). *Challenges to creating an Arabian Gulf knowledge economy', Education for a Knowledge Society in Arabian Gulf Countries (International Perspectives on Education and Society,* 24, 1-33. Emerald Group Publishing Limited.

Zaher, A. & Damaj, I. (2018). Extending STEM Education to Engineering Programs at the Undergraduate College Level. *International Journal of Engineering Pedagogy*, *8*(3), 4-16.

Zeidler, D. (2016). STEM education: A deficit framework for the twenty first century? A sociocultural socio scientific response. *Cultural Studies of Science Education, 11*(1), 11.

PART 2

CHAPTER 3

Contextualization of Interdisciplinary Science Teaching

Lara Abdallah

ABSTRACT

In this chapter we argue that teaching approaches can incorporate different aspects to allow students to recognize the purpose and the value of learning. However, connecting the different discipline topics create deep understanding of the multifaceted aspects of learning. This in turn requires a contextualization of content to link knowledge to various experiences at home, school and the wider community.

WHERE DO WE STAND?

All schools educate students to outperform standardized tests and score well, but very few educate and teach students the essential skills and knowledge to become successful candidates in the workplace. All students can acquire knowledge but what we are looking forward to is have enthusiastic candidates who are eager to learn. Education reformation is calling for bold, creative and engaging approaches that places students at the center of the learning journey. This era is witnessing many pedagogical methods; teacher-driven lectures, explicit teaching and online learning are some of the current pedagogies (Dolby & Rahman, 2008).

When it comes to teaching approaches, there is neither a right strategy nor a wrong strategy. However, there is a huge desirability to adapt pedagogical approaches that allow students to lead their own learning, to enjoy building their own future, to resolve their own personal and social problems, and to connect their classroom experiences to real life contexts. Correspondingly, scholars want to have students collaborating together to explore the way the real world works (Harada et al., 2008, Bell, 2010, Munakata & Vaidya, 2015, Krajcik, 2015) and connecting knowledge to their own context (Wyatt, 2016, Rathburn, 2015, Smith, 2010).

Science education provides meaningful explanation for the natural phenomena. It witnessed a huge shift due to the establishment of NGSS framework. The NGSS framework focuses on the performance of students to design solutions and to interpret phenomena. NGSS is not a traditional curriculum due to the inclusion of performance expectation within each standard (NRC, 2014). The instructional approach appropriate with NGSS structure has to promote investigations, collaboration and engagement. Thus, finding the best experiential inquiry approach to teach science content is considered a challenge (Krajcik, 2015, Varma et al., 2009, Amador & Miles, 2016). NGSS premise is achievable only when teaching is relevant to students' lives. Students have to conduct experiments and plan investigations to be consistent with NGSS structure (Krajcik & Czerniak, 2013). NGSS outlines the practices that need to happen to achieve the content knowledge. So, if those experiences are not part of the science class, then NGSS is not taught properly. Regardless of how well NGSS is structured, the teaching approach can either allow its beauty to shine or to wither. Becoming an NGSS classroom is a journey similar to reaching the peak of a mountain (Shelton, 2015).

The classroom successes are dependent on the joy in the desire of the students to overcome various obstacles that might hinder their academic achievements. Fulfilling NGSS vision is similar to climbing a mountain (Shelton, 2015). Several elements have to be embedded in the teaching approaches to allow students to explore the NGSS content through investigations. The classroom culture is expected to be a combination of emotional and intellectual bonds. Thus, students' minds, hands and hearts work together to master the performance expectations outlined in NGSS. Students welcome hands-on and minds-on challenging tasks. Students enjoy doing much more than simply listening (Shelton, 2015). Strategies of inquiry and experiential learning allow students to live the real experience of scientific ideas (Abrams et al., 2008, Luft et al., 2008, Windschitl, 2008, NRC, 2012).

The coming generation is supposed to be prepared for universal shifts that have an ever-accelerating pace. Thus, the learning approach has to empower students to fulfill the demands of the society. For that reason, the mastery of content knowledge is not enough. Graduates have to be skillful, creative and motivated. The dynamics of the new era awakened science educators to realize the importance of multiple perception and lines of reasoning while understanding the connections among various disciplines. The changes in the global economy are calling for school reform to focus on nurturing the skills, knowledge and competencies of students. The competitive marketplace demands high intellectual employees proficient in literacy, numeracy and technical skills. Moreover, the demographic characteristics of students have caused educators to re-examine the instructional approaches and teaching methodologies. Many theories were developed to outline new teaching and learning approaches that challenge students, such as contextual teaching and learning (CTL) and the interdisciplinary approach.

CTL is grounded by many sound educational theories and themes such as knowledge-based constructivism, effort-based learning incremental theory of intelligence, socialization, situated learning and distributed learning. CTL has interdisciplinary and contextual nature attributes. Learning is extended across disciplines so that students gain real life perspectives. Collective groups of objectives from different disciplines represent real life problems. CTL is based on connecting knowledge and experiences. The learning activities are designed to be conducted. The learning opportunities are designed based on real life themes that students experience in their everyday living. Those designed activities motivate students to work autonomously to construct their own learning. CTL provides a constructivist model. It is a teaching and learning approach that requires relating subject matter content to real life situations. Connecting the knowledge to applications in their daily lives motivates students to become engaged in the hard work that learning requires. CTL approach allows students to find meaning in the learning process. Berns and Erickson (2001) states that integrating learning subjects in appropriate context enable students to use the acquired knowledge and skills in applicable context.

Problem Statement

Fostering students' skills and competencies is a global demand. Major concerns about the capabilities of the coming generation are discussed widely among all educators. In particular, critical thinking, communication and collaboration skills are a must to have.

Schools are required to provide students with valuable learning experiences to acquire skills and competencies to fit in the workplace. Science education is designed to provide students with hands-on experiences to build their scientific inquiry skills. According to recent reports, the instructional strategies and the teacher capabilities are major components for a rich k-12 science classroom (Weiss et al., 2003, Banilower et al., 2013).

Engaging students in projects foster deeper connections to learn science. Thus, there is a growing demand to design authentic and purposeful teaching and learning approaches in the field of science education (NGSS Lead States, 2013, Luft et al., 2008, Chiapetta & Koballa, 2010). In conclusion, science education is continuously looking for an ongoing initiative to shift towards student-centered pedagogical practices to deliver NGSS framework. Meaningful experiences which foster positive ideas improve students' perceptions of science education (Barmby et al., 2008, Koch, 2010). The development of creativity and scientific exploration skills come as a result of the experiences of students in science classroom (Munakata & Vaidya, 2015).

Significance of the Study

The aim of the study is to investigate the influence of contextual teaching and learning approach in an interdisciplinary course on recognizing the purpose and the value of learning. Contextualization aims to connect academic skills to specific meaningful content focused on concrete applications to stimulate the interest of students. Rathburn (2015) states that contextualization connects the content to students' past experiences and future careers or interests. Only through effective implementation of NGSS framework, science teachers can succeed to equip students with the required skills and competencies. The setup of the learning environment shows the groundwork for guiding pedagogical strategies to deliver NGSS framework. Thus, focusing on the style of the learning experiences may inspire science teachers (Mayers & Koballa, 2013, Minger & Simpson, 2006, Hong & Lin, 2011).

Nowadays, students are discouraged to take part of their learning process due to the lack of relevance demonstrations to connect course materials to their lives. The philosophy of contextualization is based on the relevance of the content of learning with regards to real life to help students recognize the purpose and value of basic skills development (Rathburn, 2015, Premadasa and Bhatia, 2013, Krajcik & Sutherland, 2010). This research points the various possibilities and perception in order to continue researching the combined impact of contextualized teaching and learning method in an interdisciplinary course.

LITERATURE REVIEW AND THEORETICAL FRAMEWORK

This section presents the conceptual framework, theoretical framework, and the literature review. The conceptual framework discusses contextualization, contextualized approach, and interdisciplinary structure. The theoretical framework focus on learning theories and CTL, Smith model, and NGSS roadmap. At the end, the review of literature presents the recent studies which focus on the effectiveness of CTL approaches.

Theoretical Framework

Learning theories and CTL, Smith model, and NGSS roadmap underpin this study.

Learning Theories and CTL

The theory of learning by doing is promoted by many scholars (Confucius, Socrates, Aristotle). Learning happens when students actively engage in learning experiences relevant to their real life (Dewey, 1938). In this ever-changing era, "active engagement" is becoming a multidimensional phrase. Teachers and parents are not the only providers of knowledge. Students can easily obtain all answers in few seconds using internet resources (Schlechty, 2011). Thus, effective teachers have to design authentic learning experiences to motivate students to take part in their own learning journey. Students are truly engaged when they find the tasks at hand meaningful and valuable for them (Schlechty, 2011). An engaged student is an attentive, persistent and committed learner. His/her work does not have any limitations. It rather extends beyond the walls of the classroom. Project based learning is an engaging approach which provides students with opportunities to connect knowledge to real life experiences (Hallerman & Larmer, 2011). Thus, students devote their time to find resources independently in order to accomplish the task. A student feels that a project is a personal product that represents his understanding level, personality and skill set. Many scholars state that students feel a great deal of ownership of what they create (Hallermann & Larmer, 2011, Koch, 2010).

Students feel attracted to work when their opinions count. Thus, asking students to design projects leads to an intrinsic motivation and independency in achieving the set targets (Hallerman & Larmer, 2011).

Student engagement is the degree of interest that students possess to learn and progress in their education. Engagement in learning is defined as the persistence to do a task as responding to motives and values they have (Schlechty, 2011). When students are curious about a topic, they persist with tasks even if they are difficult or frustrating. For instance, some tasks such as projects open the door for students to create models using different skills. Thus, contextualizations allow students to link academic content to the knowledge acquired from various experiences at school, house or the wider community (Wyatt, 2016, Rathburn, 2015).

Students express their work creatively when the learning opportunities allow them to collaborate, communicate and reflect. Hence, they become enthusiastically eager to show their talents and personal potential. Assigning projects encourage students to become more creative and engaged (Scarbrough et al., 2004). A passionate student tries different paths to come up with the best solution. Ownership in learning process is developed as a result of defining and resolving issues in a contextual setup (Helle et al., 2006).

Smith Model

Smith model (2010) specifies the six elements of contextual teaching and learning method: (a) students are enthusiastic and actively engaged, (b) students find learning relevant to them, (c) students learn collaboratively and from each other, (d) learning is connected to real life issues, (e) prior experiences and diverse life contexts of students are fundamental to learning, and (f) teacher facilitates student learning.

NGSS Roadmap

The Next Generation Science Standards (NGSS) were created in 2011 based on the US National Research Council's Framework called "A framework for K-12 Science Education: Practices, Crosscutting Concepts, and Core Ideas (Framework)". The framework identifies the content knowledge that all k-12 students should acquire. In addition, it provides a clear vision for the "grade-appropriate proficiency in planning and carrying out investigations, analyzing and interpreting data, constructing explanations, designing solutions, and obtaining, evaluating, and communicating information (DCI arrangements NGSS, 2013: p.3). The framework includes the set of standards as well as the recent research about teaching science.

Many American associations collaboratively worked to produce the NGS framework, such as the National Science Teacher Association (NSTA), The American Association for the Advancement of Science (AAAS), and the National Research Council (NRC).) Eighty-person group worked for two years to create the science standards (NGSS Lead States 2013). Many scientists, educators and philosophers supported the group by providing continuous feedback on the released drafts.

The uniqueness of NGSS standards is due to the integration between the different standards within each standard. The dimensions within each standard are: scientific and engineering practices, crosscutting concepts, and disciplinary core ideas. Those dimensions that form each standard allow it to stand on its own. The four disciplinary science areas identified in NGSS framework are life sciences, physical sciences, earth and space sciences, and engineering, technology and applications of sciences.

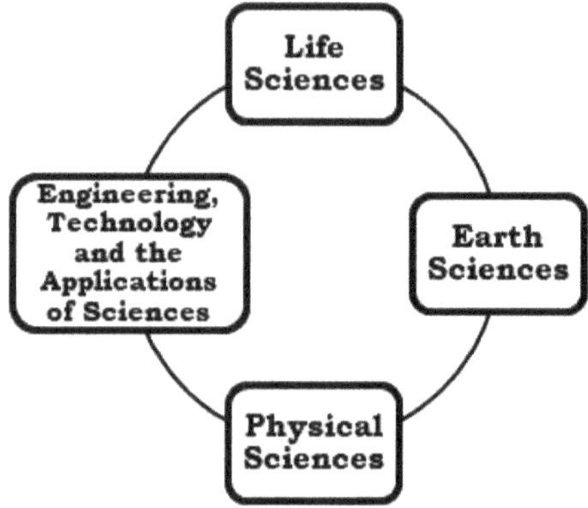

Figure 3: Four Disciplinary Science Areas

The four disciplinary areas were designed based on the most current research (National Academies, 2011). Within each disciplinary area, the framework identifies seven crosscutting concepts and eight scientific and engineering practices. Those practices have to be embedded in context with the core ideas.

The NGSS is designed to improve science education as well as student achievement. Thus, the teaching methodology has to allow students to discover scientifically and to explore logically. Significantly, many educators (Peterson, 2010, Chiapetta & Koballa, 2010, Luft et al., 2008) stress on the vital role of the teaching approach to deliver the NGSS content. Hence, the well-articulated standards delivered using a student-centered approach can lead to an improvement in teaching science (Duschl et al., 2008, Hong & Lin, 2011, Koballa, 2011). NGSS framework provides the context for students to become scientifically literate if they implemented properly.

NGSS outlines student performance expectations. The expectations clarify to educators what students are supposed to do and know. Such aspect makes NGSS more than just a curriculum overview. Moreover, it focuses on applying practices to content knowledge. NGSS is different from other science curricula because its standards focus on embedding application practices within (NGSS Lead States, 2013). NGSS focuses equally on practice, core ideas and crosscutting concepts.

Finally, the engineering design is integrated to raise the significance of engineering and technology in science education. The skills required to address the major engineering workplace challenges are clearly defined. NGSS concisely outlines the expectations for students' skills and knowledge (Porter et al., 2011). These standards provide students with deep understanding of science to face the challenges of the current world. NGSS points out that global environmental issues are big human made problems (Wysession, 2013). The stated standards are argumentative and interrogative to help students understand how human activities affect the environment. Students are expected to be engaged in arguments from evidence to increase their environmental awareness.

Although NGSS is cohesive; yet it is complex. Its complexity and interconnectedness produces an intensified learning experience in science classrooms. NGSS is mastered only when students experiment, collect data then generate arguments to address the scientific issues. NGSS provides the roadmap for a successful science learning experiences (Golden et al., 2012).

Conceptual Framework

Contextualization

Contextualization is defined by many educators, Mazzeo, Rab and Alssid (2003) state that contextualization is an instructional strategy designed to focus teaching and learning on concrete applications in a specific context. Wyatt (2016) defines contextualization as the process of linking students' academic content to prior knowledge acquired from various experiences from community, school, and house. Contextualization enhances the retention of information (Persin, 2011; Stone et al., 2006; Boroch et al., 2007). Contextualization is creating explicit connections between the different discipline areas and workplaces. Wyatt (2016) considers contextualization as an essential process in the teaching and learning of culturally and linguistically diverse students. Interdisciplinary learning and active student-centered learning are basic components of contextualization. Contextualization contributes to effective learning through extending the subject content beyond the borders of the classroom to include life applications. Persin (2011) states that contextualization is an approach designed to link content knowledge to a specific context that is of interest to the students. Contextualization has to be one of the components of any teaching model that explores the intersection of culture and education (Wyatt, 2016, Persin, 2011, Dermel & Coskin, 2010).

Many studies were conducted to evaluate the impact of contextualization on the engagement, participation and achievement of students. Scholars have had different conclusions. Some concluded that contextualization has a positive impact (Rathburn, 2015, Perin, 2011), while others reported an insignificant impact (Beswick, 2011)

Contextualized Approach

Our focus in this paper will be on contextualized approach where teachers integrate literacy into science and mathematics classes and increase students' sensitivity to their real-life situations. Persin (2011) defines contextualization as an instructional approach that creates explicit connections between different discipline areas. Contextualized basic skill instruction involves the teaching of subject objectives specifying its application in science, history or statistics. Contextualized approach is applicable to any curriculum and classroom (Wyatt, 2016) due to its focus on allowing students to bring their knowledge and personal experience to the learning process. It focuses on teaching academic skills rather than the subject matter. Despite the fact that some implicit content learning is required, students are exposed to subject-area material in the course of practicing basic skills. Rathburn (2015) states that although Contextualized learning is sound pedagogically and theoretically, mixed conclusions were obtained by scholars about its impact. Persin (2011) examined twenty-seven studies which all concluded that contextualized instruction has little impact on improving content knowledge.

Interdisciplinary Approach

Interdisciplinary approach uses knowledge coming from various disciplines for one specific purpose. Nikitina (2006) describes three interdisciplinary strategies: (a) contextualization, (b) conceptualization, and (c) problem-centering. Interdisciplinary teaching is the conceptual integration of one conception common to different disciplines (Demirel & Coskun, 2010, Thomas, 2000). This integration allows students to recognize the intersection among disciplines and deepen their analytical skills. Derimel and Coskin (2010) insist on the importance of interdisciplinary teaching approach to develop students' skills of research, discovery, synthesis, and curiosity.

An interdisciplinary course is designed to demonstrate the relationship among topics discussed in various subjects. For instance, students learn one topic and see its usage in mathematics, science, language, arts and social studies. Scholars (Repko, 2008) define interdisciplinary teaching as the presentation of one topic using meaningful disciplinary themes. The connections provide a tangible benefit to students as they understand the aim of learning abstract theories to resolve real-world dilemmas. Demirel and Coskin (2010) concluded that using interdisciplinary teaching approach revealed a significant difference in favor of a strong enhancement of students' skills and competencies.

METHODOLOGY

This study used a case study approach whereby three teachers were first studied then compared. Case study is an empirical research method used to unfold facts related to one specific topic in its specific context. Audio records and field notes from class observations were analyzed to identify the activity setting within the learning environment. Teachers' interviews and document analysis of lesson planning provided answers to understand the structure of the lesson. All documents were coded using open coding. Those codes were analyzed using axial coding process. Three questions were guiding the axial coding process:

1. Goals: What was the teacher trying to accomplish from the designed interdisciplinary course based on NGSS framework?
2. Process: How was the teacher using CTL in an interdisciplinary course?
3. Link: How was the teacher linking NGSS standards to the students' experiences?

Open-ended interview forms were used to document the views of the students and teachers on the influence of CTL in an interdisciplinary course on recognition of the purpose and value of learning. Three teachers and sixty students were interviewed. Every teacher was asked six open-ended questions. Each student was asked three open-ended questions. The opinions of two experts validated the interview forms.

Context of the Study

This study was conducted in a highly diverse private institution in Dubai, United Arab Emirates (UAE). Within the institution, the students came from forty different countries. Asians (63.4%) and European (11%) have the highest representations. Although there are clear differences among students whether cultural or linguistic, one common aspect is the deep sense of belonging to the multicultural society in UAE.

Participants

Three teachers using CTL approach to deliver interdisciplinary courses designed based on NGSS framework were similar in personal and professional background. They had taught for six years in the same school where the study was conducted, participated in same professional development and were in their mid-30s.

Data Sources

Teachers audio-recorded themselves as they planned for their lessons. The audio-recordings were transcribed for analysis. Class observation was used to capture the realities in all the subjects taught in middle school especially math, science, language, arts, and social studies. The method for class observation was non-structured observation and ethnographic. Cerda (2012) considers field notes from class observations a way to understand the dynamics of the teaching and learning process. Six classes were audio recorded to reveal insights about dialogues, relations, and class activities. Real world examples such as case studies, videos, scenarios, and newspapers were used to illustrate the approaches from various scientific disciplines along with mathematical skills and thinking.

Findings, Discussion, and Conclusion

The main interest of the present study was to explore the views of the students and teachers on the influence of CTL approach in an interdisciplinary course on recognition of the purpose and value of learning.

Key Findings

Three themes emerged as a result of the analyzed data. The three themes are: (1) Mindset of teachers and their role in classroom set, (2) Establish conceptual association, (3) Associations between disciplines, (4) Problem solving/Logical thinking skills, and (5) Scientific Study.

Mindset of teachers and their role in classroom set. Based on the planned lessons, observed classes and interview, teachers were able to design projects suitable for their students' abilities aligned with NGSS core ideas. The use of project tasks were an opportunity for teachers to provide students with hands-on experience to explore the NGSS content. The teachers' mindset allowed the implementation of CTL to be successful (Rogers et al., 2011, Harada et al., 2008, Markham, 2011). In class, they were acting as facilitators of knowledge rather than providers of information. They lead the instructions while reflecting the NGSS performance expectations to empower students to learn (NRC, 2014, Molotsky, 2011, Johnson, 2004, NGSS Lead States, 2013). The science classroom served as a venue for a diversity of practices to cater for the needs of all students.

The workload and the amount of time required to design projects were the main barriers as reported by the teachers. Yet, the three teachers believe that the time invested to design project tasks is rewarding due to witnessing huge improvements in mastering NGSS concepts. This draws parallels with the literature review about the impact of CTL approach on students' achievements in classes (Persin, 2011, Wyatt, 2016, Rathburn, 2015, Smith, 2010).

Students declared and showed their willingness to collaborate together to achieve learning targets. Their responses confirmed their enthusiasm to reflect about their own experiences and to apply the mastered knowledge to resolve real life issues. The related literature (Schlecty, 2011, Krajcik, 2015, Munakata & Vaidya, 2015, Wagner, 2010) indicate that students enjoy connecting learning to real world context and reflecting to express their preferences.

Establish conceptual association. Students confessed their ability to connect topics and concepts without the help of teachers after she demonstrated the topic using CTL. S1: "All topics are associated either directly or indirectly. Conceptualization made everything seem more meaningful for me. S 14: "The hands-on activities show the relationship between interdisciplinary connections."

Associations between disciplines. Students were convinced that there is a lack of association between disciplines. At the end of this interdisciplinary course, they related various topics with everyday events. S 49: "It was interesting for me to see the teacher able to connect math, science and history. Every subject seemed one independent entity. Many real-world issues are connected to the objectives of the course and to one another as well.

Problem solving/logical thinking skills. Students were aware of the elements of the problem-solving process such as analyzing and interpreting the causes and effects. They all expressed their desire to use logical thinking skills to solve complicated events rather than limit the usage of skills to daily life and numerical courses. S 19: "Any problem can be studied, examined and resolved if we use all what we studied collectively."

Scientific Study. The course load affected the desire for conducting researches in a negative way. Students were eager to conduct scientific studies in any topic they were interested in but there was a lack of time. S 43: "We studied historical events in relation with scientific theories. There are a lot of issues to research about to have a deeper understanding but there was no time."

Recommendation

Based on the findings from this case study, the establishment of a consistent NGSS learning environment culture sets high expectations for students to outperform. The concept of CTL approach supported by NGSS performance expectations provides foundation for successful student-centered learning environment.

CTL approach motivates teachers to incorporate many factors into designed tasks to support all students and to develop their skills further. Such a holistic approach engages everyone in the learning process regardless of their diverse needs (Harada et al., 2008, Helle et al., 2006, Krajcik, 2015).

Further Research

This study invites the potential of deeper investigations to understand the curriculum mapping design of the implemented lessons. A thorough analysis of the planning process can accurately reveal the criteria of selecting and designing project tasks. The planning process empowers teachers to design authentic projects for their students **(Tan & Leong,** 2014, **Makori & Onderi, 2013, Cullen et al., 2012, Snehi, 2011, Blumberg & Pontinggia, 2011)**

This learning group resembled a unique community of learners. Also, they provided an opportunity to understand the perceptions of both teachers and students about CTL approach. Involving elementary and high school students can provide new insights about the investigated topic.

Limitation

The sample size is one of the limitations of this study. The participants consisted of 60 students which represented all the first-year students in the private institution. The study could be extended to include all students enrolled in this interdisciplinary course for more than one semester. Another limitation is the span of the class observations. Observing more science classes provides an insight to figure out all the details related to connecting the different elements of the NGSS dimensions.

Conclusion

The designed Interdisciplinary course affects students' logical thinking skills in a positive way. Students were able to understand the common objectives of topics they were used to discuss independently in different disciplines. Dermil and Coskin (2010) recommend to give more attention to the wide benefits of interdisciplinary teaching.

The CTL process created authentic experiences for students to understand the problem statement to form hypothesizes and to display their scientific and analytical reasoning abilities. However, CTL approach needs time for planning as well as for the implementation. Teachers mainly differ in their perceptive taking, design of learning contexts, and views of students' background experiences.

Students faced difficulties in bringing the data obtained together. The use of various skills and techniques was new to them. In addition, teachers expressed the lack of proper guidance to design appropriate tasks to deliver the objectives of the interdisciplinary course. Hence, professional development was a needed component to train teachers on how to design learning opportunities to facilitate the learning process.

As a conclusion, the learning environment demonstrated the efficiency of the designed instructional strategies to implement the CTL in an interdisciplinary course of study. Students were indulged enthusiastically in exploring the scientific content through collaborative project tasks. Summing up the results, students held general willingness in working collaboratively to discover and learn. In connection to the literature reviewed, students become engaged when tasks are assigned to them.

REFERENCES

Abrams, E., Southerland, S. A., & Silva, P. C. (2008). *Inquiry in the classroom: Realities and opportunities.* International Academy Publishing.

Amador, J. A., & Miles, L. (2016). Live from Boone Lake: Interdisciplinary Problem-Based Learning Meets Public Science Writing. *Journal of College Science Teaching, 45*(6), 36-42.

Banilower, E. R., Smith, P. S., Weiss, I. R., Malzahn, K. A., Campbell, K. M., & Weis, A. M. (2013). *Report of the 2012 national survey of science and mathematics education.* Horizon Research.

Barmby, P., Kind, M., & Jones, K. (2008). Examining changing attitudes in secondary school science. *International Journal of Science Education, 30*(8), 1075-1093.

Bell, S. (2010). Project-based learning for the 21st century: Skills for the future. *Clearing House: A Journal of Educational Strategies, Issues and Ideas, 83*(2), 39-43.

Blumberg, P. & Pontiggia, L. (2011). Benchmarking the Degree of Implementation of Learner-Centered Approaches. *Innovation High Education*, 36, 189-202

Boroch, D., Fillpot, J., Hope, L., Johnstone, R., Mery, P., Serban, A., & Gabriner, R. S. (2007). *Basic skills as a foundation for student success in California community colleges.* Center for Student Success, Research and Planning Group, Chancellor's Office, California Community Colleges. Retrieved from http://css.rpgroup.org

Cerda, M.G. (2012). *Teachers exploring partnership education and ways to transform the K–12 educational system.* Saybrook University.

Chiappetta, E. L., & Koballa, T. R. (2010). *Science instruction in the middle and secondary schools: Developing fundamental knowledge and skills* (7th ed.). Allyn & Bacon.

Cullen, R., Harris, M., Hill, R. (2012). *The Learner-Centered Curriculum: Design and Implementation.* Jossey-Bass

DEMİREL, M. and COŞKUN, Y.D. (2010). Case study on interdisciplinary teaching approach supported by project-based learning. *The International Journal of Research in Teacher Education, 1*(2), 28-53.

Dewey, J., 1938. *Experience and education.* Macmillan.

Dolby, N., & Rahman, A. (2008). Research in international education. *Review of Educational Research, 78*(3), 676-726.

Duschl, R., Shouse, A., & Schweingruber, H. (2008). What research says about K-8 science learning and teaching. Principal, 16-22. Retrieved from https://www.naesp.org/resources/2/Principal/2007/N-Dp16.pdf

Golden, B., Grooms, J., Sampson, V., & Oliveri, R. (2012). Generating arguments about climate change. *Science Scope, 35(7)*, 26-35.

Hallerman, S., Larmer, J. (2011). PBL in the elementary grades: step-by-step guidance, tools and tips for standards- focused K-5 projects. Buck Institute for Education

Harada, V. H., Kirio, C., & Yamamoto, S. (2008). Project-based learning: Rigor and relevance in high schools. *Library Media Connection, 26*(6), 14-20.

Helle, L., Tynjala, P., & Olkinuora, E. (2006). Project-based learning in postsecondary education theory, practice and rubber sling shots. *Higher Education, 51*(2), 287-314.

Hong, Z. R., & Lin, H. S. (2011). An investigation of students' personality traits and attitudes toward science. *International Journal of Science Education, 33*(7), 1001-1028.

Koballa, T. (2011). *Framework for the affective domain in science education.* Retrieved from http://serc.carleton.edu/NAGTWorkshops/affective/framework.html

Koch, J. (2010). *Science stories: Science methods for elementary and middle school teachers* (4th ed.). Wadsworth.

Krajcik, J. S., & Sutherland, L. M. (2010). Supporting students in developing literacy in science. *Science Teacher*, 328(5977), 456-459. http://doi:10.1126/science.1182593

Krajcik, J. S., and C. Czerniak. (2013). *Teaching science in elementary and middle school classrooms: A project-based approach*, 4th edition. London: Routledge.

Krajcik, J. (2015). PROJECT-BASED SCIENCE. *Science Teacher, 82*(1), 25-27.

Luft, J., Bell, R. L., & Gess-Newsome, J. (2008). *Science as inquiry in the secondary setting.* National Science Teachers Association Press.

Makori, A., & Onderi, H. (2013). Evaluation of Secondary School Principals' Views on the Use of Untrained Teachers in Lesson Delivery in a Free Secondary Education System Era in Kenya. *Journal of Education and Practice*, *4*(24), 119-133.

Markham, T. (2011). Project Based Learning: A bridge just far enough. *Teacher Librarian, 39*(2), 38-42.

Mayers, R. L., & Koballa, T. (2013). *Mathematics connections: Common Core State Standards-Mathematics and the NGSS.* National Science Teachers Association Council of State Science Supervisors.

Mazzeo, C., Rab, S. Y., & Alssid, J. L. (2003). *Building bridges to college and careers: Contextualized basic skills programs at community colleges.* Workforce Strategy Center. Retrieved: http://www.workforcestrategy.org/images/pdfs/publications/Contextualized_basic_ed_report.pdf

Minger, M. A., & Simpson, P. (2006). The impact of a standards-based science course for preservice elementary teachers on teacher attitudes toward science teaching. *Journal of Elementary Science Education, 18*(2), 46-60.

Molotsky, G.J. (2011). A case study of the impact of a reformed science curriculum on student attitudes and learning in a secondary physics classroom.

Munakata, M., & Vaidya, A. (2015). Using Project- and Theme-Based Learning to Encourage Creativity in Science. *Journal of College Science Teaching, 45*(2), 48-53.

National Academies (2011). Report offers new framework to guide K-12 science education, calls for shift in the way science is taught in U.S. Retrieved from http://www8.nationalacademies.org/onpinews/newsitem.aspx?RecordID=13165

National Research Council (NRC) (2014). *Developing assessments for the Next Generation Science Standards*. National Academies Press.

National Research Council (NRC) (2012). *A framework for K-12 science education: Practices, crosscutting concepts, and core Ideas.* National Academies Press.

Next Generation Science Standards Lead States (NGSS Lead States) (2013). *Next Generation Science Standards: For states, by states.* National Academies Press.

NGSS Lead States. (2013). *Next Generation Science Standards: For states, by states.* National Academies Press.

Perin, D. (2011). Facilitating Student Learning through Contextualization: A Review of Evidence. *Community College Review, 39*(3), 268-295.

Peterson, P. E. (2010). *Saving schools: From Horace Mann to virtual learning.* Belknap Press of Harvard University Press.

Porter, A., McMaken, J., Hwang, J., & Yang, R. (2011). The new U.S. intended curriculum. *Educational Researcher, 40*(3), 103-116.

Premadasa, K., & Bhatia, K. (2013). Real life applications in mathematics: What do students prefer? *International Journal for the Scholarship of Teaching and Learning* 7(2), Article 20. Available at: http://digitalcommons.georgiasouthern.edu/ijsotl/vol7/iss2/20

Rathburn, M.K. (2015). Building Connections Through Contextualized Learning in an Undergraduate Course on Scientific and Mathematical Literacy. *International Journal for the Scholarship of Teaching and Learning, 9*(1), 11.

Repko, A. F. (2008. Assessing interdisciplinary learning outcomes. Retrieved August 20, 2009 from: http://www.uta.edu/ints/faculty/REPKO_Outcomes_AEQ.pdf

Rogers, M. A. P., Cross, D. I., Gresalfi, M. S., Trauth-Nare, A. E., & Buck, G. A. (2011). First year implementation of a project-based learning approach: The need for addressing teachers' orientations in the era of reform. *International Journal of Science and Mathematics Education, 9*(4), 893-917.

Scarbrough, H., Bresnen, M., Edelman, L., Laurent, S., Newell, S., & Swan, J. (2004). The processes of project-based learning. *Management Learning, 35*(4), 491-506.

Schlechty, P. (2011). Engaging students: The next level of working on the work. Jossey-Bass

Shelton, T. (2015). Climbing the NGSS Mountain. Science Teacher, *82*(9), 65-66.

Smith, B.P. (2010). Instructional Strategies in Family and Consumer Sciences: Implementing the Contextual Teaching and Learning Pedagogical Model. *Journal of Family & Consumer Sciences Education, 28*(1).

Snehi, N. (2011). *Improving Teaching Learning Process in Schools: A Challenge for the 21st Century*. New Delhi Publishers.

Stone, J. R., III, Alfeld, C., Pearson, D., Lewis, M. V., & Jensen, S. (2006). *Building academic skills in context: Testing the value of enhanced math learning in CTE* (Final study). St. Paul, MN: National Research Center for Career and Technical Education. Retrieved from http://136.165.122.102/UserFiles/File/Math-in-CTE/MathLearningFinalStudy.pdf

Tan, A., & Leong, W. (2014). 'Mapping Curriculum Innovation in STEM Schools to Assessment Requirements: Tensions and Dilemmas', *Theory into Practice*, 53(1), 11-17

Thomas, J.W. (2000). A Review of Research on Project Based Learning. Retrieved from: http://www.bobpearlman.org/BestPractices/PBL_Research.pdf

Varma, T., Volkmann, M., and Hanuscin, D. (2009). Preservice elementary teachers' perceptions of their understanding of inquiry and inquiry-based science pedagogy: Influence of an elementary science education methods course and a science field experience. *Journal of Elementary Science Education*, 21(4), 1–22.

Wagner, T. (2012). Graduating all students innovation-ready. *32*(1). Retrieved from http://www.edweek.org/ew/articles/2012/08/14/01wagner.h32.html?utm_source=fb&utm_medium=rss&utm_campaign=mrss

Weiss, I. R., Pasley, J.D., Smith, P.S., Banilower, E.R., & Heck, D.J. (2003). *Looking inside the classroom: A study of K-12 mathematics and science education in the United States*. Horizon Research.

Windschitl, M. (2008). What is inquiry? A framework for thinking about authentic scientific practice in the classroom. In J. Luft, R. L. Bell, & J. Gess-Newsome (Eds.), *Science as inquiry in the secondary setting* (pp. 1-20). NSTA Press.

Wyatt, M. (2016). "Are they becoming more reflective and/or efficacious?" A conceptual model mapping how teachers' self-efficacy beliefs might grow. *Educational Review*, *68*(1), 114-137.

Wysession, M. (2013). The next generation science standards and the earth and space sciences. *Science Scope,* 5(08), 17-23.

CHAPTER 4

STEM Education Policy Development

Elaine Al Quraan

ABSTRACT

STEM education has several benefits to the students and tutors, particularly, in higher education level. Therefore, the aim of this research is to give a comprehensive discussion regarding education and economic advantaged of STEM programs. The goal will be achieved by benchmarking the already established and successful K-12 STEM programs and propose a related policy to the United Arab Emirates. Countries that are currently using the STEM education have high economic growth rate compared to those with other education systems. Hence, the chapter explains the role of policymaker in promoting STEM education. It also focuses on the the reforms introduced by federal government in an attempt to create a more unified vision of STEM education to promote the awareness of the importance of incorporating the four main disciplines in the national education system. Additionally, the study discusses how the STEM education offers the students an opportunity to become more innovative and flexible to survive in the highly competitive business environment in the contemporary world.

INTRODUCTION

Currently, only a few students are willing to enroll in careers related to science, technology, engineering, and mathematics (STEM) worldwide. More importantly, the issue arises due to insufficient exposure of public to the benefits of STEM industry and ignorance of the process of engineering design found in the K-12 program. Serious measures should be formulated to facilitate the understanding the importance of pursuing careers in the STEM education system. The chapter offers relevant discussion, recommendations, and benefits that individual can obtain from international STEM education systems. Similarly, it will be crucial to focus on international policies as they fundamental in the creating the best STEM education system with practices to boost its development and growth in UAE educational context. The result will provide a set of guidelines obtained from the evaluation of various valid sources discussing the STEM education system in different countries that have adopted this system. It is imperative to offer a suitable tool for disseminating material related to STEM in the contexts of UAE.

The chapter provides insight into the importance of offering the policymakers and educators the most common practices applied by the STEM practitioners. Therefore, the chapter also aims to develop pedagogical policy and offers important suggestions that should be considered for effective implementation of STEM education in UAE. The chapter attempts to present this through a study by analyzing related documents and the critical review of pertinent literature. The research study is fundamental because it gives teachers approaches to the instructional practices including project-based learning and inquiry based strategies, which have played a crucial role in promoting a comprehensive exploration of different learning opportunities in other countries such as India. The findings will be elementary in building and designing an effective STEM program that integrates skills, knowledge, and resources in UAE. The study will also create a new impression that highlights the importance of high-quality K-12 programs to allow the policymakers and teachers to access most excellent STEM models. More importantly, the outcome of this research will generate a substantial interest amongst STEM practitioners, especially, because there are numerous studies that show that science plays a vital role in steering the strategic planning towards making an informed decision in response to the economic, technical knowledge, and skills demands in UAE. For example, other countries such as India are increasingly raising awareness of the importance of large-scale educational policies in the handling complexity that encountered implementation of the STEM education. UAE government has increased use of human capital and taking significant steps towards becoming the best in terms of economic development and growth. Therefore, the study considers the following research questions to broaden as well as deepen the comprehension UAE STEM education practices and background:

1. What are the barriers to successful implementation of STEM education in the UAE?
2. What are the responsibilities of legislators and of policymakers in promoting the K-12 setting?
3. What changes should UAE ministry of education consider to implement STEM education successfully?

The study will provide a comprehensive literature review, discussion and recommendation as well as conclusion to address the above objectives. Therefore, the study will follow a systematic flow designed to provide adequate answers to the above research questions.

LITERATURE REVIEW

There are numerous benefits associated with consistent use of integrated education system (IES) results. Some studies point out that application of IES offers crucial opportunities for more optimal and flexible learning experiences for the students. Also, STEM is considered as a learner-centered system due to its ability to improve the problem-solving as well as critical-thinking skills. Therefore, it is clear that encouraging the STEM education has important implications for the student in UAE, as it will allow them to become best problem solvers and critical thinkers as well as innovators. Several scholars demonstrate that an interdisciplinary system composed of science and mathematics play key role in developing as well as sustaining student's interest in some forms of schooling. Students become more motivated to learn and apply new concepts and ideas when exposed to flexible STEM learning environment.

According to the National Academy of Engineering and the National Research Council, there are several important benefits related to the use of STEM education system. For instance, integrating the engineering in diverse school settings improves the performance in both mathematics and science. In turn, students improve their understanding of the importance of engineering principles in different real-life situations. They can also persistently use the acquired broad technical knowledge to overcome challenges in a highly competitive business environment of the 21st century. The implications can, therefore, be used to create a relevant conceptual framework to guide ministry of education in formulation and implementing policies that support K-12 program in successfully. The research emphasizes the practical implications of STEM education, particularly, in UAE education system. A thorough description of the benefits of STEM education can help the policymakers and educators to formulate a solid model for innovative and flexible education that could benefit all the stakeholders involved.

Pertinent Issue Affecting Implementation of STEM Education

Sinay, Jaipal-Jamani, Nahornick and Douglin (2016) provide crucial information about the perception of teachers regarding STEM programs (Sinay et al. 2016 p.218). According to the source, 90% of the teachers lack understanding of the problem and knowledge of how to integrate the different subjects, disciplines, and areas when teaching. Also, the authors denote that 27% of the teachers do not understand engineering design processes and their implementation. At the same time, 55% of the tutors do not have skills related to STEM education, and 32% have no clue where to get resources related required for developing STEM career (Sinay et al. 2016 p.218). Roehrig, Moore, Wang, and Park (2012) support this finding by accerting that 23% of the teachers in UAE do not have strong knowledge of the present STEM subjects, and although 84% are confident about the ability or effectiveness of K-12, 40% wonder about the degree of skills needed to achieve the intended objectives (Roehrig, Moore, Wang, and Park 2012 p.218). Therefore, it is evident that there is a considerable discrepancy regarding teaching practice in UAE. Barrows and Tamblyn suggest that there a need for integrating inquiry and problem-based learning framework as a solution to the challenge (1980 p.221). The approach is crucial because it involves student engagement, driving question and fosters collaboration among teachers. Johnson (2012) confirms the effectiveness of the framework as analysis of six STEM schools conducted in the same years demonstrates that effective strategies and practices of engagement tutors and students are important for meeting the anticipated goal of the programs (p. 172).

Ejiwale (2013) provides various barriers that prevent proper execution of STEM education. According to Ejiwale (2013), STEM faces considerable challenges due to shortage and poor preparation of qualified teachers. Training teacher is fundamental for helping students attain high academic standard. However, most classes in UAE are under-prepared because of poor training. A survey conducted in 2013 showed that 28% of teachers in public school in UAE who teach science in K-5 lack major science knowledge and hands-on skills (Ejiwale 2013 p.229). Further, the author shows that out-of-field tutors teach 40% of the mathematic classes. Additionally, UAE is characterized by insufficient investment in teachers' professional development for strong knowledge bases. The issue contributes to poor performance of learners (Mahil 2016 p.128). The problem worsens due to poor inspiration and preparation of learners, and several studies have shown that only 20% of students are satisfied with K-12, citing lack of solid teacher proficiency which contributes to increased hardship in STEM subject (Mahil 2016 p.128). Ejiwale (2013) infers that proficiency of educators is fundamental for inspiration and motivation of students to learn STEM disciplines. Therefore, it is evident that teachers are key components for successful implementation of STEM education in UAE schools. More importantly, the author emphasizes the importance of professional development programs for teachers.

Delen, Kuzey, and Uyar (2013) extend Ejiwale's (2013) discussion by focusing on other areas that prevent proper implementation of STEM K-12. According to them, lack of connection between the teachers and students in a wide variety of ways is one of the primary challenges facing STEM programs (Delen, et al., 2013 p.118). Current research shows that project-based learning illustrates that tutors can increase students interest in K-12 by involving them in solving authentic problems and creating real integrating approach (Capraro, Capraro, and Morgan 2013 p.173). Another issue arises from the lack of support from the ministry of education (MoE). The authors hold that it is imperative for the agency to think and structure the business of education differently by considering school goals and method of engaging learners. Delen, Kuzey, and Uyar (2013) argue that teacher expertise in STEM schools is important which implies a need for the MoE to increase funding to secure and create instructor who understand how to teach subjects such as mathematics and science. More importantly, it is fundamental to focus on research collaboration across all the STEM fields as this one of the contributors that has led to poor development of skills in giving students a proper sense of purpose and direction for making effective choices related to their learning process. The scholars contend that it is fundamental to devise teaching approach through collaboration of educators to enhance connectivity and foster information sharing among all stakeholders (Green and Sanderson 2017 p.1771). What the source implies here is that effort should be made to increase partnership between and among industry personnel to bring the current gaps in STEM education. Further, they presume that it is imperative to take into account the similarities and differences that exist in the different K-12 disciplines as well as understand how to integrate them (Traphagen and Traill, 2014 p.198). By understanding the distinct practices of science and mathematics, educators would ensure a more effective implementation of STEM principles that would foster students' natural ability to learn efficiently.

According to Ceylan and Ozdilek (2015), there are real problems that emerge during the dissemination of science education. The authors further contend that it is fundamental to comprehend the interrelations of different disciplines and other issues to enhance STEM education (Ceylan and Ozdilek 2015 p.13). For example, the article notes that it is imperative to ensure integration of knowledge and skills to better instruct the learners to allow them to understand complex perspectives and concepts (Harris, 2017 p.522). The author argues that there is the need to relate concepts and ideas to facilities comprehension of the material taught in class. Thomas and Watters (2015) confirm the importance of integration approach put forward by Ceylan and Ozdilek (2015), as the source holds that there are numerous ways for integrating science and mathematics effectively into the curriculum to allow the educators to achieve the intended goals. Most importantly, the Thomas and Watters (2015) agree with Ceylan and Ozdilek (2015) arguing that knowledge and skills are important for instructors to present the content of learning in relevant and meaningful ways. Additionally, some of the things the readers quickly note about the information put forward by these scholars are that it is crucial for researchers and education policymakers to explore teachers' knowledge and skills, especially, those aspects related to multidisciplinary factors of STEM (Makhmasi et al. 2012 p.6). Further, the references are important as they offer fundamental insight into the core factors that should be considered to engage students in the classroom and support their learning experience effectively.

While scholars such as Ceylan and Ozdilek (2015) base their argument on their understanding problem associated with interrelation of STEM fields, others like Egarievwe (2015) outline critical aspect of STEM education using a theoretical approach such as integrated knowledge framework. Egarievwe (2015) demonstrates that the framework is important in assisting learners to make the relevant connection between the various STEM fields and allow them to participate in inquiry-based investigations. The article is important for this study because it also explains how cognitive and learning progression models can be used to promote student education in STEM programs (Egarievwe 2015 p.119). An interpretation of the source reveals that it is important for the educators to demonstrate a high level of coherence rather than focus on each practice or discipline separately. This implies that educators in STEM programs play a crucial role in promoting cross-disciplinary instruction which is essential in helping students build and maintain integrated STEM learning experiences. The approach is key for promoting the relevance of STEM education, emphasizing teachers' distinct role in providing high-quality instruction in the discipline in which they are proficient. Good teachers are expected to demonstrate a proper knowledge of the specificity of STEM standards and how they work in practice.

Recent Reforms Around the World That Put Emphasis on STEM

In the United States, both local and national policy makers have lobbied for efforts to increase the stock of college graduates in the STEM fields. The reasoning behind their efforts is that STEM fields are the main promoters of innovation and as a result hold significant consequences for long run economic growth and individual welfare (Grotzer, et al., 2015). On February 17, 2009, the United States President Barack Obama signed into law the American Recovery and Reinvestment Act of 2009 which was inclusive of a 787 billion dollars' legislative package that was hailed as an economic stimulus. This legislation includes 2.5 billion dollars in addition to federal funding for the National Science Foundation, including new funding for the STEM education programs (Kesidou and Koppal, 2014). This legislation is in support to the recent federal efforts, including the America COMPETES Act of 2007, to enhance the federal support for STEM education initiatives. Numerous STEM changes have been proposed since then and considered for use in several statewide efforts across the United States. Most of these changes in STEM propose that the funding should be used to improve the personal practice of science for all students at all grade levels in the United States (U.S. Department of Education, Office of Education Technology, 2014).

In Australia, the Education Ministers in an effort to support the development of skills in cross disciplinary, creative and critical thinking, digital technologies and problem-solving which are critical in all occupations in the 21st century, they established that these objectives lie at the core of the STEM field (Marginson, 2013). The government of Australia has invested in enhancing STEM education and has initiated significant activities across the country in schools and education systems, by universities and industry to boost student engagement and success in STEM and to support teachers to improve student outcome. Student engagement in STEM is affected by many factors (Drew, 2015). These are underlined by the views of the broader community about the relevance of STEM and this has pushed the Australian government to make reforms in the approach to the teaching and learning of STEM from the early years and continues as the students go through the schooling process. This is connected to the way the industry is articulating the significance of STEM-related skills that extend beyond traditional STEM occupations (Psacharopoulos and Patrinos*, 2014). Universities in Australia have put policies in place that have an influence on student choices in the senior secondary years. The primary purpose of the strategy adopted by the Australian Education Ministry is to build on a range of reforms and activities that are already on the ground as it aims at better coordinating this effort and sharpening the focus on the key areas where a collaborative action will deliver improvements to STEM education (Potter, 2013).

A survey conducted between the years 2000 and 2009 showed that the UK had dramatically fallen behind in science and math. With regards to this survey, the UK planned to implement reforms at the end of the summer of 2012 when children were returning to school with the aim of increasing their literacy in science and math (Smith, 2013). The ministry of education in the UK understands that a comprehensive foundation in STEM education is necessary to ensure that the UK remains at the forefront of science and technology in the coming decades. With the number of students specializing in higher STEM education increasing in the UK, the government understands the significance to maintain a broader perspective (Ball, 2014). The UK also wishes to stay in the forefront of science and technology and this has led them to prioritize the efforts to encourage all their students as well as the general population to have the understanding of scientific principles (Green, 2012). Raising the level of STEM education has the potential to endow citizens with the capacity to evaluate ideas from a scientific perspective and as a buffer against anti-scientific and pseudo-science ideas. The current education system in the UK that was implemented in 2014 aims at improving the standards of education and develops a strong curriculum and teaching system that encourages independent and critical thinking from an early age (Blanden and Machin, 2015).

STEM Practices in the UAE

Different scholars, such as Benkari (2013), Suchan (2014), and Edarabia (2015), provide details about the current development of STEM education in UAE. According to Benkari (2013), UAE education policymakers are recognizing the need to consider students learning perceptions and many schools are willing to introduce properly supported STEM programs in their teaching practices. Benkari (2013), Suchan (2014) and Edarabia (2015) argue in favor of fostering the existing hand-on approach to learning in UAE as a way of motivating and encouraging the learner to demonstrate their potential for internalizing new ideas. An interpretation of these sources reveals that there is considerable raise in awareness regarding the advantages of STEM education in UAE (Benkari 2013 p. 215; Suchan 2014 p.288; Edarabia 2015 p.112). Further, the three references demonstrate that there is a need to ensure national curriculum integrating a well-maintained K-12 program.

While the above authors delve in creating a picture of what is happening about the integration of STEM education in UAE, Swaid (2015) take a different perspective which is providing details about challenges the educators face in the country. The writer holds that lack of resources and teacher preparedness are primary problems. In particular, the source notes that there is a need for educators to acquire professional training as one of the approaches for achieving the intended goals of K-12 program. The scholar explains that the respective subjects are not taught holistically, implying that their separate exploration has not yielded the expected results regarding academic performance (Swaid, 2015 p.277). The technique will lead to improvement of knowledge about science. A clear analysis of the source shows that the author is trying to demonstrate a need to shift from old perspective related to education to modern methods of STEM programs in order to yield the anticipated objectives in UAE schools. The article is important for this study as it provided new concepts and ideas that curriculum developers and Ministry of Education should consider fostering student performance on the subject included in the K-12 (Swaid 2015 p.277).

Further, authors such as Carr Ronald, Lynch Bennett, and Johannes Strobel (2012) perform research to detail other viable ways for improving K-12 program. The scholars propose a descriptive framework as a method of meeting the integrated STEM education goals. The study demonstrates that STEM program takes a multidimensional space in K-12 education landscape because it involves a range of experience with a varied degree of connection. The interpretation of the data given reveals that each variant of integrated STEM program suggests the need for different strategies and techniques, consideration of challenges related to implementation, available resource, and possible outcomes (Carr, Lynch and Johannes 2012 p.229). More importantly, the source is crucial for this study because it offers insight into the various parameters that MoE must consider for successful implementation of K-12 program. The study outlines how the teachers, administrators, and curriculum developers can use the framework as well as how it can be productive in meaningful discussion of K-12 education. The framework is also important in examining and comparing feature of programs that have characteristics of integrated STEM. At the same time, it is a fundamental approach for developing as well as testing hypotheses about relationships among crucial factors and variants of integrated STEM education in UAE.

Role of Policymakers in Promoting STEM Education

Cavanagh (2008) details the roles of policymakers in promoting STEM education and the aspects they are currently considering. According to the source, leaders at all levels need to improve the various parameters of K-12 education in STEM. One of the various factors that policymakers consider is targeting different areas such as teacher training and education standards which are fundamental for attaining educational outcomes. Further, the writer demonstrates that 63% of the policymakers are focusing on promoting the relevance of STEM education, emphasizing teachers' distinct role in providing high-quality instruction in the discipline in which they are proficient (Cavanagh 2008 p.124). Corpley (2015) confirms Cavanagh's (2008) information regarding the connection of policymaker and improved K-12 education. According to the scholar educators, administrators and ministry of education are policies related to the establishment of a supportive system of assessment, accountability, and transparency in communicating different educational standards and principles to stakeholders in the field. Corpley (2015) confirms Cavanagh's (2008) information regarding that a policymaker should make significant stride to understand the distinct practices of science and mathematics to ensure a more effective implementation of STEM principles to foster students' natural ability to learn efficiently. Also, it is clear that policymakers need to reconsider the implementation of certain strategies through which teachers would have adequate time to explore the complex scientific and mathematical concepts (Corpley 2015 p.370).

Paudel (2009) provides several important recommendations that policymakers can focus on to support K-12 programs in UAE effectively. According to the writer, state, local, and national level policymaker can promote science to the level of mathematics subject using assessment system that foster understanding and learning (Paudel 2009 p.593). At the same time, they can invest in focused, sustained, and coherent set of support for the program to reach effective ways. For instance, teachers should be allowed to pursue professional learning activities to address their teaching needs. Additionally, the source recommends that federal agencies can support collaborative research to disentangle the effect of school practices from the selection of students, take into account the underlying contextual variables, and enable longitudinal examination of student outcomes (Paudel 2009 p.593). The author implies that there is a need to tie federal funding to the development of the robust approach and drive strategic research agenda to allow a full response to the current barrier preventing successful implementation of STEM programs in UAE. Some scholars like Sallee and Flood (2012) take a different approach to that of Paudel (2009). The article identifies key elements that policymakers should implement to improve K-12. The authors argue that it is fundamental to develop a coherent curriculum and set of standard that focuses on the most important topics and builds knowledge over time. More importantly, MoE and education administrators can ensure teachers understand current STEM content as well as how to teach, since most tutors are underprepared. Also, there should be a support system to oversee accountability and assessment to promote effective dissemination of concepts and ideas (Paudel 2009 p.593). At the same time, policymakers need to reconsider the implementation of certain strategies through which teachers would have adequate time to explore the complex scientific and mathematical concepts.

Additionally, National Governors Association (2008) holds that state policymaker has a role in establishing effective funding statute to set standard academic requirement and standards efficiently aligned across the K-12 level. For example, it is imperative to develop certification and preparation requirements that can support the program. The association also holds that legislators and governor, as well as state policymakers seeking to create effective STEM agendas, must understand the importance of communication. The author recommends that governors should lead to the establishment of communication strategies to engage the public regarding the availability of STEM careers and the urgency of improving the program (National Governors Association 2008 p.12). One thing that is National Governors Association (2008) and Sallee and Flood (2012) agree with, is that policymakers need to target different improvements, as one of the significant goals that should be pursued in the respective context is related to the development of a coherent set of educational standards. Having rigorous and comprehensive STEM standards is fundamental to achieving specific educational outcomes (Sallee and Flood 2012 p.673).

CRITICAL DISCUSSION OF STEM ISSUES (CHALLENGES AND RECOMMENDATIONS)

The information presented above partly demonstrates the responsibilties of legislations and policymakers in promoting STEM education system, particularly, in the K-12 setting and identify several important areas of focus to improve the program. Different scholars conquer that it is imperative to develop an effective policy that targets professional development of STEM teachers in UAE through training and establishment of education standards (Bicer et al. 2015 p.71). Further, it is fundamental to the education to create a framework that can be used to govern and promote assessment, accountability, and transparency aimed at overcoming the current barriers facing STEM education to foster students' natural ability to acquire knowledge and skills. However, although these are crucial areas that can improve STEM education in AUE, there is lack of elementary components to achieved successful execution such as a policy that governs the STEM implementation in the UAE context. Paudel (2009) and Bilgin, Karakuyu, and Ay (2015) support the idea that it is imperative to develop a robust approach to drive strategic agenda that will allow a full response to the current barrier preventing successful implementation of STEM programs in UAE. As such, the approach supports the strategy proposed and put forward by Johnson (2012): partial immersion, minimal immersion, STEM supplemental approach, collaborative partnership and infrastructure focusing instruction, and curriculum.

STEM partial immersion method is a non-traditional approach where school experience and related skills integrated into the curriculum. The policy should be developed to govern how stakeholders collaborate across the K-12 disciplines: students, staff, teachers, and community. The plan should allow the parties to partner with one another leading to the establishment of new STEM opportunity classes under the guidance of a leading team. The approach should also focus on instruction where classes should include inquiry or project-bases technique to allow the students to solve real world problems. The proposal put forward by Sinay, Jaipal-Jamani, Nahornick, and Douglin (2016) and Egarievwe (2015) support the importance of this method as they argue in favor of using the inquiry-based framework to enable stakeholders to work together to achieve the intended K-12 goals. Additionally, the policy should take into account curriculum by focusing on overseeing that courses taught in UAE STEM education are aligned with international academic standards to allow natural progression of students from subject to subject. STEM minimal immersion is important to approach because it is directly related to the problem-based learning, supplemental to the adopted curriculum. The policy should ensure that K-12 education includes separate STEM units completed at the end of the school year. It could also govern how to learning institutions can encompass short units offered by the industry to provide students with sufficient experience and develop required workforce skills (Breiner et al. 2012 p.9). The crucial infrastructure for achieving the goal includes establishing a school support from the administrators to the STEM tutors. The plan can also provide guideline related to how the provision of essential material and resources to teachers to allow them effectively teach STEM. The administrators can serve as an important party for the successful implementation of the proposed policies because they can encourage collaboration of teachers as well as the integration of the STEM in all the classes. Finally, successful STEM implementation in the UAE requires the development of a policy that governs collaborative partnerships, for example, the higher education is a fundamental body that plays an important role when it comes to preparing K-12 teachers and students for their STEM careers (Capraro and Nite 2014 p.310). It is imperative to the adopted strategy to recognise that higher education institutions in UAE are crucial for developing qualified teachers successfully which means the need to change the current institutional practices, teaching methods, and candidate recruiting procedures (Gehrke and Kezar, 2017 p.273).

a) Organizational and Formalized Efforts to Measure the Impact of STEM on National Development Properly

The current body of literature gives a clear and comprehensive view of the various barriers that prevent successful implementation of the STEM education, particularly, in UAE. Most importantly, most teachers do not understand the useful approach that can be used to teach K-12 education such as problem and inquiry-based methods and lack requires skills and knowledge. Apparently, UAE is also facing other challenges such as shortage and poor preparation of qualified teachers, insufficient investment in teachers' professional development and poor inspiration and preparation of students. Further, there is lack of coherent connection between teachers and students, support from the ministry of education, research collaboration across all the STEM fields, reduced comprehension of the interrelations of different disciplines (Capraro and Han, 2014, p. 9). Therefore, successful implementation of K-12 education in UAE requires organizational and formalized efforts to measure the impact of STEM on National development properly.

The approach calls for the need to restrain secondary school teachers to teach STEM subjects in an integrated manner such that students learn in an active and interdisciplinary manner. According to Traphagen and Traill (2014), administrators, government and education institutions can measure the impact of STEM education properly for national development through establishment and sustenance of cross-sector partnership where the cultivators of STEM education can assess the shift of resources and gaps to evaluate connectedness on learning experiences. Concerned national institutions can determine if the program produces the intended benefits by focusing on its collective goals and community needs as well as identify areas that need creating pathways for further engagement, learning, and development of careers. The first effort that education management institutions can put forward to assess the impact of STEM education is to establish common measurement designed to allow the investigation of questions about interconnections between teachers' professional development and student learning (Claymier 2014 p.5). More importantly, the programs need to examine the impact that teacher practice has on the student education process. Therefore, it is imperative for organizations such as MoE to develop a common measurement system for the K-12 plan focused on relevant, leverage, and interconnected variables that will allow empirical understanding of the effect of STEM education in UAE.

Furthermore, the formalized effort should assess the shift in priority for research and evaluation as well as summative data to measure the impact of the designed approach to optimizing the student learning process, collaboration among the different parties and effectiveness as well as benefits of K-12. Additionally, it is fundamental to create an assessment plan that will help to measure the project goals and to provide prompt feedback instead of just documenting changes (Corlu, Capraro, and Capraro 2014 p.76). The strategy should empower scholar to perform research studies to develop conceptualizations based on strong theoretical foundations. Organizations that address complex social problems should extend this practice to the development of common measure program to understand and describe the organizational changes needed to foster the effective implementation of STEM education.

b) Identification of Metrics to be Measured Regularly

Data garnered from the literature review under pertinent issues facing the STEM education demonstrates that one of the primary challenges is lack of teachers' knowledge for teaching K-12 education disciplines. Therefore, the most important metric that should regularly be measured is content knowledge and ability to disseminate the material effectively. The indicators would focus on assessing of tutors understand of how to integrate content in a way that reflects comprehension of the students (Han, Capraro, and Capraro 2015 p.1089). Another crucial indicator is the participation of teachers in the K-12 specific professional development activities. The metric can be important in measuring the participation of teachers in research-based, and high-quality professional development in STEM education focused on continued development of their capacity to teach, address classroom challenges and allow the establishment of sustainable opportunity for reaching the desired goals. Also, it is elementary to measure the contribution of instructional leaders in creating a condition that supports the implementation of STEM education. Additionally, UAE can use assessment to measure the core practices and concepts of K-12 followed by analysis of examination of consistency based on the set standards (Kunberger 2013 p.259). Therefore, curriculum leadership and teachers should identify the instructional methods suitable for STEM education as well as project-based learning. At the same time, it would be important to use student assessment and self-reflection, which can demonstrate their level of understanding and perception regarding teachers' effectiveness in teaching STEM. Moreover, school administrators can focus on observing learners' approaches to the underlying problems and measure variables related to collaboration among all parties, questioning data, and provide critical thinking.

c) Teacher Training and Certification to Deliver STEM Curriculum

Ejiwale (2013) demonstrates that teachers are key components for successful implementation of STEM education in UAE schools. More importantly, the author emphasizes the importance of professional development programs to equip and allow teachers embrace the programs. Therefore, successful implementation of K-12 education in UAE requires teacher training and certification to deliver STEM curriculum. Instructors need training on different areas of STEM, reasons for its establishment, how the program works and its possible economic, community, and student impact. In addition, professional development sessions should include theory, program, and classroom establishment instruction to produce well-prepared teachers and become certified educators, which require another different session for completing lesson plan as well as other important processes to assess teacher content knowledge (Alyammahi, Zaki, Barada and Al-Hammadi 2016 p.960). Training can also encompass face-to-face workshops, synchronous and asynchronous activities, post professional development and certification. The strategy is fundamental for ensuring that teachers deployed in the various public schools in UAE are qualified to handle the STEM subject taking into account need for integrating their similarities and differences to ensure efficient dissemination of the material to students. Certification of teachers to occur after successful completion of the examination task once the person has met the requirements, which should correspond to the respective level of course, offered in the curriculum.

d) Curriculum Development: Adaptation of International Standard STEM Curriculum and Developing Local Materials

Benkari (2013), Suchan (2014), and Edarabia (2015) agree that one of the challenges facing STEM in UAE is the poor formulation of content and lack of teacher understanding of the material in the current curriculum. Another issue arises from the lack of support from the Ministry of Education. Therefore, MoE should adopt international standard STEM curriculum and curriculum developers should create local materials that meet the international standards/learning outcomes. The approach is important because it allows the developers of curriculum comprehend effective methods of including practices and core ideas explicitly from technology and engineering. Also, international standards can influence approaches taken to teach science and integration of students in across the STEM disciplines effectively

(Mcdonald 2016 p.533). Similarly, these international standards can foster educator self-efficacy as they lay structure that UAE can adapt to oversee the effective transfer of knowledge as well as understanding to the learners. Finally, considering international education standard can help UAE administrators adapt the fundamental conceptual framework for the development of teacher professional knowledge and skills for effective dissemination of material to ensure intended student performance.

e) Monitor and Improve STEM Training

There is need to monitor K-12 training and collect data to see whether or not it has the desired impact. The strategy requires gathering information about the current system before implementing change for comparison of the performance of the program later. The technique allows the administrators and government stay aware of the progress that the country is making and identify areas that need changes. Monitoring provides a way of achieving continued improvement in the adopted education standard. The process should focus on the areas such as development planning, implementation of school improvement strategies, continue the professional development of teachers to ensure quality teaching to enable the students to attain their learning potential (Munakata and Vaidya 2015 p.48). Additionally, it allows inclusion of a needed area of improvement in the future development plan to promote continuous progress. Therefore, monitoring enables the establishment of priorities in the prospected training strategies (Green and Sanderson 2017 p.1772). As a result, the school can produce quality and standard report with details of attained goals and progress in relation to the national and international education principles. The goal can be achieved through teacher self-evaluation. Similarly, it may be imperative to focus on analyzing classroom materials and documents such as lessons and forward plans and homework. More importantly, it is crucial to interview the teaching staff about the learning process and classroom method used (Muller 2015 p.411). The process follows the analysis of data collected to see areas that need improvement.

f) Raising Awareness and Effective Communication with the Community and Citizens on Career Opportunities

According to the National Governors Association (2008), a few UAE students are considering STEM careers that will guarantee them economic prosperity. Primarily, the problem related to the fact that many families and students have no comprehension of the importance of STEM skills to the country's economic development in the future. Therefore, just a number of people are enrolling in the available career opportunities. Further, UAE education institutions have not been conveying information about the need for STEM knowledge in the future workforce and that the occupations in the STEM industry are stable and well paying (Parker and Lazaros 2014 p.24; Rennie, Venville, and Wallace 2012 p.182). As such, there is a need to raise awareness and communicate effectively with the community and citizens on career opportunities. Educators and government should cultivate interest and enthusiasm in UAE STEM industry and education broadly through effective campaign statewide to ensure increased public awareness of the need for STEM education for improvement of the economy (Aswad, Vidican, and Samulewicz 2011 p.559). They need to inspire excitement about career opportunities in the STEM and correct misconception about the industry. At the same time, while the overarching vision of the awareness and communication program should be tied to the country's broader STEM agenda, the philanthropic community and private sector should shape, finance, and carry the campaign by providing substantial advocacy and financial investment. They can also contribute positively to the objective because they have positive media attention and affiliation with marketing institutions. Successful communication will oversee that families throughout the region understand what STEM is about and the reason it is fundamental to the economy and individuals. Most importantly, the players can demonstrate what UAE's stem careers apparently look like and define as well as differentiate strategies for a distinct audience such as females (Pasha-Zaidi and Afari 2016 p.1231).

g) Sectors Interconnectedness to Enhance the Investment and Success in STEM Education

According to Delen, Kuzey, and Uyar (2013) funding STEM education in UAE is a challenge that requires sectors interconnectedness to enhance the investment and success in STEM education. Private and public schools should work together with philanthropists to prepare the program for successful implementation (Traphagen and Traill 2014 p.198). Private and public organizations should co-operate to invest collectively in their perceived interest for K-12 in UAE. The institutions should collaborate in financing higher education programs for the development of professional STEM educators. They should work together to develop a solid policy for provisions and introduce college scholarships to students who pursue coursework in the four core subjects. Furthermore, STEM funding increment in government budget is important for improving the foundations of STEM education (Traphagen and Traill 2014 p.198; Raiyn and Tilchin, 2016 p.124). The enactment of new policies is intended to improve teacher quality in different STEM fields. It is only after demonstrating and linking the utility of STEM education to businesses within identified economic sectors that businesses and economic sectors proactively demand effective STEM education.

Conclusion

In achieving the vision set in STEM education, policymakers and educators need to consider various existing challenges and opportunities. The collaboration of multiple stakeholders in the education field is necessary to foster the application of STEM education in different spheres of social and business life. One of the challenges identified in the process is related to the promotion of equitable access to STEM teaching and learning experiences. It has been reported that disadvantaged students, particularly those attending high-poverty schools, have no access to high performing and functional teachers in the core subjects of science and mathematics (Suchman 2014). The disparities in providing equitable access to STEM education may additionally challenge policy makers to enhance STEM teaching and learning practices across the United States. Another challenge that has been identified is the establishment and promotion of engaged communities of practice. To overcome this challenge, policymakers need to focus on developing and enhancing the conditions for meeting the needs of local communities (Ralston, Hieb, & Rivoli, 2013). This usually takes place through facilitating collaborative learning. Also, lesson activities can be redesigned and modified to meet the evolving needs of stakeholders in the education field. The fear of failure should be directly addressed within the education system, which would enable stakeholders to understand the implications of STEM education.

Indeed, the importance of STEM in the UAE is understood in broad terms by government and academia. Government, economy or industry does not understand the details of why STEM is important. Further, considerable problems are impeding successful implementation of STEM education in UAE. Most importantly, teachers do not understand the useful approach that can be used to teach K-12 education such as problem and inquiry-based methods and lack required skills and knowledge. Apparently, UAE is also facing other challenges such as shortage and poor preparation of qualified teachers, insufficient investment in teachers' professional development, and poor inspiration and preparation of students. Therefore, there is a need to help facilitate an understanding of the linkages between STEM competencies with general learning and university education and fundamental economic requirements. In other words, few understand the strong linkages between K-12 and learning ability with sustainable competitiveness which means there is a need to develop a policy to government implementation of K-12. As such, it is important to work with individual sectors, plan a competitive direction, identify the required skills to sustainably succeed in that direction and communicate these linkages to industry and academia and the government. Further, it is important to establish organizational and formalized efforts to measure the impact of STEM on National development properly, ensure effective teacher training, certification to deliver STEM curriculum efficiently, and MoE should adopt international standard STEM curriculum and curriculum developers should create local materials that meet the international standards/learning outcomes. Similarly, government and education institution should raise awareness and communicate with the community and citizens on career opportunities. Additionally, there is a national requirement to retrain secondary school teachers to teach STEM subjects in an integrated manner such that students learn in an active and interdisciplinary manner.

REFERENCES

Almekhlafi, A.G. & Almeqdadi, F.A. (2010). Teachers' perceptions of technology integration in the United Arab Emirates school classrooms. *Journal of Educational Technology & Society*, *13*(1), 165-175.

Alyammahi, S., Zaki, R., Barada, H. & Al-Hammadi, Y. (2016). Overcoming the challenges in K-12 STEM education. In *Global Engineering Education Conference (EDUCON), IEEE* (pp. 951-960). IEEE.

Aswad, N.G., Vidican, G. & Samulewicz, D. (2011). Creating a knowledge-based economy in the United Arab Emirates: realising the unfulfilled potential of women in the science, technology and engineering fields. *European journal of engineering education*, *36*(6), 559-570.

Barrows, H.S. (1985). *How to design a problem-based curriculum for the preclinical years*. Springer Pub Co.

Barrows, H.S. & Tamblyn, R.M., 1980. *Problem-based learning: An approach to medical education*. Springer Publishing Company.

Benkari, N. (2013). The 'sustainability' paradigm in architectural education in UAE. *Procedia-Social and Behavioral Sciences*. 102, 601-610.

Bicer, A., Navruz, B., Capraro, R.M., Capraro, M.M., Oner, T.A. & Boedeker, P. (2015). STEM schools vs. non-STEM schools: Comparing students' mathematics growth rate on high-stakes test performance." *International Journal of New Trends in Education and Their Implications*, *6*(1), 138-150.

Bilgin, I., Karakuyu, Y. & Ay, Y. (2015). The Effects of Project Based Learning on Undergraduate Students' Achievement and Self-Efficacy Beliefs Towards Science Teaching. *Eurasia Journal of Mathematics, Science & Technology Education*, *11*(3), 469-477.

Breiner, J.M., Harkness, S.S., Johnson, C.C. & Koehler, C.M. (2012). What is STEM? A discussion about conceptions of STEM in education and partnerships." *School Science and Mathematics*, *112*(1), 3-11.

Capraro, M.M. & Nite, S.B. (2014). Stem integration in mathematics standards. *Middle Grades Research Journal*, *9*(3), 1-10.

Capraro, R.M., Capraro, M.M. & Morgan, J.R. (2013). STEM project-based learning. *An Integrated Science, Technology, Engineering, and Mathematics (STEM) Approach*.

Capraro, R.M. and Han, S. (2014). STEM: the education frontier to Meet 21st century challenges. *Middle Grades Research Journal*, *9*(3), 10-15.

Carr, Ronald L., Lynch D. Bennett, and Johannes Strobel. (2012). Engineering in the K-12 STEM Standards of the 50 US States: An Analysis of Presence and Extent. *Journal of Engineering Education* 101(3), 539-564.

Cavanagh, S. (2008). Where is the 'T' in STEM? *Education Week*. 27(30), 17-19.

Ceylan, S. & Ozdilek, Z. (2015). Improving a sample lesson plan for secondary science courses within the STEM education. *Procedia-Social and Behavioral Sciences*. 177, 223-228.

Claymier, B. (2014). Teaching 21st century skills through an integrated STEM approach. *Children's Technology & Engineering, 18*(4), 5-10.

Corlu, M.S., Capraro, R.M. & Capraro, M.M. (2014). Introducing STEM education: implications for educating our teachers for the age of innovation. *Egitim ve Bilim, 39*(171). 74-85.

Corpley, D. (2015). 'Teaching engineers to think creatively: Barriers and challenges in STEM disciplines.' In *the Routledge International Handbook of Research on Teaching Thinking. Routledge International Handbooks* (pp. 402-410). Routledge Taylor & Francis Group.

Delen, D., Kuzey, C. & Uyar, A. (2013). *Successful K-12 STEM education: Identifying effective approaches in science, technology, engineering, and mathematics*. National Academies Press.

Egarievwe, S. U. 2015. Vertical education enhancement-A model for enhancing STEM education and research. *Procedia-Social and Behavioral Sciences*. 177, 336-344.

Ejiwale, J.A. (2013). Barriers to successful implementation of STEM education. *Journal of Education and Learning (EduLearn), 7*(2), 63-74.

Gehrke, S. & Kezar, A. (2017). The Roles of STEM Faculty Communities of Practice in Institutional and Departmental Reform in Higher Education. *American Educational Research Journal, 54*(5), 803-833.

Green, A. & Sanderson, D. (2017). The Roots of STEM Achievement: An Analysis of Persistence and Attainment in STEM Majors. *The American Economist, 63*(1), 79-93.

Han, S., Capraro, R.M. & Capraro, M.M. (2016). How science, technology, engineering, and mathematics project-based learning affects high-need students in the US. *Learning and Individual Differences, 51*, 157-166.

Harris, S.Y. (2017). Undergraduates' assessment of Science, Technology, Engineering and Mathematics (STEM) information literacy instruction. *IFLA Journal*, 43(2), 171-186.

Johnson, C.C. (2012). Implementation of STEM education policy: Challenges, progress, and lessons learned. *School Science and Mathematics, 112*(1), 45-55.

Kunberger, T. (2013). Revising a design course from a lecture approach to a project-based learning approach. *European journal of engineering education, 38*(3), 254-267.

Mahil, S. (2016), April. Fostering STEM+ education: Improve design thinking skills. In *Global Engineering Education Conference (EDUCON), 2016 IEEE* (pp. 125-129). IEEE.

Makhmasi, S., Zaki, R., Barada, H. & Al-Hammadi, Y. (2012, October). Factors influencing STEM teachers' effectiveness in the UAE. In *Frontiers in Education Conference (FIE), 2012* (pp. 1-6). IEEE.

McDonald, C.V. (2016). STEM Education: A review of the contribution of the disciplines of science, technology, engineering and mathematics. *Science Education International, 27*(4), 530-569.

Muller, J. (2015). The future of knowledge and skills in science and technology higher education. *Higher Education, 70*(3): 409-416.

National Governors Association. (2008). Promoting STEM education: A communications toolkit.

Parker, J. and Lazaros, E.J. (2014). Teaching 21st century skills and STEM concepts in the elementary classroom. *Children's Technology & Engineering, 18*(4): 24-27.

Pasha-Zaidi, N. & Afari, E. (2016). Gender in STEM Education: An Exploratory Study of Student Perceptions of Math and Science Instructors in the United Arab Emirates. *International Journal of Science and Mathematics Education, 14*(7), 1215-1231.

Paudel, N.R. (2009). A critical account of policy implementation theories: status and reconsideration. *Nepalese Journal of Public Policy and Governance, 25*(2), 36-54.

Raiyn, J. & Tilchin, O. (2016). The Self-Formation of Collaborative Groups in a Problem Based Learning Environment. *Journal of Education and Practice, 7*(26), 120-126.

Rennie, L., Venville, G. & Wallace, J. eds. (2012). *Integrating science, technology, engineering, and mathematics: Issues, reflections, and ways forward.* Routledge.

Roehrig, G.H., Moore, T.J., Wang, H.H. & Park, M.S. (2012). Is adding the E enough? Investigating the impact of K-12 engineering standards on the implementation of STEM integration. *School Science and Mathematics, 112*(1), 31-44.

Sallee, M.W. & Flood, J.T. (2012). Using qualitative research to bridge research, policy, and practice. *Theory into practice, 51*(2), 137-144.

Sinay, E., Jaipal-Jamani, K., Nahornick, A. & Douglin, M. (2016). *STEM teaching and learning in the Toronto District School Board: towards a strong theoretical foundation and scaling up from initial implementation of the K-12 STEM strategy. Research Series I* Research Report. (15/16), 16.

Suchman, E. L. (2014). Changing the academic culture to improve undergraduate STEM education. *Trends in Microbiology*. 22(12), 657-659.

Swaid, S. I. (2015). Bringing computational thinking to STEM education. *Procedia Manufacturing*. 3, 3657-3662.

Thomas, B. Watters, J. J. (2015). Perspectives on Australian, Indian and Malaysian approaches to STEM education. *International Journal of Educational Development*. 45, 42-53.

Traphagen, K. & Traill, S. (2014). *How cross-sector collaborations are advancing STEM learning*. Noyce Foundation.

Vaidya, A. (2015). Using Project-and Theme-Based Learning to Encourage Creativity in Science. *Journal of College Science Teaching*, *45*(2), 48-53.

CHAPTER 5

Teaching and Assessing Creativity in STEAM Education

Areej ElSayary

ABSTRACT

Science, Technology, Engineering, and Mathematics STEM has become an international focus of paramount significance. Through educational reform, the United Arab Emirates UAE government has stated national strategic measures in the Vision 2021 to raise students' attainment in TIMSS & PISA standardized assessments as well as promotes STEM education. Furthermore, developing STEM talents within Emirati students is one of the main purposes of the Science, Technology, and Innovation STI Policy (2015). Adding art to STEM has a positive impact on students' attitudes, motivations and interests that leads to developing creativity skills (Yakman, 2007). The purpose of this chapter is investigating the factors that affect teaching and assessing students' creativity. A mixed method design is used to answer the research questions. The study was conducted in a private school in the UAE. The participants are science, technology, language art, and mathematics teachers (n=30). The results of the study emphasized that motivation, cognition, and metacognition set as factors affecting students' creativity in STEAM classes. A balance between formative and summative assessment should be considered to shift the focus from raising students' attainment in standardized assessment to developing their creativity skills.

INTRODUCTION

There is a distinct gap between the way students learn in the classrooms and the way they are assessed. This is because most teachers do not use the depth of knowledge required to focus on students' cognitive levels. The standardized assessments of math and science (such as TIMSS and PISA) are designed according to the cognitive domains: knowledge, application and reasoning. In addition, the questions included in the reasoning domain is to assess the students' skills in dealing with the real-life applications and performance tasks where the students should, reason, reflect, explain, and find the solutions of the problems. However, the learning practices don't match the students' assessments where teachers feel the tension between developing students' creativity and between preparing them to perform well in the fact-based assessments (Beghetto, 2015). Students experience the skills of each subject solely when learning separate subjects. In other words, students are not able to transfer what they have learned in different situations. However, in the STEAM class, the students experience the essence of the skills of all subjects that are intertwined together in order to produce new and unique ideas. The STEAM program has been implemented in a private school in Dubai, United Arab Emirates from grade 1 – 8 which is used as a case study for this chapter. The purpose of this study is to examine teachers' perceptions about the factors that affect teaching and assessing creativity and to recommend ways to fill the gap between students' learning and the way they are assessed. The factors that affect creativity: motivation, cognitive process (convergent and divergent) and metacognitive process (Tan et al., 2014; Beghetto & Kaufman, 2010; Smith & Smith, 2010) will be used as a conceptual framework that guided this study using a mixed method design with multiple tools. A questionnaire with open- and close-ended items is used to measure teachers' perceptions about the factors that affect STEAM creativity. An observation is conducted in a duration of three weeks to explain and explore how STEAM education fosters students' creativity.

The following questions are used to fulfill the aim of the study:

> In what ways are motivation, cognitive, and metacognitive set as factors that affect teaching and assessing creativity?
>
> 1) What are the teachers' perceptions about the factors that affect teaching and assessing creativity?
> 2) To what extent does STEAM education fosters student creativity?

According to previous study of Sternberg (2015) that motivation, cognitive and metacognitive process foster creativity, the hypothesis of teachers' perceptions is that they believe that motivation, cognitive, and metacognitive set as factors affect creativity. However, they will differ in teaching creativity based on the subject taught. The hypothesis of the second question of the study that the STEAM education is fostering students' creativity due to the use of the cognitive and metacognitive process that increase their intrinsic motivation (Corply, 2015).

The STEAM program has been implemented in a private school in Dubai, United Arab Emirates from grade 1 – 8 which is used as a case study for this research. The purpose of this study is to examine teachers' perceptions about the factors that affect teaching and assessing creativity and to recommend ways to fill the gap between students' learning and the way they are assessed. The factors that affect creativity: motivation, cognitive process (convergent and divergent) and metacognitive process (Tan et al., 2014; Beghetto & Kaufman, 2010; Smith & Smith, 2010) will be used as a conceptual framework that guided this study using a mixed method design with multiple tools. A questionnaire with open- and closed-ended items is used to measure teachers' perceptions about the factors that affect students' creativity. An observation is conducted in a duration of three weeks to explain and explore how the STEAM education fosters students' creativity.

Factors that affect developing creativity such as: motivation, cognition, and metacognition (Runco, 1987; Sternberg, 1985; Tardif & Sternberg, 1988) are used as a conceptual framework to guide this study as presented in the figure below.

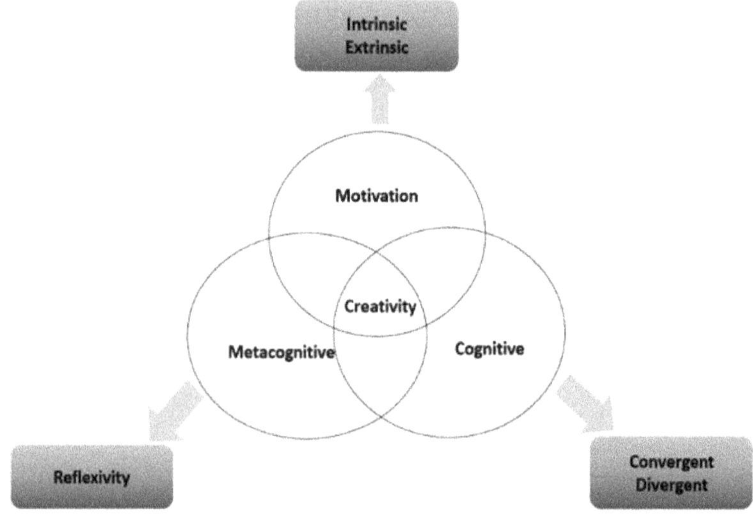

Figure 1: the conceptual framework used to guide this study (Runco, 1987; Sternberg, 1985; Tardif & Sternberg, 1988).

The STEAM education is the fusion of the disciplines; science, technology, engineering, art and mathematics which is considered to be an essential paradigm for creative teaching and learning (Corpely, 2015). In STEAM classes students develop their cognitive and metacognitive thinking and are intrinsically motivated to finish their tasks. Furthermore, adding "A" to STEM can enhance students' creativity and has a positive impact on students' attitudes and interests (Yakman, 2007). The United Arab Emirates reinforce the importance of STEM education. The ADEC (Abu Dhabi Education Council) aims to develop 21st Century skills through enhancing creative thinking skills among students (ALQubaisi, 2014).

Literature Review

Creativity has been defined as the interaction between the field, domain and individuals (Csikszentmihalyi, 1999). Guliford (1950) stated that it is significantly important to teach students how to think in order to produce creative products. Kaufman and Beghetto (2009) proposed a framework of the 4C model (Mini-c, Little-c, Pro-c, and Big-C) of creativity that enable people to understand the scale used to measure creativity. The Little-c creativity is focusing on everyday activities such as the creative actions where the non-experts may participate in. People who scored high in the Torrance test considered to be in the Little-c even the students who learn new concepts or make a new metaphor also seen as Little-c (Kaufman and Beghetto, 2009). As a result, Kaufman and Beghetto (2009) designed a new category to inherit in the learning process called Mini-c. It focuses on the personal and developmental aspects of creativity and known as transformative learning (Runco, 1996; and Vygotsky, 1967). The Mini-C highlights the importance of innovative interpretations of experiences and actions made by learners where it is important indicators on how to assess, monitor, and develop creativity (Wang and Greenwood, 2013). This model of creativity is in alignment with the Vygotskian conception of cognitive and creative development as all learners use their working memory in organizing and transforming the input information by using the existing knowledge (Kaufman and Beghetto, 2009). Pro-c creativity is known as professional expertise as it represents a development progression in the little-c but not reached yet the Big-C. The Pro-c level of creativity is implied in any one who attained a professional experience in any creative area. The Pro-c model is consistent with the acquisition approach of creativity (Ericsson, 1996; Ericsson, Roring & Nandagopal, 2007). Finally, the Big-C model is known as eminent accomplishments. People who are considered to be in the Big-C creativity are winners of prestigious awards or being included in encyclopedia. Everyone starts by the mini-c of creativity and rare people jump to the pro-c. The second step of the mini-c is the little-c and from this level there are two transitions. The first is informal preparation to the pro-c level of creativity and the second end by the reflection. In the pro-c level there are also two paths, the first one when the people remain creative in their professional lives. The other path is the peak of creativity where people develop and fertile their creativity to reach to the Big-C level. The following chart shows the 4C model of creativity.

Mini-c creativity has been defined as product and process of learning, that shows a balance of novelty and assessment (Cachia & Punie, 2010). Creativity does not exist outside of a particular subject area, but it is shaped or defined partially by the subject area (Blamires & Peterson, 2014). It has different forms from one subject to the other based on the skills required to master the subject knowledge and innovate a unique idea. However, the integration between disciplines allows to easily connect information to produce a meaningful product especially by experiencing the flavor of the skills used to master the subjects. It is the shift from the mini-c level to the little-c level of creativity. Further shift to the pro-c level can be reached when dealing with more specific areas of domains, projects or problems.

STEAM Education and Creativity

The aim of STEM education is to prepare an innovative and creative generation that focuses on the technical skills. Adding "A" to STEM is sparking the interchange between convergent and divergent thinking (Yakman, 2007). According to Yakman (2007) who has announced a framework used for teaching integrated subjects, stated that the art in STEAM is considered to be design art, language, sociology, philosophy, psychology, and history.

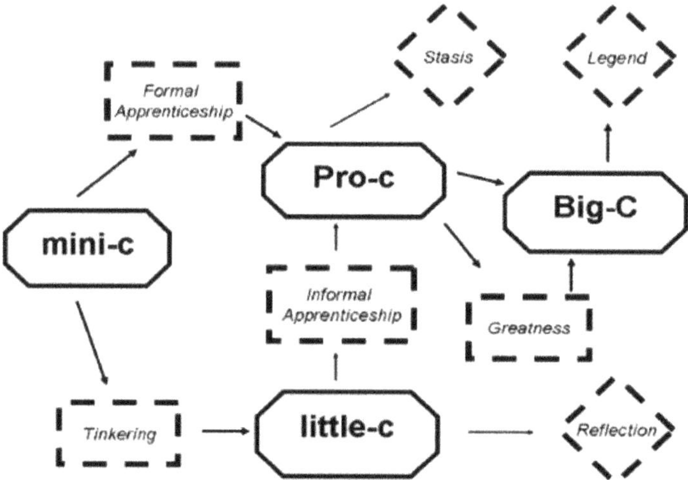

Figure 2: the 4C model of creativity (Kaufman and Beghetto, 2009).

However, the focus in this study is on the language art. Corply (2015) mentioned three elements that enhance creativity. First, students should have the opportunity to be engaged in creativity through learning integrated courses, problems or projects. Second, the importance of the positive encouragement to students who were engaged in creative tasks. Finally, students should be rewarded for completing and producing creative products. Furthermore, Sternberg (2007) stated twelve strategies that used to drive the habit of creativity. They are essential to develop students' creativity however, there are still many areas required the convergence approach. These strategies imply in involving students in open-ended projects where they need to redefine problems and make good choices. Students should be encouraged to ask questions and analyze assumptions and not only to accept problems given to them. Students need to be taught how to generate ideas and to have a team-spirit for persuading others about their ideas and think how to use the best idea and justify it in their group activities. As a result, extrinsic and intrinsic motivation are important factors in students' learning. Students need to be users of information, for example, try to find connection between a concept in biology and mechanical engineering when creating hinges and relating this to parts of the body. Challenging students in a task given to find obstacles and have the opportunity to fail and try again. They need to learn how to assess risk and judge whether this risk is acceptable or no. Allowing students to deal with haziness and think independently by giving them ill-structured problems instead of well-structured steps of project or problem. Build students' self-efficacy by requiring creativity as an assessable component of project work. Another strategy is by helping students to find what excites them through a real- world projects so they will be able to find their desired field. The importance of pushing students to the extent of their ability within their comfort zone allowing flexibility in assessments where each student will be assessed according to his or her limit. Finally, STEAM educators need to role model creativity.

Factors that Affect Creativity

Many psychologists and educators stated that the creative thinking improves students' motivation, metacognition, and interpersonal and intrapersonal skills in addition to the ability to write creatively, solve problems, and interpret scientific process (Beghetto & Kaufman, 2010; Plucker, et al., 2004; Smith & Smith, 2010; Torrance, 1959). Recently, policymakers and stakeholders are increasingly paying attention to the students' scores in the standardized assessments with ignorance on teaching students' creative thinking (Beghetto & Kaufman, 2010). The main objective is to think constructively by teaching students how to think as the success of this leads to the creative products of students' learning (Guliford, 1950). Torrance (1959) indicated that creative thinking enhances personality development, information acquisition, and success in future career. Vygotsky (1967) described creative thinking as exercise of imagination which is essential for students' future. It is interestingly important to note that the working memory has an essential role for the rehearsal and practice of the cognitive and creative thinking in order to transfer information to the long-term memory otherwise it will stay in short-term memory and causes loss of information (Long, 2011). Ofsted (2010) reported that students' motivation, progress and attainment have been improved by creative approaches to learning through allowing students to question, explore, challenge ideas, reflect and evaluate their learning.

The intrinsic and extrinsic motivation are two important types of motivation. Relying on one type of them may result of not completing the task. The intrinsic motivation is derived by the learning goal where the excitement and enjoyment of learning occur especially in the unpleasant or difficult task (Long, 2011). The extrinsic motivation is derived by the performance goal where the target here is to get a perfect product or reward from learning (Kaufman et al., 2015). A result from a dominant research indicated that the intrinsically motivated learners are driven by the curiosity, interest and desire to learn (Deci & Ryan, 2000). A positive relationship between the intrinsic (learning oriented) and extrinsic (performance prove) has been found (Gong, Huang & Farh, 2009; Hirst, Van Knippenberg & Zhou, 2009; To et al., 2012). Kaufman et al. (2015) stated that the extrinsic motivation may not reduce the creativity as suspected however there is a relationship between extrinsic and intrinsic motivation.

Creativity is considered to be a cognitive ability that should be developed across the lifespan. The learning through multidisciplinary encourage the convergent thinking. Sternberg (2006) stated that creativity requires the interconnection of knowledge, cognitive abilities, ways of thinking, personality and motivation. It starts by the knowledge that is considered to be input information where the working memory takes the role to influence the ability to think divergently (think about many solutions) and convergent (focus on one way) while solving problems (Long, 2011). According to Guliford (1959), convergent thinking is to come up with a single answer of well-structured problem. However, creativity foster the divergent thinking that focus on innovative ways of thinking. Art inspire divergent thinking among different disciplines by shifting students' thinking from convergent thinking.

It is significantly important to provide students the opportunity to reflect their learning process before interpreting their own views (Wang & Greenwood, 2013). One of the things that helped educators to nurture students' creativity is through listening to their point of views. Types of formative assessment is considered to foster students' metacognitive skills where they are able to evaluate their work, reflect, write reports, maintain portfolios, and make presentations (Tan et al., 2014). Infusing creative thinking into science, technology, engineering, art and mathematics is not only enhancing students' creativity but also their academic achievements.

The advanced TIMSS assessments (Mullis & Martin, 2015) provides specific information in preparing students to pursue careers of STEM fields as well as creating a reference point to ensure the quality of students' learning. Earl (2013) suggested a balance of three types of assessments: assessment of, for and as learning in a pyramid shape. The assessment of learning (summative and standardized assessments) is considered to be the least and at the top. The assessment as learning is the most and at the bottom where the assessment for learning is in the middle. In order to foster students' creativity, there should be a balance in assessing students' learning in terms of process and products, unexpected outcomes, subject knowledge, authentic tasks, standard tests (Cheng, 2015). Assessment for learning (formative assessment) is essential to successful of teaching and learning creativity where questioning, reflection and evaluation take place. The Assessment Reform Group (2002) proposed that assessment for learning is characterized by cyclical process where teachers gather data about students' information and skills through observing, questioning, monitoring their work, and gathering feedback. This gives indication to teachers about their teaching practices and for students to improve their work. Black and William (2006) suggested that open questioning and dialogue, feedback, and peer and self-assessment are forms of formative assessments which inherit in the cognitive process. Furthermore, students need to know their goals and how to judge their quality for self-assessment to be successful (Cheng, 2015) which is the metacognitive process.

Research Method

The study implemented over a period of three weeks in a private school in United Arab Emirates, Dubai. The study highlights the gap between teaching strategies and assessments. There are two paths used in this study. First research question focuses on the teachers' perceptions which is measured using the questionnaires in order to explain and explore their perceptions about teaching and assessing creativity. Second question focuses on fostering students' creativity using observation tool with a rubric and field notes that are conducted for the science, technology, math, English language, and STEAM classes.

A mixed method is implemented to address the research question of the study. The type of mixed method used is concurrent transformative method with the features of the embedded design (Creswell, 2009). Both data were collected concurrently however the big status is for the quantitative data where the qualitative data is nested and merged within it (Creswell, 2009). Morse (1991) noted that the qualitative data is nested in the quantitative data in order to describe aspects of quantitative data that cannot be quantified. The results of both data are integrated.

The population is the large group to which the results are generalized (Johnson & Christensen, 2012). The participants of the study are grade 1 – 8 teachers (N = 45). The characteristic of the population is that all the teachers are teaching science, technology, language art, or mathematics. However, the purposive sample is selected from the population because the main aim is to select the teachers who taught STEAM education, projects or cross-curricular link in their teaching strategies. As a result, the sample selected is n = 30.

Two instruments have been used; Teachers' questionnaire and observation to fulfil the research questions of the study. The teachers' questionnaire is designed to address the first question of the study: What are the teachers' perceptions about the factors that affect teaching and assessing creativity? The questionnaire started by demographic information. It is categorized according to the factors that affect creativity: motivation, cognitive (convergent and divergent) and metacognitive. According to Johnson and Christensen (2012) the questionnaire type is called intra-method mixing questionnaire where each category consists of closed- and open-ended questionnaire. the responses of the closed-ended items with rating scale. The first category is the motivation rated according to Likert-scale: strongly disagree, disagree, uncertain, agree, strongly agree. The cognitive and metacognitive categories are measured based on 5-point rating scale: very often, often, sometimes, seldom, and never. The second part of each category is the open-ended questionnaire to allow further clarification of the teachers' perceptions.

The observation tool is conducted for confirmatory and exploratory purposes (Johnson & Christensen, 2012) to measure how STEAM education foster students' creativity. It consists of rubric based on the categories of the factors that affect creativity to collect data quantitatively which is analyzed into frequencies and percentages. In addition, field notes are used to describe the results that cannot be quantified. The role of observer in this study is a participant-as-observer. The participant-as-observer is one of the useful styles of observation as the researcher is allowed to take a mix of insider and outsider roles (Johnson & Christensen, 2012). The observation is conducted in a duration of three weeks.

The pragmatism philosophy is reinforcing the importance of combining and integrating between qualitative and quantitative data (Johnson & Christensen, 2012). The duration of the study is three weeks where the teachers received the questionnaire at the beginning of the three weeks and collected after two weeks of the study. Teachers' permissions were taken prior the study for ethical consideration and all data have been kept confidentially. The observation is conducted during the three weeks of the study. The data collected quantitatively and qualitatively were merged in the light of the three factors that affect creativity: motivation, cognitive, and metacognitive dimensions.

Results

The results represented of the two different tools are merged and integrated according to the conceptual framework used.

Results of teachers' questionnaire is interpreted into three different categories: motivation, cognitive, and metacognitive.

Motivation

This section is consisted of closed and open-ended-questions. The responses of the closed-ended questions are based on five points Likert-scale. The highest mean is 4.73 shows that students like praising their efforts in any task. The lowest mean is 3.47 where the students get worried if they cannot solve problems. The open-ended question in this section is asking teachers to explain how they guide and facilitate students' learning in order to be self-directed learners. The teachers' responses differ however, most teachers responded that the students learn from each other and learn from their mistakes when they are working in small groups. Students feel responsible when they take the role of teaching in flipped classrooms. Many teachers stated that they are guiding students to prepare lessons by giving them the objectives and guidelines where they prepare and present the lesson. In addition, giving students different tasks in the same project within the team make them responsible, increase their interests and work collaboratively to produce the product. Using technology in the learning process increase students' ability to be motivated and self-directed learners.

Cognitive Thinking

This section has been divided into two parts the first one is the convergent thinking and second is the divergent thinking. Each of them consists of closed- and open-ended questions. The closed-ended items have a five-rating scale. The highest mean in the convergent thinking is 3.93 shows that students ask questions and then investigate their own questions. The lowest mean is 3 where students rarely set up data table when they do activities. The highest mean in the divergent thinking is 3.9 shows how often students do concept maps, mind maps, or webs and doing tasks that require generating multiple ideas and find possible solutions that are specific in nature. The lowest mean is 2.2 shows that students do tasks that require designing projects, activities, or experiments. Furthermore, they discuss "why" if the task doesn't appear as it was predicted. The open-ended questions in cognitive thinking are asking teachers to explain in what ways do experiments, problem solving, or projects are related to creativity, what type of problems used to develop students' creativity, and what can make each subject creative? The teachers' responses differ according to their specializations. The English teachers stated that when giving students the opportunity to analyze characters through the development of actions and events in a novel or poem, these help them to create their own stories and poems and, in its turn, will lead to analyze problems and consider the constraints. The math teachers explain how the problem-solving help students to understand the concepts easily and understand the meanings behind learning difficult concepts. Science teachers explained that the problems they are facing give students the opportunities for discovering and learning new concepts. The scientific inquiry approach not only help to reach their goals but also help to go beyond learning and extend their information to new concepts. The STEAM teachers described students' high interactions in solving the problems together while planning for their projects and in generating ideas for solving problems and focus again on one solution. In STEAM classes, students generate new and unique ideas when they are working on ill-structured problems.

Metacognitive Thinking

This section included closed- and open-ended questions. In this section the teachers answered questions that measure their perception about students' reflection and improvement of their work. The highest mean is 3.73 where students maintaining a portfolio about their work. The lowest mean is 2.4 where students rarely do long-term products or projects. The open-ended question is asking teachers how the discussions and reflections incorporated in concepts are higher-order thinking? All the teachers stated that students in these processes can freely express their opinions, think of more proof to sustain them, raise their confidence, and help them in being capable to create more different notions. Students who think, judge, and reflect on their work can think deeply and improve their way of thinking. Students have the chance to express, generate new and unique ideas and connect them to real-life and to other disciplines.

Findings of Observation

The observation is conducted to confirm and explain the results of the questionnaire. In addition, it is used for exploratory purposes in order to identify any unexpected results that might affect creativity. The observation has been conducted in the duration of three weeks. The classes observed are science, technology, language art, mathematics and STEAM. The observation has been categorized according to the conceptual framework of the study. The results analyzed based on the rubric developed from the (KHDA) Knowledge and Human Development Authority framework (2015). In addition, there is field notes used by the observer to cite any unexpected results or gaps that might occur.

Motivation Dimension

This dimension includes two items to measure: students' interactions, collaboration and communication skills; and attitudes to learning. The results differ according to the subject observed. The highest percentage is in STEAM classes with 90% in first item and 93% in second item. However, the lowest score is in the math classes where the first item is 30% and the second item is 26.6%. The second highest percentage is in the technology classes where the first item is 70% and the second item is 73%. It is noticeable the difference between the science and English classes. The students' interactions, collaboration, and communication skills in English is 53% while in science is 60%. The results of attitudes to learning in English is 66% while in science is 53%. The figure below shows the difference between the subjects in motivation dimension.

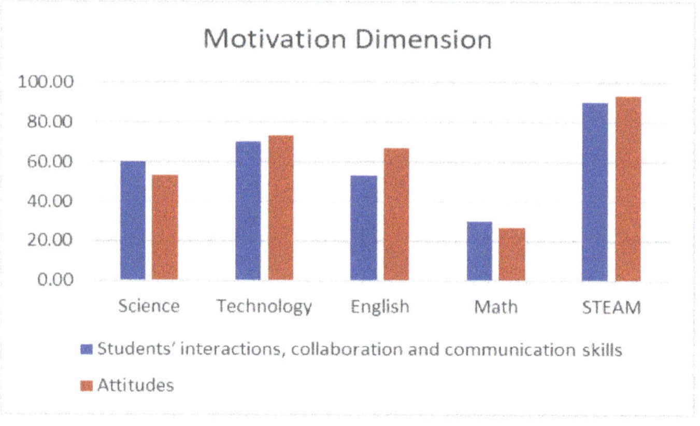

Figure 3: The percentage of the classes observed per subjects in motivation.

Cognitive Dimension

The cognitive dimension also includes two items to measure. The first item is enquiry, research, critical thinking and use of technologies and the second item is application of learning to the world and making connections between areas of learning. Again, the highest percentage is in the STEAM classes where the first and second item in this dimension are 96%. The lowest score is in the math classes where the results are 46% of the first item and 30% of the second item. The second highest score is in the technology which is 76% in the first item and 70% in the second item. Surprisingly that science and English differ in their percentage where the first item in science is 53% and the English is 46%. However, the second item in science is 50% while in English is 56%. The figure below shows the difference between the subjects' percentages.

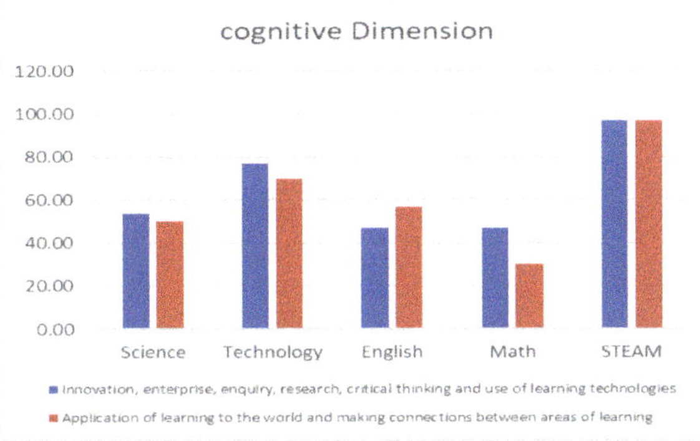

Figure 4: The percentage of the classes observed per subjects in cognition.

Metacognitive Dimension

The metacognitive dimension has two items: students' engagement in and responsibility for their own learning, and interactions that includes the use of questioning and dialogue. The highest results occurred in this category is the STEAM classes where the two items of metacognitive dimension are 96%.

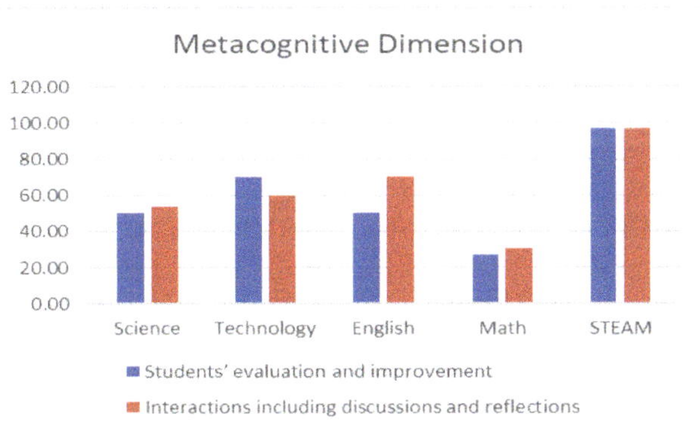

Figure 5: The percentage of the classes observed per subjects in metacognition.

The lowest result is also the math classes where the first item is 26% and the second item is 30%. The second highest percentage is in the technology classes where the first item is 70% and the second item is 60%. The science and English classes differ in their percentages per items where the first item in science and English are the same 50% while the second item in science is 53% and English is 70%. It has been noticeable in the observation that there is flipping between science and English classes in the three factors that affect creativity.

Findings and Discussion

This study set out with the aim of exploring and explaining the factors that affect teaching and assessing creativity, teachers' perceptions, and how STEAM education foster creativity in a private school in Dubai. The factors that affect creativity (motivation, cognitive, and metacognitive process) are used as a conceptual framework to guide the study. The data collected from the teachers' questionnaire and the observation is merged and integrated in the light of the conceptual framework.

The highest response of teachers' perceptions about motivation stated that students like praising their efforts in any task which implies in the role of extrinsic motivation. In addition to intrinsic motivation about how students enjoy doing experiments, activities or projects. This is compatible with the results that mentioned positive relationship between intrinsic and extrinsic motivation (Gong et al., 2009; Hirst et al., 2009; To et al., 2012). Regarding the observation, it has been mentioned that the highest percentage of students' motivation is in STEAM classes where students interact and collaborate effectively in a wide range of learning situations and communicate their learning to achieve goals. In addition, they showed very positive and responsible attitude, demonstrated self-reliance, and flourish in critical feedback. Technology came in the second rank of students' motivation after STEAM classes. Surprising results have been shown in science and English classes where there is flip between the observation results in science and English. In other words, students' interaction, collaboration, and communication was higher in science classes however their attitudes were higher in English classes than science. On contrary, students' show low motivation in math classes.

In this category there is two dimensions, convergent and divergent thinking. Regarding the convergent thinking, teachers responded that students ask their own questions and investigate them. They design their own activities, experiments, or projects. In addition, they make observations and write conclusions about what they have observed. Teachers' stated that students show low performance in setting up data tables which is compatible by the observation results that math classes have the lowest percentage in students' cognitive process. A study of Bolden et al. (2009) mentioned that teachers' have difficulty in encouraging and assessing students' creativity in math. In divergent thinking, teachers noted that most students do tasks that require generating ideas, do concept maps and mind maps. In addition, most of students explain, and provide further information where they can make connections with different areas. The low responses of teachers' perceptions that students do not complete tasks that require designing activities, experiments, or projects and they get worried if it did not appear to work as predicted. However, this doesn't appear in the observation of all classes. The highest percentage was in the STEAM classes where students used convergent and divergent thinking in the cognitive process as they ask questions, try to investigate them, design projects, define problems, generate ideas, and create models and prototypes. Sternberg (2007) emphasizes that the cognitive abilities and way of thinking are essential to foster creativity. The technology got the second rank in the cognitive process while in science and English there is again flip in the results. In science classes students show higher percentage than English in innovation, enterprise, enquiry, critical thinking, and use of technologies. However, in English classes the percentage was higher in making connections to the real-world and areas of learning. This proves what Evans (2013) noted that language art sparks the interplay between the convergent and divergent thinking.

The teachers' mentioned that students are able to express their opinions, think deeply in reflecting and improving their work. Wang and Greenwood (2013) emphasizes the importance of the reflection of students before putting their own opinions about their work. This has been shown clearly during the STEAM classes and some of the English classes. They are able to connect their ideas to the real-life and to other disciplines. The observation showed that the highest percentage of the metacognition is also in the STEAM classes. Surprising results were shown in technology and English classes where flipping between the results of the metacognition items occurred. English classes were higher in students' interactions when involved in discussions and reflections while technology classes were higher than English in students' self-evaluation and improvement. The science classes got a close percentage to the English and technology while math classes got the lowest percentage.

Conclusion and Recommendations

Motivation, cognition, and metacognition set as factors affecting creativity (Beghetto & Kaufman, 2010; Plucker et al., 2004). Creativity is not receiving attention in teacher education programs (Davies et al., 2004). Stakeholders, educators and teachers need to understand creativity, its value, the factors that affect it, and the reason behind including it in the curriculum rather than giving great attention to students' scores of the standardized assessments. It is essential to increase teachers' awareness of identifying creative thinking, attitudes and dispositions (Long & Plucker, 2015). This will lead to raising of students' scores in the standardized assessments that focus on the use of cognitive and metacognitive skills. Earl (2013) suggested a balance between the three types of assessments: assessment as, for, and of learning. The creative process focuses on the use of assessments for learning in increasing students' creativity through the use of cognitive and metacognitive process that are driven by motivation. Tan et al. (2014) emphasized the strong relation between the use of formative assessments in learning process that enhance students' creativity. Motivation, cognitive, and metacognitive process considered to be the shift from mini-c to little-c of creativity that lead to pro-c and in term prepare students for the Big-C (Kaufman & Beghetto, 2009). STEAM education is considered to foster students' creativity as students experience the flavor of skills acquired from all subjects to complete their projects. Surprising results were the flipping between science and English art in the motivation and cognition while in metacognition the flipping was between science and technology. The math classes showed low percentage in fostering students' creativity. This is compatible with the study of Bolden et al. (2009) which indicated that teachers found difficulty in teaching and assessing creativity in math subject. Furthermore, adding "A" to STEM flourish the cognitive process that increase students' creativity. Motivation, cognition, and metacognition implies in the twelve strategies of Sternberg (2007) that are used to drive the habit of creativity through STEAM classes. The benefit of STEAM education that it deepens students' understanding by integrating contents, broaden it by exposing them to STEAM contexts, and increase their interests toward STEAM fields (Roehrig et al., 2012). Further research should focus on the nature of each subject of STEAM education, the relation between subjects and how using the skills of all subjects benefit students in STEAM classes. In addition to exploring the effectiveness of STEAM on students' scores in assessments.

REFERENCES

Al Qubaisi, A (2014). Abu Dhabi Education Council [online].[Accessed 18 October 2014] Available at: https://www.adec.ac.ae/en/MediaCenter/News/Pages/AD-students-explore-spacecraft.aspx

Assessment Reform Group (2002). *Assessment for Learning: 10 principles research-based principles to guide classroom practice.* Assessment Reform Group.

Blamires, M. and Peterson, A. (2014). Can creativity be assessed? Towards an evidence-informed framework for assessing and planning progress in creativity. *Cambridge Journal of Education*, 44(2), 147-162.

Beghetto, R. A. (2015). Teaching creative thinking in K12 schools: lingering challenges and new opportunities, pp 201-211. In *the Routledge International Handbook of Research on Teaching Thinking.* Routledge International Handbooks. Routledge Taylor & Francis Group.

Beghetto, R. and Kaufman, J. (2010). *Nurturing creativity in the classroom.* Cambridge University Press.

Black, P. & Wiliam, D. (2015). Developing the theory of formative assessment. *Educational Assessment, Evaluation and Accountability*, [online] 21(5), 5-31. Available at: https://kclpure.kcl.ac.uk/portal/files/9119063/Black2009_Developing_the_theory_of_formative_assessment.pdf [Accessed 1 Dec. 2015].

Bolden, D., Harries, T. and Newton, D. (2009). Pre-service primary teachers' conceptions of creativity in mathematics. *Educational Studies in Mathematics*, 73(2), 143-157.

Cachia, R. and Ferrari, A. (2010). *Creativity in schools.* Luxembourg: Publications Office.

Cheng, V. (2015). Consensual Assessment of Creativity in Teaching Design by Supportive Peers-Its Validity, Practicality, and Benefit. *J Creat Behav*, p.n/a-n/a.

Corpley, D. (2015). *Teaching engineers to think creatively: barriers and challenges in STEM disciplines,* pp. 402-410. In *the Routledge International Handbook of Research on Teaching Thinking.* Routledge International Handbooks. Routledge Taylor & Francis Group.

Creswell, J. (2009). *Research design.* Sage Publications.

Csikszentmihalyi, M. (1999) *Implications of a Systems Perspective for the Study of Creativity*, in R. J. Sternberg (ed.) *Handbook of Creativity.* Cambridge University Press. 313-335.

Davies, D., Howe, A. & McMahon, K. (2004). Challenging primary trainees' views of creativity in the curriculum through a school-based directed task. *Science Teacher Education*, 41, 2-3.

Deci, E. L., & Ryan, R. M. (2000). The 'what' and 'why' of goal pursuits: Human needs and the self-determination of behaviour. Psychological Inquiry, 11, 227-268.

Earl, L. (2013). *Assessment as Learning: Using Classroom assessment to maximize student learning*. 2nd ed. Corwin.

El Sayary, A. Forawi, S. & Mansour, N. (2015). *Teaching thinking in STEM subjects: STEM education and problem-based learning*, pp. 357- 368, in: *The Routledge International Handbook of Research on Teaching Thinking*. Routledge International Handbooks. Routledge Taylor & Francis Group.

Ericsson, K. (1996). *The road to excellence*. Lawrence Erlbaum Associates.

Ericsson, K., Roring, R. & Nandagopal, K. (2007). Giftedness and evidence for reproducibly superior performance: an account based on the expert performance framework. *High Ability Studies*, 18(1), 3-56.

Evans, N (2013). *Language Diversity as a Resource for Understanding Cultural Evolution*. In Peter J. Richerson and Morten H. Christiansen (ed.), *Cultural Evolution: Society, Technology, Language, and Religion*, MIT Press, Cambridge, pp. 233-268.

Gong, Y., Huang, J. & Farh, J. (2009). Employee Learning Orientation, Transformational Leadership, and Employee Creativity: The Mediating Role of Employee Creative Self-Efficacy. *Academy of Management Journal, 52*(4), 765-778.

Guilford, J. P. (1950). Creativity. *American Psychologist, 5*, 444-454.

Guilford, J.P. (1959). *Traits of creativity in Creativity and its Cultivation*. 142-161. Harper and Row

Hirst, G., Van Knippenberg, D. & Zhou, J. (2009). A Cross-Level Perspective on Employee Creativity: Goal Orientation, Team Learning Behavior, and Individual Creativity. *Academy of Management Journal*, 52(2), 280-293.

Johnson, B. & Christensen, L. (2012). Educational research. Sage Publications.

Kaufman, J. & Beghetto, R. (2009). Beyond big and little: The four-c model of creativity. *Review of General Psychology*, 13(1), pp.1-12.

Kaufman, J., Palmon, R. & Royston, R. (2015). *What we want impacts how we create: creativity, motivation and goals*, pp. 181- 190, in: *The Routledge International Handbook of Research on Teaching Thinking*. Routledge International Handbooks. Routledge Taylor & Francis Group.

Long, M. Wood, C. Littleton, K. Passenger T. & Sheehy K. (2011). *The Psychology of Education*. Routledge Taylor & Francis Group.

Long, H. & Plucker, J. (2015). *Assessing creative thinking: practical applications*, pp. 315- 329, in: *The Routledge International Handbook of Research on Teaching Thinking*. Routledge International Handbooks. Routledge Taylor & Francis Group.

Morse, J. (1991). *Qualitative nursing research*. Sage Publications.

Mullis, I. and Martin, M. (2015). *Timss 2015 Assessment Frameworks*.

Ofsted (2010). *Learning: Creative Approaches that Raise Standards*. Ofsted.

Plucker, J. A., & Makel, M. C. (2010). Assessment of creativity. In J. C. Kaufman & R. J. Sternberg (Eds.), Cambridge handbook of creativity (pp. 48–73). Cambridge University Press.

Roehrig, G., Michlin, M., Schmitt, L., MacNabb, C. & Dubinsky, J. (2012). Teaching Neuroscience to Science Teachers: Facilitating the Translation of Inquiry-Based Teaching Instruction to the Classroom. *Cell Biology Education, 11*(4), 413-424.

Runco, M. (1987). Interrater agreement on a socially valid measure of students' creativity. *Psychological Reports, 61*(3), 1009-1010.

Runco, M. A. (1996). *Personal creativity: Definition and developmental issues.* New Directions for Child Development, 72, 3–30

Smith, J. K., & Smith, L. F. (2010). Educational creativity. In J. C. Kaufman & R. J. Sternberg (Eds.), Cambridge handbook of creativity (pp. 250–264). Cambridge University Press

Sternberg, R. (1985). Implicit theories of intelligence, creativity, and wisdom. *Journal of Personality and Social Psychology, 49*(3), 607-627.

Sternberg, R. (2006). The Nature of Creativity. *Creativity Research Journal*, 18(1), 87-98.

Sternberg, R. (2007). Creativity as a habit. In A-G. Tan ed. Creativity: A Handbook for Teachers (pp. 3-25). World Scientific.

Sternberg, R. (2015). Teaching for creativity: The sounds of silence. *Psychology of Aesthetics, Creativity, and the Arts, 9*(2), pp.115-117.

Tan, J., Caleon, I., Jonathan, C. & Koh, E. (2014). a dialogic framework for assessing collective creativity in computer-supported collaborative problem-solving task. *Research and Practice in Technology Enhanced Learning, 9*(3), 411-437.

Tardif, T. Z., & Sternberg, R. J. (1988). *What do we know about creativity?* In R. J. Sternberg (Ed.), *The nature of creativity* (pp. 429-440). Cambridge University Press.

To, M., Fisher, C., Ashkanasy, N. a& Rowe, P. (2012). Within-person relationships between mood and creativity. *Journal of Applied Psychology, 97*(3), 599-612.

Torrance, E. (1962). *Guiding creative talent.* Prentice-Hall.

Vygotsky, L. (1967). Play and Its Role in the Mental Development of the Child. *Journal of Russian and East European Psychology, 5*(3), 6-18.

Wang, B. & Greenwood, K. (2013). Chinese students' perceptions of their creativity and their perceptions of Western students' creativity. *Educational Psychology, 33*(5), 628-643.

Yakman, G, (2007). *STΣ@M Education: an overview of creating a model of integrative education. Pupils Attitudes Towards Technology*. 2008 Annual Proceedings.

CHAPTER 6

STEAM Education Implementation Roadmap

Noura Assaf

ABSTRACT

The purpose of this chapter is to present a brief overview of the implementation of STEAM education in schools in developed countries such as China, Australia, United Kingdom and United States of America and to provide a roadmap of its implementation in the context of the United Arab Emirates. The research study in this chapter adopts a qualitative approach whereby purpose sampling of secondary data is collected, compiled and analyzed. Themes are generated after coding the content: implementation of STEAM, challenges related to STEAM application and implementation and requirements for success implementation. For the purpose of ensuring proper integration of STEAM in UAE educational system, a roadmap is proposed with policy drafting recommendations, such as curriculum reform, technology integration, teacher professional development and financial funds.

INTRODUCTION

Globally, national education programs are emphasizing the understanding of knowledge and practices related to mathematics and science, as well as that of engineering and technology (Kelley & Knowles, 2016). The Next Generation Science Standards (NGSS) implemented in the United States includes engineering practices and design as one of the core and primary element of science education (Kang, 2019). Keeping with the practices of the world, the United Kingdom (UK) has also developed an educational policy agenda for the promotion of integrating Science, Technology, Engineering, and Mathematics (STEM) practices, both inside and outside of schools (Kang, 2019). Henriksen (2017) cited the findings of the Institute for the Future who has estimated that the jobs which 85% of today's K-12 learners will be doing by 2030 have not been identified yet. Moreover, he highlights that this finding has wide implications, especially towards the demand of a workforce which is both adequately prepared to effectively respond to real-life problems. Dell'Erba

(2019) added that these findings also led academic scholars and practitioners to argue the inclusion of 'arts' in the STEM policy.

Inclusion of 'arts' in the current STEM policy can have wide implications for student achievement and teaching practices; it can encourage students to creatively solve real-life problems while building upon the existing STEM approaches (Dell'Erba, 2019).Considering the constantly changing demands of the workforce, policymakers in the developed countries of world are increasingly emphasizing on including the aspect of arts in the STEM policy to develop and implement Science, Technology, Engineering, Arts and Mathematics/Applied Mathematics (STEAM) programs (Dell'Erba, 2019; Henriksen, 2017). For the purposes of this research, STEAM education is characterized as an approach to teaching in which the students will have the potential of demonstrating critical thinking, innovation, and problem-solving skills. The integration of arts is used as an instructional approach in STEAM education, and for promoting inquiry-based and experiential learning (Dell'Erba, 2019). Finally, Shaer, Shibl, and Zakzak (2019) highlighted that the Government of the United Arab Emirates (UAE) must develop policies which will efficiently allocate both human and financial resources for the implementation of STEAM education at both the local and national level. Evidence based reforms on STEAM can enable improvements in the education system of the country. A report published by the Education Commission of States argued that investments in the implementation of STEM Education are more likely to fail if they are not supported by adequate and reliable funding on an annual basis (Goetz & Zinth, 2016). Thus, changes in education policy from STEM to STEAM will bring reforms in current procedure, which might positively affect the UAE development as well.

The sustainable development of UAE is currently hindered by three issues: the pressure of economic transformation, lack of high-level talents in the workforce, and difficulties in reforming education at a national level (Shaer, Shibl, & Zakzak, 2019). Considering these issues, there exist a high-level need of innovative and practical talents in the workforce with enhanced skills and knowledge; however, the current educational policies and standards are not aligned with this goal (Shaer, Shibl & Zakzak, 2019). These problems can be solved through developing and implementing a new model of "talent cultivation" which can synchronize with the development needs of UAE. Thus, based on this focus on promotion (Shaer, Shibl, & Zakzak, 2019), a further step is urgently required for the innovation of education and advancing current teaching practices. Guo, Xu, and Wang (2018) highlighted that STEAM education can be of significant assistance for the highlighted focus as this policy provided new tools and strategies. Moreover, they have shown that over the years STEAM education has been widely adopted due to its advanced teaching methods, a teaching system of multidisciplinary integration, and a unique curriculum concept. Due to STEAM education's feature of multidisciplinary fusion, it cultivates the engineering literacy and innovative capabilities of students (Guo, Xu & Wang, 2018). Further, the implementation of STEAM education standards and policies at the national level is also the vision of Mohammed Bin Rashid Al Maktoum's Global Initiatives (Shaer,

Zakzak, & Shibl, 2019). Therefore, this chapter is based on comprehensive review of related journal studies, texts, reforms of STEAM materials whereby it explores a roadmap for STEAM education implementation in the education sector in the UAE, specifically addressing this roadmap to be adopted in elementary private and public schools throughout all Emirates. Although this context is associated with all education sectors (schools and university), the implementation of STEAM will lead to powering students with necessary skills and creativity that will ultimately lead to educational and economic growth.

Purpose of the Study

The aim of this research is to review the implementation of STEAM education in schools in developed countries and provide a change in roadmap of its implementation in the UAE elementary schools, both public and private for which STEAM education standards and policies proposal draft might be based on. In order to address the main purpose of this chapter, the following research question will be addressed:

- What are the major features of STEAM education implementation roadmap for which a STEAM draft policy might be based on in UAE?

To achieve this aim, the research will complete the following objectives:

- To understand STEAM education initiatives implemented in developed countries including the UK, USA, China and Australia.
- To explore major features of a STEAM education implementation roadmap for which a STEAM draft policy might be based on in UAE.
- To explore the in-progress implementation of STEAM education in UAE and compare with the framework on which implementation initiatives was carried out in developed countries.

Theoretical Framework and Literature Review

This section of the research shall focus on establishing the link between variables of the research objective. Factors affecting STEAM implementation shall be evaluated to determine the link of each on education system. The theoretical framework, thus, consists of implementation models which translates policies into action. The theoretical framework of this research is grounded on the Behavior Change Wheel (BCW). Michie, Van Stralen, and West (2011) define behavior change interventions as coordinated set of activities with the purpose and intention of changing specified behavioral patterns in any individual or group of individuals. The determination of the broad approach to be adopted is primarily the first step in the process of designing change interventions and then work is conducted on the specifics of the intervention(s). Once this step is completed, the specific components of the intervention are identified (Atkins, Michie, & West 2014; Michie, Van Stralen, & West, 2011).This literature review will, therefore, discuss the policies and their interventional components that are to be translated into action while remaining in the framework of the Behavior Change Wheel.

Theoretical Framework

Behavior Change Wheel

The "behavior change wheel" that arises from the transtheoretical model of the change admits the existence of four, five or six stages, in the form of a (circular) wheel. Thus, people go through the different stages of the wheel, as if they were sliding through them (Atkins, Michie, & West 2014). The circular shape of the wheel reflects a reality that a change is not linear: in any process of change, the person revolves around the process several times before reaching a stable change. First, these stages of change are usually represented by a wheel to symbolize the fact that the person "spins" several times around the process before achieving a stable change. On the other hand, the model considers relapses as a normal event in the process of change, and subjects are often even told that each relapse "takes another step toward improvement." This is not intended to encourage relapse, but to motivate the subject to continue with his process of change (Atkins, Michie, & West, 2014).

However, before focusing on behavioral change, the psychologist must face an even more complex challenge: motivational change (Atkins, Michie, & West, 2014). This step must always precede the intervention itself because it is easier for those who are willing to change. It is important to know in what motivational state the patient is to know if it is feasible to ask for a change in behavior or if he is still in a premature phase.

The behavior change wheel model can be evidenced from Figure 1 (Hendriks et al., 2013). Exploring the implication of this model on the selected research problem, it can be demonstrated that individuals can show a change in their behavior as per three important elements, which can include opportunity, capability and motivation.

The Transdisciplinary Essence of Bibliological-Informative Theory

The complexity of the world in which we live forces us to value interconnected phenomena. Current physical, biological, social and psychological situations do not act but interact with each other. The description of the world and current phenomena demands a new way of valuing it from a broader perspective, with a new way of thinking that demands finding a new paradigm capable of interpreting current reality.

This brings us to the transdisciplinary conception (Brier, 2014; Lacono, Vaidyanathan, & Vrieze 2015).

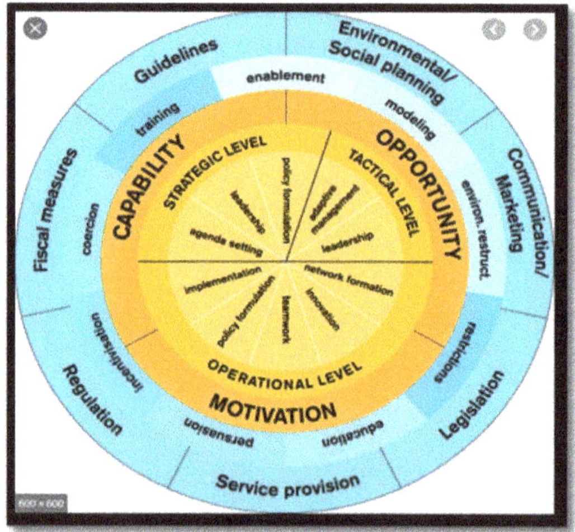

Figure 1: Behavior Change Wheel Model (Hendriks et al., 2013)

When making a brief characterization of the set of elements, it can be demonstrated that conditioned the appearance of the informative bibliological phenomenon, and we would have to start it from the history of human communication itself, that is, language and writing. Although universal culture does not escape these precepts and the transition from oral to written culture produced revolutions that influenced the behavior of man as a social being, today's transformations of technology show a nascent digital culture that leads to new approaches in the treatment of communications and information (Brier, 2014). All this generates that at present there is talk of "informative phenomenon" as a set of manifestations that have arisen and that characterize in a very peculiar way the incomparable world of current information (Lacono, Vaidyanathan, & Vrieze 2015). This is giving rise in the quantification of information, its storage and communication with others so that effective use of knowledge can be done. This theory has created a rise in technological transformations, which is applicable in a variety of fields, and this can also be applied in the education sector of the UAE.

Integrative Theory

Integrative or integrated theories are those that are constructed "from the combination of parts or ideas taken from a variety of existing explanations" (Chemers, 2014). These new theories are consistent with the new criminological approach that understands the criminal act as a multicausal event. This criminology theory shows that changes are arising in criminal offences of individuals with the changes evidenced in culture and social aspects of life.

The influence of social learning has the potential to affect the acts of individuals. Throughout its evolution, this theory integrates different aspects such as that of the delinquent subculture, inequality of opportunity, social learning, differential association, tension, control, labeling and rational choice (Borghi et al., 2013). This demonstrates the fact that social learning of individuals can affect their decision-making processes, their ideas as well as acts. Hence, social learning in educational systems of the UAE should be focused as the theory shows positive social learning can impacts positively. Integrative theory can be incorporated in theoretical framework of the study as it directs social learning attributes, which can cover up STEAM education policy.

Finally, the theoretical framework referenced the policy context for STEAM Education and the important policy areas on which the UAE Government should work to ensure the effective implementation of STEAM, at both public and private levels. There exists a lack of research on policy areas for the implementation of UAE and those which are available are rarely supported by empirical research.

Literature Review

Policy Context of STEAM Education

STEAM education is grounded on the concept of innovation where education is the key component for developing critical thinkers to lead the innovation economy (English, 2017). Clarke (2019) asserts that the perceived need for linking arts with science as one of the key workforce and education policy concerns is one of the primary reason policymakers are emphasizing STEAM more than STEM. STEM to STEAM is the term associated with this shift of focus with the primary purpose of workforce education to instill skills, creativity and flexibility in the nee workforce. This is achieved through arts integration and aligning it with the academic core already developed within the education system (Clarke, 2019). The author explored STEAM education within the context of research and highlighted that it is viewed as a key tool for the promotion of innovation between and within different disciplines including arts, sciences and humanities. Graduates' not possessing innovative and creative spirits are a commonly held view among policymakers and are believed to have an impact on economic development (Escude, Hooper, & Vossoughi, 2016). They added that this view has further exaggerated the shift to STEAM approach in many developed countries, especially including Japan, China and the USA. Adherents of STEAM education contend that this approach has the potential of fostering critical thinking, personal growth, creativity and team-working skills which can positively influence workforce opportunities to contribute effectively to societal needs while gaining employment (Clarke, 2019; English, 2017).

Policies for STEAM Education

Shaer, Shibl, and Zakzak (2019) identify the current state of STEAM Education in UAE private schools. The authors reported that though some private schools have implemented STEAM Education for early year education, its implementation is not guided by policies or standards. Subsequently, certain policy-focus areas have been recommended by these authors; however, they have not been viewed through a critical lens. Within the theoretical framework of this research, the key policies suggested in their report will be critically analyzed in the light of peer-reviewed literature for identifying the most high-priority areas on which the UAE Government should work. They have also discussed several policy-related challenges for the implementation of STEAM education in the UAE. These included the integration of curricula, capacity of teachers, allocation of resources especially financial, incorporation of technology, and balancing the requirements of the curricular with that of STEAM objectives. Shaer, Shibl, and Zakzak (2019) ended up their research by discussing several policy-related interventions for these identified challenges; however, this research explores the three most crucial areas: curricular integration, professional development of teachers (capacity of teachers), and allocation of resources.

Curricular Integration: Shaer, Shibl, and Zakzak (2019) recommended the implementation of an interdisciplinary STEAM guide and framework for providing the schools and educators tools for effectively integrate STEAM education with curricula, and to measure the success of integration. Dell'Erba (2019) provided certain policy considerations for STEAM Education including access, finance, and statewide coordination. The education department of South Carolina implemented an implementation guide and continuum for STEAM Education in liaison with relevant governmental educational authorities (South Carolina Education Department, 2014). The purpose of this framework was to achieve statewide consistency and to provide guidance for STEAM Education at all educational levels. Dell'Erba (2019) also cited the case study of the education department of Georgia Department which developed similar frameworks that also included reflection and self-assessment tools for the educators. These tools ensure that educators can measure the quality and progress of STEAM implementation or use these tools for the planning of quality implementation (Dell'Erba, 2019). The author further highlighted the importance of understanding the STEAM educational practices to achieve curricular integration in relation to STEAM education. These include: leveraging of concepts from one or more STEM frameworks for the creation of purposeful artwork, embedding 'intention' during the design process, focusing on outcomes that have an aesthetic and personal meaning, conducting open-exploration in the context of both art and science;, achieving communication about the process as well as the outcome, and iteration through several prototypes, drafts and/or models (Dell'Erba, 2019). Finally, Chu, Martin, and Park (2018) added that governments have to ensure that these core teaching practices of STEAM education are effectively followed to enable its effective implementation.

Allocation of STEAM Resources: Shaer, Zakzak, and Shibl (2019) highlighted that the UAE Government must develop policies which will efficiently allocate both human and financial resources for the implementation of STEAM education at both the local and national level. A report published by the Education Commission of States argued that investments in the implementation of STEM Education are more likely to fail if they are not supported by adequate and reliable funding on an annual basis (Goetz & Zinth, 2016). Clarke (2019) studied the case of the UK and Ireland and discussed the ways through which governments of both these countries have allocated significant funds for the development and implementation of STEAM Education in all schools. The Education Commission of States details the steps taken by the US government to implement STEM in Utah. The report highlighted that the most important step taken is the development of legislation to assure adequate and reliable funding on an annual basis (Goetz & Zinth, 2016).Though the case study of UTAH represents the implementation of STEM Education, the same policies can be used for the implementation of STEAM. The report also highlighted that the development and maintenance of a STEM Centre are not possible without legislative funding (Goetz & Zinth, 2016).

Professional Development of STEAM Teachers: Despite an increase in interest in STEAM education, teachers still find it challenging to integrate and adopt this approach into their subject matter (Henriksen, 2017). Herro and Quigley (2016) added that this might be due to the fact STEAM implementation has often been viewed through the narrower lens understanding it as a simple integration of arts into the sciences. The authors further added that the acronym of STEAM might also cause confusion to some as they might consider it as just plugging an 'a' into STEM. Bahrum, Ibrahim, and Wahid (2017) added that some consider STEAM Education as the sole amalgamation of 'arts' with 'science' which can have consequences. Some teachers are uncertain about the inclusion of an 'a' into STEM and others might lack the artistic capabilities required to effectively achieve arts integration (Henriksen, 2017). Teachers who have artistic capabilities might be uncertain about how to incorporate the arts into STEM. This calls for the need of STEAM with an inclusive approach that encompasses various disciplines and provides entry points to teachers across various contexts (Hong, 2017). Henriksen (2017) added that this view of STEAM has multiple aspects; it focused on the integration of arts but also on creativity, interdisciplinary, project-centered thinking, and real-world learning.

Hong, Jho, and Song (2016) advocated that the capabilities of STEAM educators should be strengthened through constructing learning and research communities beyond individual-level training programs. Hong (2017) detailed the implementation of policies related to STEAM Education in Korea and highlighted that a "STEAM Bridge Center" model was developed, which promotes collaborative research between teachers and academic scholars. This results in the improvement of teacher's capabilities regarding arts-integration and the integration of STEAM in their educational practices. Clarke (2019), however, highlighted that governments have a major role to play in the development of such programs and 'future strategies' and 'roadmaps should focus on this aspect.

Finding and Discussion

This chapter referred to previous studies (secondary data) in order to address the research problem. As findings of the critical analysis of the research topic, all chosen studies were put side by side in Table 2 and compared against each other as this set-up allows to draw on the conclusion about STEAM implementation, challenges and success criteria through which implementation becomes possible.

In order to address the first objective related to understand the STEAM education initiatives in different developed countries, it has been assessed that implementation of STEAM initiatives started from elementary level schools. This has positively affected lesson planning of schools in other countries. Whereas, the in-progress implementation of STEAM education in the UAE shows that it is only limited to early years and STEAM policies should be developed by the government.

Referring to Table 1, the implementation of the STEAM-related educational modules is beneficial; however, there were numerous problems. All four of the selected in the above table, range from the years 2013 to 2017 and in all of them similar authors exist, thus, linking the researches together and making them significant for the analysis. It is significant because the challenges identified during the research done in the year 2013, challenges relating to the STEAM application were identified to be teacher's experience, availability of finance, lack of organized structure to implement STEAM modules.

The gap was filled in 2014 by taking responses from the experienced teachers; however, the problem persisted. Although the teachers agreed with the benefits, they claimed that experience could not aid them in getting around unsynchronized course material developed in the country or financial funds required to implement these modules. The authors concluded that in addition to financial support, the teachers required a clear policy on the kind of material available for the students. The research conducted in 2017 supported the fact, that in order to manage the benefits from STEAM modules in true essence, financial support is required for reorganizing the curriculum. The difference in the material available through textbooks and modules generated results confused students.

Final research selected for the comparison is also performed in 2017; nevertheless; it is different from the prior researches because it researched the application of STEAM modules in schools. After policy related to course material had been put into effect, the answers to a survey from experienced teachers aid us to conclude that if the UAE is also to apply the STEAM technique, the course material of the textbook as well as modules implementation, would have to be facilitated by drafting policy. In order to ensure a proper implementation and policy drafting recommendations of STEAM education in UAE a roadmap showcased in Figure 2 is proposed based on the research literature review and document analysis sections.

Table 1: Comparison between old and subsequent research

Older research	Subsequent research	Comparison
Hye Jeong, Jiwon, and Jung Bog, (2013) noted that South Korean teachers implemented the STEAM education system in their schools. Because they thought that the system's implementation should generate interest for arts, math's, science and engineering. Nevertheless, the system is difficult and time-consuming moreover; its successful application also requires significant experience as well as a change to the curriculum. Changing the curriculum is not only difficult but is not possible without financial back-up.	Byun, Park, and Sim (2016) observed that the majority of experienced Korean teachers thought positively about the STEAM framework. Authors found out that the system can permeate learning into students; however, significant change to current educational curriculum structure is required. Moreover, the system can also generate interest in the students; however, there are problems that are causing this program to fail in generating benefits. These problems include a lack of developed educational curriculum structure in line with the framework required for STEAM. In addition, financial support is also required because a significant investment might be needed.	The research conducted in 2013 revealed that teacher's experience, as well as finance, is required to conduct the successful implementation of STEAM education system. These points were covered when research performed in 2016 by authors including several as well as the ones, who were part of the research in 2013. They covered the gap related to the experienced teachers nevertheless; it was found that unless significant financial support is available. As well as a major change to the curriculum is not made, the STEAM cannot be implemented for generating effective results.
Soo-min, Tae-sang, and (2014) conducted research about STEAM education modules implementation in middle schools. The authors found that though teachers had implemented the modules in the school and the interest related to it is very high. However, problems came out when the students were unable to connect the textbook to the results of STEAM education. The reason is that the curriculum had not been re-organized to facilitate the benefits of STEAM education system. In addition to the development of curriculum, paving down of methods related to dissemination of STEAM is also important.	Bolger and Kim (2017) concluded that drafting of policy related to STEAM education in South Korea began in the year 2009. It had been found that with careful instructions related to the embedment of STEAM modules into the curricula of the Science, Arts, Math, Engineering and Technology aided the benefits management among the students. The authors engaged 119 elementary level teachers, who were experienced to conduct lessons based on STEAM. In the post-policy application, the teachers' commitment had improved significantly, and their lesson planning had become a lot easier after clarification of the curriculum.	Author Kim was part of the research conducted on the application of STEAM in 2014 and 2017. In both instances, elementary school teachers participated and responded to the survey related to a STEAM application. The research conducted in 2014 found that teachers believed in the benefits of teaching STEAM subjects through modules. However, these benefits cannot be felt unless significant changes are made to the curriculum. The problem is corroborated because the research performed in 2017 focused on gathering data from elementary school teachers. where upgraded course material had been applied, teachers found it a lot easier to apply STEAM-related modules, thereby earning students' interests with simple clarification of the policy related to the subject matter.

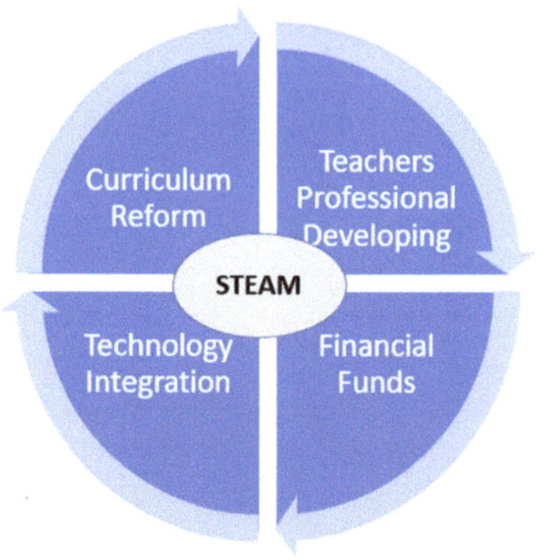

Figure 2: STEAM roadmap in UAE

The roadmap scheme is nonlinear, reflecting the reality of the map as one can always go back and reflect in order to improve the areas mentioned. However, for an effective implementation all four components need to be well addressed. Starting by a clear definition of STEAM for all stakeholders and subject involved in order to ensure consistency. Moreover, a reform in the curriculum is needed in order to integrate the STEAM objectives within the curriculum, explore all its discipline by recognizing arts as one of its subjects in order to gain knowledge and cognitive skills and upgrade course materials by developing collaborative plans. These materials should be in line with the STEAM modules so that the students do not get befuddled with the contrasting test results and course material of the textbook. It will be more efficient if this integration was adopted as the national curricula as it will ensure consistency in delivery across all schools. Ensuring appropriate professional development for teachers that allows them to explore STEAM as an interdisciplinary learning and delivers STEAM through its best recommended practices set by experts and understands the importance of creativity within this framework. It is suggested as well to integrate a section on STEAM in the teacher licensing exam set by the country. To add, according to Atkins, Michie, and West (2014), the roadmap and policies are going to be implemented through teachers; thus, their behavior is another essential ingredient for ensuring that STEAM modules are applied. Therefore, psychological training will be given to teachers in order to motivate them, change their mindset and be able to accept the following STEAM frame as lots of teachers are afraid to try anything that is out of their comfort zone and that is why the BCW is very important. In addition, the technology subject in STEAM is highly important in order to cope with the fourth industrial revolution. This can only happen through preparing teachers to use and handle advanced technology such as simulations, virtual reality and much more in a way that stimulate STEAM education. It will be best if all teachers can be International Computer Driving License (ICDL) certified in order to ensure that the basic usage of a computer and other devices are met. Finally, nothing can be planned and executed without proper funding. Allocating these resources ensure the equity and allocation of STEAM resources to all students. These could be funded by the Ministry itself, the schools without increasing its tuition fees in an abnormal way, increment needs to be monitored/set by the ministry or even organizing a charity within the country that ensure the funds for this purpose.

Conclusion

In conclusion, this chapter has reviewed the implementation of STEAM education in schools in developed countries whereby it has showed that most of the schools do not have a proper roadmap and policies that guides and assure the effectiveness of STEAM implementation. Throughout the literature review and analysis, it can be concluded that for the implementation of the STEAM modules in the UAE, the government must pen policies. Because education is not an individual's work, it rather shapes the future of the society, and therefore, the education course material should have to be made as per the requirements of the UAE industry. Moreover, stakeholders should be involved in drafting STEAM related policies beside the ministry of education as they are the primary line of contact with students. Since children then later in their life become part of the industry, they must be using their skills and therefore it is suggested that industry practitioner's perspective need to be heard when drafting the policy. These skills require evaluation of the data and gaining an insight into their decisions. These skills shall be developed through educational knowledge which is taught from the course books available in the schools. According to Shaer, Shibl, and Zakzak (2019), UAE private schools had implemented the STEAM techniques for giving education to early year students. However, in the absence of the proper policies identifying the correct application of the structure or the kind of change in students to be expected, it is impossible to measure the change being wrought in students due to these policies. Therefore, in order to integrate the STEAM modules into the UAE education system, clear policies must be drawn, and roadmaps need to be followed so the appliers can measure the change this can only happen through commitment, change in the people (parents, students, teacher and society) mindset and behavior in terms of opportunity, capability and opportunity that are offered through STEAM implementation and that are highlighted through the BCW. Finally, defining STEAM for all stakeholders, curriculum reform and integration, teachers training, advanced technology training and support and financial findings are the road that leads to effective STEAM implementation as well as they constitute core standards required to draft a STEAM policy. Implementation and monitoring process of the following proposed roadmap will be studied in further research.

REFERENCES

Atkins, L., Michie, S., & West, R. (2014). *The Behaviour Change Wheel*. Zaltbommel, Netherlands: Van Haren Publishing.

Bahrum, S., Wahid, N., & Ibrahim, N. (2017). Integration of STEM Education in Malaysia and Why to STEAM. *International Journal of Academic Research in Business and Social Sciences*, *7*(6). https://doi.org/10.6007/ijarbss/v7-i6/3027

Borghi, A. M., Caligiore, Daniele, Thill, S., Ziemke, T., & Baldassarre, G. (2013). Theories and computational models of affordance and mirror systems: An integrative review. *Neuroscience & Biobehavioral Reviews*, *37*(3), 491–521. https://doi.org/10.1016/j.neubiorev.2013.01.012

Brier, S. (2014). The Transdisciplinary View of Information Theory from a Cybersemiotic Perspective. *Studies in History and Philosophy of Science*, 23–49. https://doi.org/10.1007/978-94-007-6973-1_2

Byun, S., Park, HyunJu, Sim, J., Han, H.-S., & Baek, Y. S. (2016). Teachers' Perceptions and Practices of STEAM Education in South Korea. *EURASIA Journal of Mathematics, Science and Technology Education*, *12*(7), 1739–1753. https://doi.org/10.12973/eurasia.2016.1531a

Chemers, M. (2014). *An Integrative Theory of Leadership*. Abingdon, United Kingdom: Taylor & Francis. https://doi.org/10.4324/9781315805726

Chu, H.-E., Martin, S. N., & Park, J. (2018). A Theoretical Framework for Developing an Intercultural STEAM Program for Australian and Korean Students to Enhance Science Teaching and Learning. *International Journal of Science and Mathematics Education*, *17*(7), 1251–1266. https://doi.org/10.1007/s10763-018-9922-y

Clarke, M. (2019). Chapter 15: STEM to STEAM: Policy and Practice. In t*he STEAM Revolution* (First edition, Vol. 1, pp. 223–236). New York, United States: Springer Publishing. https://doi.org/10.1007/978-3-319-89818-6_15

Dell'Erba, M. (2019). *Policy Considerations for STEAM Eductaion*. Retrieved from https://files.eric.ed.gov/fulltext/ED595045.pdf

English, L. D. (2017). Advancing Elementary and Middle School STEM Education. *International Journal of Science and Mathematics Education*, *15*(S1), 5–24. https://doi.org/10.1007/s10763-017-9802-x

Escudé, M., Hooper, P. K., & Vossoughi, S. (2016). Making Through the Lens of Culture and Power: Toward Transformative Visions for Educational Equity. *Harvard Educational Review*, *86*(2), 206–232. https://doi.org/10.17763/0017-8055.86.2.206

Guo, L., Wang, X., & Xu, W. (2018). The Status Quo and Ways of STEAM Education Promoting China's Future Social Sustainable Development. *Sustainability*, *10*(12), 4417. https://doi.org/10.3390/su10124417

Hendriks, A.-M., Jansen, M. W., Gubbels, J. S., De Vries, N. K., Paulussen, T., & Kremers, S. P. (2013). Proposing a conceptual framework for integrated local public health policy, applied to childhood obesity the behavior change ball. *Implementation Science*, *8*(1). https://doi.org/10.1186/1748-5908-8-46

Henriksen, D. (2017). Creating STEAM with Design Thinking: Beyond STEM and Arts Integration. *STEAM*, *3*(1), 1–11. https://doi.org/10.5642/steam.20170301.11

Herro, D., & Quigley, C. (2016). STEAM Enacted: A Case Study of a Middle School Teacher Implementing STEAM Instructional Practices. *Journal of Computers in Mathematics and Science Teaching*, *35*(4), 319–342. Retrieved from https://learntechlib.org/primary/p/174340/

Hong, O. (2017). STEAM Education in Korea: Current Policies and Future Directions. *Science and Technology Trends: Policy Trajectories and Initiatives in STEM Education*, *1*(1), 92–102. Retrieved from https://researchgate.net/publication/328202165/

Hong, O., Jho, H., & Song, J. (2016). An Analysis of STEM/STEAM Teacher Education in Korea with a Case Study of Two Schools from a Community of Practice Perspective. *EURASIA Journal of Mathematics, Science and Technology Education*, *12*(7), 1843–1862. https://doi.org/10.12973/eurasia.2016.1538a

Hye Jeong, P., Jiwon, Lee, & Jung Bog, K. (2013). Primary Teachers' Perception Analysis on Development and Application of STEAM Education Program. *Eurasia Journal of Mathematics, Science and Technology Education*, *32*(1), 47–59. Retrieved from https://researchgate.net/publication/263623230/

Kang, N.-H. (2019). A review of the effect of integrated STEM or STEAM (science, technology, engineering, arts, and mathematics) education in South Korea. *Asia-Pacific Science Education*, *5*(1). https://doi.org/10.1186/s41029-019-0034-y

Kelley, T. R., & Knowles, J. G. (2016). A conceptual framework for integrated STEM education. *International Journal of STEM Education*, *3*(1), 1–11. https://doi.org/10.1186/s40594-016-0046-z

Kim, D., & Bolger, M. (2016). Analysis of Korean Elementary Pre-Service Teachers' Changing Attitudes About Integrated STEAM Pedagogy Through Developing Lesson Plans. *International Journal of Science and Mathematics Education*, *15*(4), 587–605. https://doi.org/10.1007/s10763-015-9709-3

Lacono, W. G., Vaidyanathan, U., & Vrieze, Scott I. (2015). The Power of Theory, Research Design, and Transdisciplinary Integration in Moving Psychopathology Forward. *Psychological Inquiry*, *26*(3), 209–230. https://doi.org/10.1080/1047840x.2015.1015367

Michie, Susan, Van Stralen, M. M., & West, R. (2011). The behaviour change wheel: A new method for characterising and designing behaviour change interventions. *Implementation Science*, *6*(1), 1–11. https://doi.org/10.1186/1748-5908-6-42

Shaer, S., Shibl, E., & Zakzak, L. (2019). *The STEAM Dilemma: Advancing Sciences in UAE Schools the Case of Dubai.* Retrieved from https://mbrsg.ae/home/research/education-policy/the-steam-dilemma-advancing-sciences-in -uae-school/

Soo-min, L., Tae-sang, L., & Yougshin, K. (2014). Analysis of Elementary School Teachers' Perception on Field Application of STEAM Education. *Journal of Science Education*, *38*(1), 133–143.

South Carolina State Department of Education. (2014). South Carolina STEAM Implementation Continuum Center for Standards, Assessment, and Accountability. Retrieved January 29, 2020, from https://csaa.wested.org/resources/south-carolina-steam-implementation-continuum

Zinth, J., & Goetz, T. (2016). *Promising Practices: A Statae Policymaker's STEM Playbook.* Retrieved from https://files.eric.ed.gov/fulltext/ED569158.pdf.

PART 3

CHAPTER 7

Metacognitive Awareness in Mathematics and Science

Lames Abdul-Hadi

ABSTRACT

Metacognition has been identified as a crucial component of effective learning of mathematics and science education. It entails students recognizing their learning processes, styles, preferences and self-efficacy. The literature review presented highlights the following major topics: the theoretical review of metacognition, the recent literature and studies on metacognitive strategies and inventories to assess metacognition, reflection and self-regulation in mathematics and physics. This is followed by how this research was translated into educational practice.

INTRODUCTION

Metacognition has been identified as a fuzzy concept due to its various definitions and dimensions (Flavell, 1981, p.37; Veenman et al., 2006) and has been investigated from various perspectives and for several purposes. Metacognition is defined as a multidimensional set of skills that includes thinking about own thinking and involves different types of knowledge that make metacognition an inspiring area for research (Lai, 2011). The literature review on metacognition points to the numerous studies (Lai, 2011; Veenman, 2012; Zohar & Barzilai, 2013; Zohar & Dori, 2012) that highlight the effect of metacognitive strategies on students' learning processes and metacognitive skills and consequently, learning outcomes. The following major parts are emphasized in the literature: definition of metacognition, trends and theoretical review, recent research on metacognitive treatments and inventories to enhance and measure metacognitive awareness, the relationship among metacognition, reflection and self-regulation in mathematics and physics education. This chapter looks at the current state of research as it relates reflection as metacognitive strategy to metacognitive awareness and self-efficacy in mathematics and physics education.

Definitions and Components of Metacognition

Cognitive psychology can be understood as the attempt to understand how people think and how they employ their basic mental abilities to recall information, analyze, and reason (Hunt & Ellis, 2004). Cognition includes various skills that can help the learners to achieve particular goals and can be identified and measured (Schraw et al., 2006). Metacognition is defined as the individuals' awareness to monitor and regulate their own cognitive processes (Hennessey, 1993). However, not all cognitive processes require metacognition (Lai, 2011). Within cognitive psychology, metacognition is mainly defined as the executive control of cognitive processes (Kuhn & Dean, 2004). Schraw and Moshman (1995, p.350) describe metacognitive theory as "a relatively systematic structure of knowledge that can be used to explain and predict a broad range of cognitive and metacognitive phenomena". Metacognitive theory focuses on cognitive characteristics of the mind, ways of thinking and levels of control and understanding of the cognitive processes (King, 1999 cited in Rahman, 2011). The cognitive system consists of "metacognitive self-instructions and the cognitive processes that are involved in the execution of those instructions" (Veenman, 2012, p.27). Moreover, within the cognitive system, both cognitive and metacognitive actions feed different goals and functions (Brown, 1987; Butler, 2006; Veenman, 2012). The attempts to conceptualize metacognition indicate the complex relationship between cognition and metacognition where metacognition is part of the cognitive system and the higher order factor that controls the cognitive system (Veenman et al., 2006).

Metacognition as a concept originally indicates the individual's knowledge and regulation of one's cognitive activities in learning processes (Schraw et al., 2006). Flavell (1976, p. 232) was the first to refer to metacognition as "one's knowledge concerning one's own cognitive processes and products, or anything related to them". This definition is applied in several ways to point to the process of thinking about our own thinking (Flavell, 1979, p.906). Later, Schraw and Dennison (1994) clarified that metacognition is linked to the ability to understand, reflect and control one's learning, while King (1999 cited in Rahman, 2011) described the attributes of metacognition as a person's ability to think about own learning processes and to identify suitable strategies to analyze and to implement what has been learned. There are several terms that are commonly associated with metacognition such as metacognitive awareness, metacognitive activities, theory of mind, metacognitive skills, judgement of learning, self-efficacy and self-regulation (Veenman et al., 2006). "Metacognitive Awareness" (Hennessey, 1999; Schraw & Dennison, 1994) as a term has been adapted in several studies and means the individuals' awareness of how they learn, how they think and engage in self-reflection. Metacognitive awareness also includes the awareness of monitoring and assessing one's cognitive processes related to further learning (Balcikanli, 2011; Schraw et al., 2006). Generally, definitions of metacognition point to the importance of learners' awareness of their own thinking to have the knowledge about and regulation of cognitive strategies used in learning and to be able to reflect on one's performance and learning experiences (Bransford et al., 2000; Flavell, 1979; Pintrich, 2002).

Researchers like Brown & DeLoache (1978) and Kluwe (1987) believe that self-regulation is a secondary component of metacognition while other researchers like Winne (1996) and Zimmerman (1995) consider self-regulation as a higher order concept of metacognition because it also includes motivational and emotional processes (Veenman et al., 2006). Another distinction is made between metacognitive knowledge and skills by referring to metacognitive knowledge as the individual's declarative knowledge about the relations between learner, task and strategy feature (Flavel, 1979), while metacognitive skills refer to the individual's procedural knowledge for regulating one's problem-solving and learning processes (Brown & DeLoache, 1978; Veenman, 2005).

Flavell (1979), Schraw and Dennison (1994), and Schraw et al. (2006) classify the metacognition components into two main components: metacognitive knowledge and metacognitive regulation. Metacognitive knowledge as defined by Flavell (1979, p.906) is the knowledge about an individual's cognition as a learner and how, when and why to apply certain strategy to improve performance and involves three factors: (1) declarative knowledge which refers to the individual's knowledge about own beliefs and perception of task structure and self-efficacy, (2) procedural knowledge which means the "knowledge about the execution of procedural skills" (Schraw & Moshman 1995, p.353), while (3) conditional knowledge means "knowing when and why to apply various cognitive actions" (Schraw & Moshman, 1995, p.353). Metacognitive regulation contains three regulatory skills comprising planning, monitoring and evaluating (Balcikanli, 2011; Schraw, 2001). (1) Planning indicates choosing appropriate strategies and resources. (2) Monitoring indicates individual's awareness of comprehension and task performance. (3) Evaluating indicates the evaluation and judgment of outcomes and effectiveness of the regulation process if it matched the task goals. The components and factors of metacognition are shown in Figure 1.

Metacognitive components are generally correlated, which indicate that the sources for all the main metacognitive schemas are in a loop (Balcikanli, 2011). One of the main questions about metacognition in research is whether it is domain specific or general. Flavell (1979, p.906) clarifies that the factors of metacognition are defined as "four classes of phenomena": metacognitive knowledge and experience, tasks and strategies. One group of researchers (Schraw, 1998; Veenman et al., 2006) is involved in the debate that an individual's cognitive skill is subject or domain related while an individual's metacognitive skills can be employed similarly in any domain. Other researchers (Davis, 2003; Thomas et al., 2008) debate that the employed metacognitive skills differ according to the different cognitive tasks. Therefore, metacognitive skills should be researched in different contexts and subjects.

Metacognition has been considered a fuzzy concept due to the interrelationship between the concept and its components, and the inconsistency in the meaning of metacognition as a concept in the field of research (Zohar & Dori, 2012; Veenman,

2012). Several frameworks were developed to categorize metacognitive components; the typology of these components is shown in Table 1 below. Researchers like Flavell (1979) and Brown (1978) distinguish between metacognitive knowledge and skills (Veenmanet al., 2006) where metacognitive skills represent the self-regulatory processes of metacognition (Zimmerman, 1995).

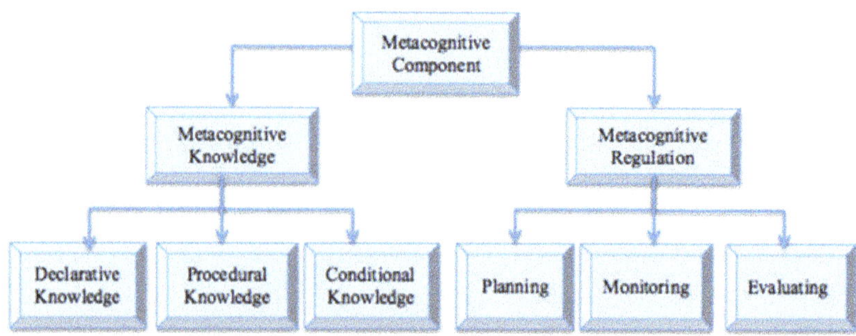

Figure 1: Metacognitive Components

Basic development of metacognitive knowledge and skills starts at an early age during early-school years and can be observed (Whitebread et al., 2009) while metacognitive skills become academically oriented in formal education when the learners need to regulate their knowledge (Veenman et al., 2006). Veenman (2012) states that during high school age the steep linear development of metacognitive skills happens. Learner's metacognitive knowledge alone is not enough to be an indicator of success in performing a task because it is affected by the students' self-efficacy, capability and motivation (Veenman, 2005). From this point of view, the recommendation is to study metacognition in specific task or domain (Veenman et al., 2006) and to consider the executive process of metacognition which is the self-regulatory processes during learning activities (Lai, 2011; Zohar & Dori, 2012). Moreover, learners may not use their metacognitive skills constructively because of the lack of knowledge of how to employ these skills (Veenman et al., 2006).

Assessment of Metacognitive Awareness

Although much effort is being channeled into creating an inventory to assess learners' metacognitive awareness, several studies discuss the reliability and validity of these inventories due to the multivariate nature of metacognition (Balcikanli, 2011; Lai, 2011; Rahman, 2011; Veenman, 2012). Another debate about metacognition arises from the difficulty to assess the metacognitive skills and knowledge because they cannot be observed directly and the available measures of metacognitive awareness is narrow in focus and decontextualized from the school context (Lai, 2011, p.2). The distinction of metacognition components demands development of a variety of assessment tools and measures (Zohar & Dori, 2012). Veenman (2005) distinguishes between two methods of assessing metacognitive skills: on-line methods where the metacognitive skills are assessed during the task like observations (Veenman & Spaans, 2005), computerized logs (Stephens & Winterbottom, 2010; Veenman et al., 2004) and thinking aloud (Afflerbach, 2000), stimulated recall (Van Hout-Wolters, 2000) while off-line methods refer to assessing metacognitive skills either before or after the task such as questionnaires (Schraw & Dennison, 1994; Thomas et al., 2008), reflection (Wiezbicki-Stevens, 2009) and interviews (Zimmerman & Martinez-Pons, 1990); this latter method mainly depends on the learners' self-reports (Veenman, 2012).

Table 1: Typology of Metacognitive Components (Lai, 2011, p.7)

Metacognitive Component	Type	Terminology	Citation
Cognitive knowledge	Knowledge about oneself as a learner and factors affecting cognition	Person and task knowledge	Flavell, 1979
		Self-appraisal	Paris & Winograd, 1990
		Epistemological understanding	Kuhn & Dean, 2004
		Declarative knowledge	Cross & Paris, 1988 Schraw et al., 2006 Schraw & Moshman, 1995
	Awareness and management of cognition, including knowledge about strategies	Procedural knowledge	Cross & Paris, 1988 Kuhn & Dean, 2004 Schraw et al., 2006
		Strategy knowledge	Flavell, 1979
	Knowledge about why and when to use a given strategy	Conditional knowledge	Schraw et al., 2006
Cognitive regulation	Identification and selection of appropriate strategies and allocation of resources	Planning	Cross & Paris, 1988 Paris & Winograd, 1990 Schraw et al., 2006 Schraw & Moshman, 1995 Whitebread et al., 2009
	Attending to and being aware of comprehension and task performance	Monitoring or regulating	Cross & Paris, 1988 Paris & Winograd, 1990 Schraw et al., 2006 Schraw & Moshman, 1995 Whitebread et al., 2009
		Cognitive experiences	Flavell, 1979
	Assessing the processes and products of one's learning, and revisiting and revising learning goals	Evaluating	Cross & Paris, 1988 Paris & Winograd, 1990 Schraw et al., 2006 Schraw & Moshman, 1995 Whitebread et al., 2009

Metacognitive Inventories

Metacognitive awareness is crucial in mathematics and physics education because it develops reasoning skills and improves learning outcomes in mathematics (DuToit, 2013; Mevatech & Fridkin, 2006) and physics (Nashon & Nielson, 2011; Zohar & Dori, 2012). Consequently, the need to assess learners' metacognitive awareness has led to several attempts to create a valid and reliable inventory (Lai, 2011; Schraw & Dennison, 1994). The following will cover some of the most used inventories in mathematics and science education.

One of the most well-known and frequently used inventories to measure metacognitive aspects using self-report items is the Metacognitive Awareness Inventory (MAI) by Schraw and Dennison (1994). The MAI was created for adults and involves fifty-two self-report items to measure the eight factors of metacognition under two broader components: metacognitive knowledge and metacognitive regulation. Another inventory is the Metacognitive Skills and Knowledge Assessment (MSA) by Desoete et al. (2001). MSA involves 160 items that assess the learners' metacognitive knowledge and skills through seven components of metacognition. Three factors of metacognitive knowledge: declarative, procedural and conditional alongside four factors of metacognitive skills: monitoring, planning, evaluation and prediction. MSA is a multi-method inventory where students' performance is compared with predictions (Ozsoy & Ataman, 2009).

The Self-Efficacy and Metacognition Learning Inventory-Science (SEMLI-S) was created by Thomas et al. (2008) based on: first, the theoretical framework that relates students learning processes in science with metacognitive science learning "orientation" (Thomas et al., 2008, p.1702) which means that the assessment and evaluation of students' metacognitive knowledge, skills and learning processes are task and context related and varies between subject domain and overtime (Veenman et al., 2006). Second, is the methodological consideration to create empirical self-report inventory to measure nature and level of students' metacognitive awareness in science classrooms as a result of the effect of certain treatment to enhance students' metacognitive awareness (Thomas et al., 2008). Although Thomas et al. (2008) agree with Veenman et al. (2006, p.7) that "general metacognition may be instructed concurrently in different learning situations and may be expected to transfer to new ones", they included items in their inventory that were related to the context to emphasize the importance of the setting where teachers and students assume their roles. The domain of learning environments research in science education is full of inventories that support the researchers' perspective, for example, the Metacognitive Orientation Learning Environment Scale-Science inventory (Thomas, 2003) and prior to that, the Constructivist Learning Environment Scale inventory (Taylor et al., 1994). All these tools include items or an introductory sentence that ask the learners to focus and to relate their reflection or self-report to domain-specific learning processes and metacognition. SEMLI-S inventory is used in several studies in the domain of biology (Aurah, 2013; Mavrikaki & Athanasiou, 2011), chemistry (Sandi-Urena et al., 2011) and physics (Nashon & Nielson, 2011; Thomas, 2011; Chantharanu-wong et al., 2012), and is also cited in several studies (Butterfield, 2012; Schererl & Tiemann, 2014). SEMLI-S consists of 30-items related to five factors then was piloted with 505 students (13-18 years old) and tested for validity and reliability. The Rasch analysis shows that SEMLI-S fits the Rasch model well. The reliability of SEMLI-S is tested by Cronbach's-Alpha and the results have reliable alpha scores (0.68 to 0.85). There is also significant correlation (at the 0.01 level) between the five subscales (Thomas et al., 2008). The SEMLI-S factor analysis indicates that it can be used for all its factors simultaneously or for each factor separately.

The majority of the inventories that assess metacognitive skills are students' self-report tools (Schraw & Dennison, 1994; Thomas et al., 2008). 'Self-report' assesses the students' perceptions of their metacognitive awareness by asking the students to focus on the nature and the level of their metacognitive knowledge and skills along with evaluating their use of cognitive and metacognitive processes. There are several empirical self-report inventories that explore students' learning and metacognition like the Learning Processes Questionnaire (LPQ) (Biggs, 1987), the Learning and Study Strategies Inventory (LASSI) (Weinstein et al., 1987), the Motivated Strategies for Learning Questionnaire (MSLQ) (Pintrich et al., 1993) and the Assessment of Cognitive Monitoring Effectiveness (ACME) (Osborne, 1998). The above inventories are based on general cognitive and metacognitive constructs rather than specific context or subject (Schraw, 2000).

Metacognition in Mathematics and Physics Education

Mathematics and science learning reap various cognitive processes involved in problem-solving, inquiry learning, reading and writing (Kuo et al., 2013; Veenman, 2012). Undeveloped problem-solving skills, poor conceptualization and difficulty in applying knowledge across disciplines are common addressed parries to learn mathematics and physics (DuToit, 2013; Kuo et al., 2013). The literature in science and mathematics education supports several definitions of metacognition as the awareness of learning processes and the learner ability to think about own thinking (DuToit, 2013; Nashon & Nielsen, 2011; Parson, 2011). Relatively little is done about metacognitive processes in reading and problem-solving (Parsons, 2011; Lai, 2011).

Metacognitive awareness empowers the students' self-efficacy and motivation to overcome any anxiety in math and physics learning (Lai, 2011; Ramirez & Beilock, 2011). Teachers find difficulties in engaging students in guided-inquiry process so to overcome this issue, students must develop metacognitive skills of "thinking about their own thinking" that are necessary to facilitate this kind of exploration (Veenman, 2012; Zohar & Dori, 2012). These metacognitive skills are developed as students engage in developing new conceptual understandings, build on prior knowledge and reflect on their own learning as they move through the phases of this guided-inquiry process. The evidence on the effect of metacognition to empower inquiry and discovery learning in science is strongly proved in physics domain (Anderson &

Nashon, 2007; Manlove et al., 2007) and correlated to problem-solving skills in physics (Nashon & Nielsen, 2011; Phang, 2010) and mathematics (DuToit, 2013; Mevarech & Fridkin, 2006).

Teaching metacognitive strategies has shown evidence of improvement in teachers' and students' engagement with subject's concepts and writing tasks (LaVaque-Manty & Evans, 2013), students' motivation and self-efficacy (Silver, 2013; Wu & Looi, 2013). Bransford et al. (2000, p.21) affirm the need to explicitly implement metacognitive strategies and to conduct action research. There is much evidence that metacognitive skills are considered the most crucial of intellectual abilities that support learning performance (Veenman et al., 2004; Zohar & Dori, 2012). General metacognition can be taught alongside any learning situations; however, task or domain related metacognition should be taught for each task or domain distinctly (Zohar & Dori, 2012). A considerable number of studies support studying metacognition related to specific domain or task like problem-solving in math (Desoete et al., 2003; Kramarski & Mevarech, 2003), science (Thomas, 2003) and physics (Phang, 2010; Nashon & Nielsen, 2011). These studies recommend linking instruction and teachers' feedback to metacognition (Balcikanli, 2011; Ku & Ho, 2010) and developing the students' metacognitive awareness in explicitly (Veenman et al., 2006).

Although the growing body of literature on metacognition underlines its importance in learning, the effects of metacognitive strategies on learning outcomes are merely reported and measured (Lai, 2011; Zohar & Dori, 2012). To ascertain the cause and effect relationship between metacognitive strategies and learning outcomes, metacognitive knowledge and skills need to be measured using a pretest-posttest design (Veenman et al., 2006; Zohar & Barzilai, 2013). Ozsoy and Ataman (2009) adopted a quasi-experimental design to find the effect of implementing structured metacognitive strategy in problem-solving activities to enhance forty-seven fifth grade students' problem-solving achievement and metacognitive skills. The study lasted nine weeks and the Metacognitive Skills and Knowledge Assessment (MSA-TR) (Desoete et al., 2001) was used. The findings indicated that the students in the experimental group significantly improved in metacognitive skills and mathematical problem-solving achievement (Ozsoy & Ataman, 2009, pp.76-78). The same findings were corroborated by other studies in the field which had also adopted the quasi-experimental design (DuToit, 2013; Mevarech & Fridkin, 2006). The importance of metacognitive awareness lies in its influential role to develop students' thinking and learning processes (Thomas et al., 2008).

Metacognition and Reflection

Reflection as one of the metacognitive strategies has been highlighted in the past decade for its influential role to empower learners' metacognitive awareness and learning processes (Hacker et al., 2009; Lai, 2011). Metacognition and reflection have been traced back to educational psychologists and early thinkers of metacognition such as James, Dewey, Vygotosky and Piaget (Slavin, 2011). Hence, 'Reflection' is defined as the individual awareness of own experience (Boud et al., 1985). Dewey (1909, in Fisher 2011, p.2) defines "reflective thinking" as "Active, persistent, and careful consideration of any belief or supposed form of knowledge in light of the grounds that support it and the further conclusions to which it tends". Later on in the 1980s, reflection started getting scholarly attention with more focus on development of a reflective pedagogy framework that points to reflection as the learners' response to their learning experience. Definitions of metacognition and reflection are usually used interchangeably in the literature. The definition of reflection is aligned with the components of metacognitive self-knowledge of an individual's awareness of his or her own strengths and weaknesses besides motivational beliefs related to learning (Pintrich, 2002).

In class practice, reflection may be in the form of "metacognitive reflection" activity (Tarricone, 2011, p.6), to reflect on certain learning experience or one's cognitive processes. Reflection sometimes represents a stage in a metacognitive schema. When learners are given the opportunities to reflect on their learning process, they can better organize and manage new information and they can recognize what learning strategies accelerate understanding (Rando, 2001; Schraw, 2000; Weinstein, 2006). Bransford et al. (2000) assert that this ability to reflect differentiates expert from novice learners. Moreover, reflection transfers experience into learning and enables learners to employ their experiences in new situations (Boud et al., 1985). Thus, the main source for constructing learning objectives, Bloom's Taxonomy, has been modified to include metacognition, with the addition of metacognitive self-knowledge, as a new category of knowledge and an influential aspect of learning (Pintrich, 2002). Several studies (Bransford et al., 2000; Flavell, 1979; Murray, 1999; Pintrich, 2002; Schraw, 2000; Weinstein, 2006) have suggested that more opportunities should be provided to learners to reflect on their learning processes to improve their metacognitive awareness and to enhance their learning by being better in regulating their knowledge.

Reflection as a learning activity requires time and focus on cognitive processing to engage in continuous internal mental dialogue of one's thinking about own thinking, to regulate one's knowledge towards applying it in unfamiliar situations or contexts and to judge assumptions or understandings when interfered with one's learning (Donnelly, 2010; Stefani et al., 2007). Empirical studies (Boud et al., 1985; Hoffman & Spatariu, 2008; Pintrich, 2002; Zohar & Dori, 2012) have demonstrated the efficiency of different types of reflection such as journals, self-reflection, metacognitive prompting and guided-reflection especially when used in specific subjects and related to certain concepts and tasks.

Reflection Methods

Mathematics and physics share a lot of characteristics in subject's nature. Students should be proficient in mathematical skills to be successful on a science inquiry where many topics in physics are applications of math concepts (Collins, 2011, p.28). Problem-solving skills require from the learners reflection, conscious and regular monitoring of their own thinking processes to recognize the effective technique and strategy to solve a problem and to regulate the required knowledge to help in solving. Reflection on problem-solving strategies and the rationale behind choosing certain techniques or reaching certain answers, enhance the learners' problem-solving skills to perform a task and to make sense of the problem. The whole process helps students to clarify their understanding and empower their self-efficacy. Students with poor metacognitive skills have difficulty in applying mathematical and physics concepts in problem-solving and subsequently, in learning. Teaching metacognitive strategies explicitly such as reflection, thinking aloud, and modeling help students with metacognitive deficits to be metacognitively aware (Kaplan et al., 2013).

The literature about metacognition in mathematics and science education indicates that writing has become a recent prerequisite in enhancing a learner's metacognition and investigation skills (Collins, 2011, pp.15-16). Scaffolding approach, in math and science writing tasks, has been identified as an effective technique to empower students' understanding in content knowledge (Yore & Treagust, 2006; Hohenshell & Hand, 2006). According to these findings, the different methods of reflection such as guided reflection, self-reflection and metacognitive prompting helps in empowering the students' learning, achievement and metacognitive awareness level (Collins, 2011; Lai, 2011; Wiezbicki-Stevens, 2009). Metacognitive prompting was developed to help in progressing the reflection from dialogic to critical (Collins, 2011; Surgenor, 2011). Guided reflection is a method of reflection that involves engagement in a series of questions that help the individual to think and explore their actions (Surgenor, 2011). A study (Parsons, 2011) in higher education aimed to explore and to investigate the effect of 'writing to learn' as metacognitive intervention in mathematics classes on the students overall achievement and self-reflection with respect to learning mathematics concluded that writing in mathematics enhanced the students' meaningful connections about content and themselves as learners and encouraged them to reflect on their learning.

The demand to stimulate students' thinking processes about mathematical matters has been addressed in the last two decades in mathematics education literature, therefore, the call for reform in mathematics education refers to reflection as a high level of cognitive thinking process and a central part of mathematics education (Kaune, 2006; Kilpatrick, 1986). Reflection has been proved as a useful metacognitive activity to enhance high school students' mathematical understanding (Parson, 2011; Tok, 2013), learners' thinking and learning processes (De Corte, 1995; Schoenfeld, 1992). White and Frederiksen (1998) conducted a study to explore how reflection, as a metacognitive intervention implemented in physics classes, can be used to improve student understanding of the connection between scientific inquiry and real world. The researchers hypothesized that the difficulties in physics learning are due to the learners' low level of metacognitive awareness rather than the lack of intellectual ability. The results confirm previous findings that reflection is not an end but a means towards a helpful metacognitive approach that facilitates student learning and understanding of the inquiry process. Davis (2003) studied the role of generic and directed reflection of 180 middle school students at the end of a unit in physics about heat flow and energy conversion. The students were asked to reflect eleven times after analyzing and critiquing a news article about the topic. The study found that students who did not reflect or whose reflections were poor in quality obtained less successful results in the final project. The findings also indicated that the general type of responses endorsed quality of reflection that is more productive.

'Metacognitive prompting' is defined as "an externally generated stimulus that either tacitly or explicitly activates reflective cognition or evokes strategy use with the objective of enhancing a learning or problem-solving objective" (Hoffman & Spatariu, 2008, p.878). Metacognitive prompting has been proved as effective in enhancing students' self-efficacy and problem-solving efficiency in mathematics and physics (Collins, 2011; Hoffman & Spatariu, 2008). Endorsing "prompted metacognitive reflections" (Collins, 2011, p.39) by asking students reflective questions as they work their way through mathematics and physics learning, triggers their reflective cognition and helps to connect their learning experience and content knowledge to unfamiliar contexts (Collins, 2011; Hoffman & Spatariu, 2008; Wiezbicki-Stevens, 2009). A quasi-experimental study (Berthold et al., 2007) on the effect of cognitive and metacognitive prompting on learning concluded that cognitive and metacognitive prompts empower students' learning especially when metacognitive and cognitive prompts are combined. Distinctive methods of metacognitive prompting have been effective in many disciplines and contexts such as mathematics (Kramarski & Gutman, 2006) and physics (Davis, 2003). Prompting has been used to get students to think, review and reflect before, in or after the lesson to deepen understanding and comprehension (Fogarty, 2006, p.8). Self-reflection questions and comments by naming and describing while learning help the students to better understand when there is any difficulty.

Metacognitive Awareness and Self-Efficacy

The relationship between metacognitive awareness, self-efficacy, self-regulation and learning processes is highlighted in the literature (Lai, 2011; Veenman et al., 2006). Metacognition does not influence learning outcomes when it is isolated but rather is related to other elements of learning theory (Veenman, 2012). The literature related to this matter (Lai, 2011; Schraw et al., 2006), indicates that students' metacognitive awareness and the metacognitive strategies that they use in learning processes are subsets of self-regulation such as self-efficacy, where the learner's self-confidence about performance and goal attainment influence the learning outcomes. A self-directed or self-regulated learner is the learner who can monitor his own progress and make modifications when needed. By reviewing self-regulated literature, a connection is highlighted between engagement and metacognition.

Many researchers (Efklides, 2011; Mason & Scrivani, 2004; Thomas et al., 2008) are interested in the complex relations between metacognitive knowledge, skills, epistemological beliefs, self-regulation and learning processes on one side and self-efficacy, motivational processes and study interests on the other side. Simultaneously, other researchers studied the relationship between affective variables such as subject and test anxiety and metacognition (Tobias & Everson, 1997; Zohar & Dori, 2012) or learning difficulties (Harris et al., 2004). Neuropsychological studies on metacognition have been narrowed to very specific metacognitive processes like feeling of knowing and judgement of learning phenomena (Metcalfe & Shimamura, 1994; Pinon et al., 2005). Those studies recommended more investigation from other perspectives such as instructional, developmental or diagnostic perspectives and the learners' monitoring, planning and reflection skills as part of metacognitive components.

However, the role of student's self-efficacy in empowering academic outcomes has been proven where students with high level of self-efficacy often persevere longer with tasks, and are more likely to set and monitor their goals (Bandura, 2006; Collins, 2011; Zimmerman & Cleary, 2006). Although gender differences seem to play a role in the level of the student's self-efficacy, the research about gender differences in self-efficacy shows inconsistent results (Zimmerman & Martinez-Pons, 1990; Jacobs et al., 2002). A study by Pajares (2003) concludes that although grade nine female students obtained better writing scores, the male students showed a higher level of self-efficacy than the female students. Another study (Zimmerman & Martinez-Pons, 1990) indicates that there is no significant difference in self-efficacy in mathematics between male and female students while Jacobs et al. (2002) concludes that female students have higher self-efficacy than males from kindergarten through grade twelve in mathematics. While metacognitive awareness plays significant role in self-regulated learning, self-efficacy is important to help the learners to have the belief that they can perform tasks and achieve goals (Zimmerman & Cleary, 2006). The learners with less successful strategies are the individuals that have low level of self-efficacy (Brown, 1987). Self-reflection on performance is the last stage of self-regulation where the learner evaluates the extent of their satisfaction about performance outcomes and it is found that self-efficacy plays a crucial role at this stage because it influences the learners' abilities to judge their task performance and goal achievement (Collins 2011, p.28). In this light, teacher's feedback to students increases their self-efficacy (Zimmerman & Cleary, 2006). What leads to the empowerment of learners' self-efficacy is the goal achievement coupled with the cognitive processing that is required to achieve the targeted goal (Collins, 2011, p.37).

Self-regulation is among the transferable skills that can be developed in children through social interactions between adults and children as concluded by Vygotsky (1987 cited in Slavin 2011). Therefore, the emphasis on metacognitive strategies to be taught explicitly in the classroom is addressed in the literature (Balcikanli, 2011; Lai, 2011) by highlighting the effect of metacognitive teaching approach to enhance learning and metacognitive skills (Veenman, 2011). Students' engagement in metacognitive thinking requires teacher's guide through well-designed metacognitive activities which demand teacher's metacognitive awareness in the first place (Ku & Ho, 2010; Rahman, 2011). Although several studies investigated the role of metacognition in teaching and learning, the cause and effect of implementing certain metacognitive strategies in specific domains is not explored deeply and openly in the literature (Lai, 2011; Zohar & Barzilai, 2013).

Flavell (1979, p.27) states "metacognition is congruent with the learners' need and desire to communicate, explain and justify thinking to organisms as well as to himself". Numerous studies in science and mathematics education (Ku & Ho, 2010; Rahman, 2011) differentiate between metacognitive strategies and skills. This body of research concludes that metacognitively aware students are more strategic and perform better than students who lack such awareness (Bransford et al., 2000). However, few studies are done about metacognitive awareness in mathematics and science education in the UAE. Haidar and Al Naqabi (2008) researched hundred and sixty-two grade eleven students' implementation of metacognitive strategies and their influence on the former's understanding of chemistry. The study debates that learners should be taught how to employ different metacognitive strategies to enhance their learning in chemistry. Similarly, Al Khatib (2010) investigated the relationship between self-efficacy, self-regulation and academic performance among 404 student-teachers using a self-report inventory to assess their motivational orientations and their use of different learning strategies. The results conclude that self-efficacy and self-regulation have positive significant effect on academic performance. Other studies' results (Al Khatib, 2010; Haidar & Al Naqabi, 2008) indicate that engaging the learners in their learning processes has a strong influence on their academic performance and the recommendations are to conduct more studies in the field.

The literature review concludes: Firstly, the focus on developing students' metacognitive skills is taking more empirical consideration than developing students' metacognitive knowledge that require more studies on students' metacognitive awareness which involves both of them. Secondly, there is a lack of empirical research that employs cause and effect approach to study the effectiveness of implementing certain metacognitive strategies such as reflection on students' metacognitive awareness and self-efficacy in mathematics and physics. Thirdly, there is an insufficient number of studies on metacognitive awareness of high school students.

REFERENCES

Afflerbach, P. (2000). Verbal reports and protocol analysis. In M. Kamil, P. Mosenthal, P. Pearson & R. Barr (eds.). *Handbook of Reading Research,* 3, 163-179.

Al Khatib, S.A. (2010). Meta-cognitive self-regulated learning and motivational beliefs as predictors of college students' performance. *International Journal for Research in Education (IJRE),* 27, 57-72.

Anderson, D., & Nashon, S. (2007). Predators of knowledge construction: Interpreting students' metacognition in an amusement park physics program. *Science Education, 91*, 298-320.

Aurah, C. (2013). The effects of self-efficacy beliefs and metacognition on academic performance: a mixed method study. *American Journal of Educational Research, 1*(8), 334-343.

Balcikanli, C. (2011). Metacognitive Awareness Inventory for Teachers (MAIT). *Education & Psychology*, vol.9(23), pp. 1309-1332.

Bandura, A. (2006). Guide for constructing self-efficacy scales. In F. Pajares & T. Urdan (eds.). *Self-efficacy beliefs of adolescents*, vol. 5. Information Age Publishing, pp. 307-337.

Berthold, K., Nückles, M., & Renkl, A. (2007). Do learning protocols support learning strategies and outcomes? The role of cognitive and metacognitive prompts. *Learning and Instruction*, vol.17(5), pp.564-577.

Biggs, J. (1987). *Students Approaches to Learning and Studying*. Australian Council for Educational Research.

Boud, D., Keogh, R. & Walker, D. (1985). 'Promoting reflection in learning: a model'. In D. Boud, R. Keogh, & D. Walker (eds.) *Reflection: Turning Experience into Learning*, Nichols.18-40.

Bransford, J., Brown, A.L. & Cocking, R. (2000). *How People Learn: Brain, Mind, Experience and School.* Expanded edn. National Academia Press.

Brown, A. (1978). Knowing when, where, and how to remember: A problem of metacognition. In R. Glaser (ed). *Advances in instructional psychology, 1.* 77-165.

Brown, A., DeLoache, J. (1978). Skills plans and self-regulation. In R. Siegler (eds.), *Children's thinking: What develops?* Lawrence Erlbaum Associates. 3-35.

Brown, A. (1987). 'Metacognition, executive control, self regulation and other more mysterious mechanisms. In F. Weinert & R. Kluwe (eds). *Metacognition, Motivation and Understanding.* Erlbaum, pp.65-116.

Butler, P. (2006). *A Review of the Literature on Portfolios and Electronic Portfolios.* from: https://www.researchgate.net/publication/239603203_A_Review_Of_The_Literature_On_Portfolios_And_Electronic_Portfolios

Butterfield, A. (2012). *Employing Metacognitive Produces in Natural Science Teaching.* M.Ed. Thesis. Stellenbosch University.

Chantharanu-wong, W., Thatthong, K., Yuenyong, C. & Thomas, G. (2012b). Exploring the metacognitive orientation of the science classrooms in a Thai context. *Procedia Social and Behavioral Sciences.* 46, 5116-5123.

Collins, T. (2011). *Science Inquiry as Knowledge Transformation: Investigating Metacognitive and Self-regulation Strategies to Assist Students in Writing about Scientific Inquiry Tasks.* Ph.D. Dissertation. Oregon State University.

Davis, E. (2003). Prompting Middle School Science Students for Reflection: Generic and Directed Prompts. *The Journal of the Learning Sciences, 12*(1), 91-142.

De Corte, E. (1995). Fostering Cognitive Growth: A Perspective from research on mathematics learning and instruction. *Educational Psychologist, 30*(1), 37-46.

Desoete, A., Roeyers, H. & Buysee, A. (2001). Metacognition and Mathematical Problem Solving in Grade 3. *Journal of Learning Disabilities, 34*(5), 435-449.

Desoete, A., Roeyers, H. & De Clercq, A. (2003). Can off-line metacognition enhance mathematical problem solving? *Journal of Educational Psychology, 95*(1), 188-200.

Donnelly, B. (2010). *Digital portfolios and learning: the students' voices.* Ed.D. Thesis. University of California Davis.

DuToit, D.S. (2013). *The effect of metacognitive intervention on learner metacognition and achievement in mathematics.* Ph.D. Thesis. University of the Free State Bloemfontein.

Efklides, A. (2011). Interactions of Metacognition with Motivation and Affect in Self-Regulated Learning: The MASRL Model. *Educational Psychologist, 46*(1), 6-25.

Fisher, A. (2011). *Critical Thinking: An Introduction.* 2nd ed. Cambridge University Press.

Flavell, J. (1976). Metacognitive aspects of problem solving, in L. Resnick (ed). *The Nature of Intelligence.* Ealbaum, 231-235.

Flavell, J. (1979). Metacognition and cognitive monitoring: A new area of cognitive-developmental inquiry. *American Psychologist, 34*(10), 906-911.

Flavell, J. (1981). *Cognitive monitoring.* In W. P. Dickson (ed). *Children's oral communication skills.* Academic Press, 35-60.

Fogarty, E. (2006). *Teachers' use of differentiated reading strategy instruction for talented, average, and struggling readers in regular and SEM-R classrooms.* Ph.D. Dissertation. University of Connecticut.

Hacker, D., Dunlosky, J. & Graesser, A. (2009). *Handbook of Metacognition in Education.* Routledge.

Haidar, A. & Al Naqabi, A. (2008). Emiratii high school students' understandings of stoichiometry and the influence of metacognition on their understanding. *Research in Science and Technological Education, 26*(2), 215-237.

Harris, K., Reid, R. & Graham, S. (2004). Self-regulation among students with LD and ADHD. In B. Wong (eds.), *Learning about learning disabilities*, 3rd ed. Elsevier Academic Press, 167–195.

Hennessey, S. (1993). Situated cognition and cognitive apprenticeship: Implications for classroom learning. *Studies in Science Education, 22*(1), 1-41.

Hennessey, M.G. (1999). *Probing the dimensions of metacognition: Implications for conceptual change teaching-learning.* Paper presented at Annual Meeting of the National Association for Research in Science Teaching (NARST), Boston, MA.

Hoffman, B. & Spatariu, A. (2008). The Influence of Self- efficacy and Metacognitive Prompting on Math Problem solving Efficiency. *Contemporary Educational Psychology, 33*(4), 875-893.

Hohenshell, L. & Hand, B. (2006). Writing-to-learn strategies in secondary school cell biology: A mixed method study. *International Journal of Science Education, 28*(2-3), 261-289.

Hunt, R., & Ellis, H. C. (2004). *Fundamentals of Cognitive Psychology.* 7th ed. Steve Rutter.

Jacobs, J., Lanza, S., Osgood, D., Eccles, J. & Wigfield, A. (2002). Changed in children's self-competence and values: gender and domain differences across grades one through twelve. *Child Development, 73*(2), 509-527.

Kaplan, M., Silver, N., LaVaque-Manty, D. & Meizlish, D. (2013). *Using Reflection and Metacognition to Improve Student Learning.* Kindle.

Kaune, C. (2006). Reflection and metacognition in mathematics education- tools for the improvement of teaching quality. *ZDM: The International Journal of Mathematics Education, 38*(4), 350-360.

Kilpatrick, J. (1986). Reflection and Recursion. In M. Carss (ed.), *Proceedings of the Fifth International Congress on Mathematical Education.* Birkhauser, 7-29.

Kluwe, R. (1987). 'Executive decisions and regulation of problem-solving behavior'. In F. Weinert, & R. Kluwe (eds.), *Metacognition, Motivation, and Understanding.* Erlbaum, 31–64.

Kramarski, B. & Gutman, M. (2006). How can self-regulated learning be supported in mathematical e-learning environments? *Journal of Computer Assisted Learning, 22*(1), 24–33.

Kramarski, B. & Mevarceh, Z. (2003). Enhancing Mathematical Reasoning in the Classroom: The Effects of Cooperative Learning and Metacognitive Training. *American Educational Research Journal, 40*(1), 281-310.

Ku, K. & Ho, I. (2010). Metacognitive strategies that enhance critical thinking. *Metacognition Learning, 5*, 251-267.

Kuhn, D. & Dean, D. (2004). A bridge between cognitive psychology and educational practice. *Theory into Practice, 43*(4), 268-273.

Kuo, E., Hull, M., Gupta, A. & Elby, A. (2013). How students blend conceptual and formal mathematical reasoning in solving physics problems. *Science Education, 97*(1), 32-57.

Lai, E.R. (2011). *Metacognition: a literature review*. Pearson. From: https://images.pearsonassessments.com/images/tmrs/Metacognition_Literature_Review_Final.pdf

LaVaque-Manty, M. & Evans, E.M. (2013). 'Implementing Metacognitive Interventions in Disciplinary Writing Classes', in M. Kaplan, N. Silver, D. LaVaque-Manty & D. Meizlish (eds). *Using Reflection and Metacognition to Improve Student Learning Across the Discipline, Across the Academy: New Pedagogies and Practices for Teaching in Higher Education.* Stylus Publishing. Ch.6.

Manlove, S., Lazonder, A.W. & De Jong, T. (2007). Software scaffolds to promote regulation during scientific inquiry learning. *Metacognition and Learning, 2*(2-3), 15-31.

Mason, L. & Scrivani, L. (2004). Enhancing students' mathematical beliefs: An intervention study. *Learning and Instruction, 14*(2),153-176.

Mavrikaki, E. & Athanasiou, K. (2011). Development and Application of an Instrument to Measure Greek Primary Education Teachers' Biology Teaching Self-Efficacy Beliefs. *Eurasia Journal of Mathematics, Science & Technology Education, 7*(3), 203-213.

Metcalfe, J., & Shimamura, A. (1994). *Metacognition: Knowing about knowing.* MIT Press.

Mevarech, Z. & Fridkin, S. (2006). The effects of IMPROVE on mathematical knowledge, mathematical reasoning and meta-cognition. *Metacognition and learning, 1*(1), 85-97.

Nashon, S. & Nielsen, W. (2011). Connecting student self-regulation, metacognition and learning to analogical thinking in a physics problem-solving discourse. *Educational Research Journal, 1*(5), 68-83.

Osborne, J.W. (1998). *Measuring Metacognition: Validation and Assessment of Cognition Monitoring Effectiveness*. Ph.D. Thesis. State University of New York.

Ozsoy, G. & Ataman, A. (2009). The effect of metacognitive strategy training on mathematical problem-solving achievement. *International Electronic Journal of Elementary Education, 1*(2), 67-82.

Pajares, F. (2003). Self-efficacy beliefs, motivation, and achievement in writing: A review of the literature. *Reading and Writing Quarterly, 19*(2),139-158.

Parsons, M. (2011). *Effects of Writing to Learn in Pre-Calculus Mathematics on Achievement and Affective Outcomes for students in a Community College Setting: A Mixed Methods Approach.* Ph.D. Thesis. Colorado State University. Fort Collins, Colorado.

Phang, F. (2010). Patterns of Physics Problem-solving and Metacognition among Secondary School Students: A Comparative Study between the UK and Malaysian Cases. *The International Journal of Interdisciplinary Social Sciences,* vol.5, pp.1833-1882.

Pinon, K., Allain, P., Kefi, M., Dubas, F. & Le Gall, D. (2005). Monitoring processes and metamemory experience in patients with dysexecutive syndrome. *Brain and Cognition,* vol.57, pp.185-188.

Pintrich, P. (2002). The role of metacognitive knowledge in learning, teaching, & assessing. *Theory into Practice*, vol.41(4), pp.220 - 227.

Pintrich, P. & Schunk, D. (2002). *Motivation in education: Theory, research and applications.* Prentice Hall Merrill.

Pintrich, P., Smith, D., Garcia, T. & McKeachie, W. (1993). Reliability and Predictive Validity of the Motivated Strategies for Learning Questionnaire (MSLQ). *Educational and Psychological Measurement,* vol.53(3), pp.801-813.

Rahman, F. (2011). *Assessment of science teachers' metacognitive awareness and its impact on the performance of students.* Ph.D. Thesis. Allama Iqbal Open University.

Ramirez, G. & Beilock, L. (2011). Writing about testing worries boosts exam performance in the classroom. *Science*, vol.331(6014), pp.211-213.

Rando, W.L. (2001). Writing teaching assessment questions for precision and reflection. *New Directions for Teaching & Learning,* vol.87, pp.77 - 83.

Sandi-Urena, S., Cooper, M. & Stevens, R. (2011). Enhancement of Metacognition Use and Awareness by Means of a Collaborative Intervention, *International Journal of Science Education*, vol.33(3), pp.323-340.

Scherer, R. & Tiemann, R. (2014). The Development of Scientific Strategy Knowledge Across Grades: A Psychometric Approach. *Sage Open*, vol.4, pp.1-14.

Schoenfeld, A. (1992). Learning to think mathematically: problem solving, metacognition and sense making in mathematics. In D.A. Groues (Ed.), *Handbook of Research on Mathematics Teaching and Learning*. Macmillan, pp. 334-370.

Schraw, G. (1998). Promoting general metacognitive awareness. *Instr Scich EducEthics*, vol.26(2). pp.113-125.

Schraw, G. (2000). 'Assessing metacognition: Implications of the Buros symposium'. In G. Schraw, & J.C. Impara (ed). *Issues in the Measurement of Metacognition.* Buros Institiue of Mental Measurements, pp.297-323.

Schraw, G. (2001). 'Promoting general metacognitive awareness'. In H.J. Hartman (ed). *Metacognition learning and instruction.* Kluwer Academic Publisher, pp.3-16.

Schraw, G. & Dennison, R. (1994). Assessing metacognitive awareness. *Contemporary Educational Psychology*, vol.19(4), pp. 460-475.

Schraw, G. & Moshman, D. (1995). Metacognitive theories. *Educational Psychology Review*, vol.7(4), pp.351-373.

Schraw, G., Crippen, K. J. & Hartley, K. (2006). Promoting self regulation in science education: Metacognition as part of a broader perspective on learning. *Research in Science Education,* vol.36(2), pp.111-139.

Silver, N. (2013). 'Reflective Pedagogies and the Metacognitive Turn in College Teaching'. In M. Kaplan, N. Silver, D. LaVaque-Manty and D. Meizlish (eds), *Using reflection and metacognition to improve student learning: Across the Disciplines, Across the Academy: New Pedagogies and Practices for Teaching in Higher Education.* Stylus Publishing. Ch.1.

Slavin, R.E. (2011). *Educational Psychology: Theory and Practice.* 10th ed. Pearson International Edition.

Stefani A, Lozano A, Peppe A., Stanzione, P., Galati, S., Tropepi, D., Pierantozzi, M., Brusa, L., Scarnati, E. & Mazzone, P. (2007). Bilateral deep brain stimulation of the pedunculopontine and subthalamic nuclei in severe Parkinson's disease. *Brain*, vol.130(6), pp.1596-1607.

Stephens, K. & Winterbottom, M. (2010). Using a learning log to support students' learning in biology lessons. *Journal of Biological Education (Society of Biology).* vol. 44(2), pp.72-80.

Surgenor, P. (2011). *Tutor, Demonstrator & Coordinator development at UCD.* UCD Teaching and Learning Resources. From, http://www.ucd.ie/t4cms/Reflective%20Practice.pdf

Tarricone, P. (2011). *The Taxonomy of Metacognition.* Psychology Press.

Taylor, P.C., Fraser, B.J. & White, L.R. (1994, April). *The revised CLES: A questionnaire for educators interested in the constructivist reform of school science and mathematics.* Paper presented at the annual meeting of the American Educational Research Association.

Thomas, G. (2003). Conceptualisation, development and validation of an instrument for investigating the metacognitive orientations of science classroom learning environments. The Metacognitive

Orientation Learning Environment Scale–Science (MOLES–S). *Learning Environment Research*, vol.6(2), pp.175–197.

Thomas, G. (2011). Metacognition in science education: Past, present and future considerations. In B. Fraser, K. Tobin & C. McRobbie (eds.), *Second international handbook of science education*. Springer. pp.131-144.

Thomas, G., Anderson, D., & Nashon, S.M. (2008). Development and validity of an instrument designed to investigate elements of science students' metacognition, self-efficacy and learning processes: The SEMLI-S. *International Journal of Science Education*, vol. 30(13), pp. 1701-1724.

Tobias, S. & Everson, H. (1997). Studying the relationship between affective and metacognitive variables. *Anxiety, Stress, and Coping*, vol.10, pp. 59–81.

Van Hout-Wolters, B. (2000). 'Assessing active self-directed learning'. In R. Simons, J. van der Linden & T. Duffy (eds.). *New Learning*. Kluwer, pp.83-101.

Veenman, M. (2005). The assessment of metacognitive skills: What can be learned from multi-method designs? In C. Artelt, and B. Moschner (eds). *Lernstrategien und Metakognition: Implikationen fur Forschung und Praxis*. Waxmann, pp.75-97.

Veenman, M. (2011). Learning to self-monitor and self-regulate. In R. Mayer & P. Alexander (eds). *Handbook of Research on Learning and Instruction*. Routledge, pp.197-218.

Veenman, M. (2012). 'Metacognition in science education: definitions, constituents, and their intricate relation with cognition'. In A. Zohar, and Dori, Y. (eds). *Metacognition in Science Education: Trends in Current Research*. Springer, Ch.2.

Veenman, M. & Spaans, M. (2005). Relation between intellectual and metacognitive skills: Age and task differences, *Learning and Individual Differences*, vol.15, pp.159-176.

Veenman, M., Wilhelm, P. & Beishuizen, J. (2004). The relation between intellectual and metacognitive skills from a developmental perspective, *Learning and Instruction*, vol.14, pp.89-109.

Veenman, M., Van Hout-Wolters, B. & Afflerbach, P. (2006). Metacognition and learning: conceptual and methodological considerations. *Metacognition Learning*. vol. 1, pp.3–14.

Weinstein, C.E. (2006). Teaching students how to become more strategic and self regulated learners. In W.J. McKeachie & Svinicki, M. (Ed.), *McKeachie's teaching tips*. Houghton Mifflin, pp. 300-317.

Weinstein, C.E., Schulte, A.C., & Palmer, D.R. (1987). *LASSI: Learning and Study Strategies Inventory*. HH Publishing.

Wiezbicki-Stevens, K. (2009). *Metacognition: Developing Self-Knowledge Through Guided Reflection*. Ed.D. Thesis. University of Massachusetts-Amherst.

White, B. & Frederiksen, J. (1998). Inquiry, Modeling, and Metacognition: Making Science Accessible to All Students. *Source Cognition and Instruction,* vol.16(1), pp.3-118.

Whitebread, D., Coltman, P., Pasternak, D., Sangster, C., Grau, V., Bingham, S., Almeqdad, Q. & Demetriou, D. (2009). The development of two observational tools for assessing metacognition and self-regulated learning in young children. *Metacognition and Learning,* vol.4(1), pp.63-85.

Winne, P. H. (1996). A metacognitive view of individual differences in self-regulated learning. *Learning and Individual Differences*, vol.8, pp. 327–353.

Wu, L. & Looi, C.K. (2013). Incorporation of Agent Prompts as Scaffolding of Reflection in an intelligent Learning Environment. *Smart Innovation, Systems and Technologies*, vol.17, pp.369-391.

Yore, L. D., & Treagust, D. F. (2006). Current realities and future possibilities: Language and science literacy empowering research and informing instruction. *International Journal of Science Education*, vol.28(2-3), pp.291-314.

Zimmerman, B. (1995). Self-regulation involves more than metacognition: A social cognitive perspective. *Educational Psychologist*, vol.30, pp.212-221.

Zimmerman, B. & Cleary, T. (2006). Adolescents' development of personal agency: The role of self-efficacy beliefs and self-regulatory skill. In F. Pajares and T. Urdan (eds). *Self-efficacy beliefs of adolescents.* Information Age Publishing, pp. 45-70.

Zimmerman, B. & Martinez-Pons, M. (1990). Students' Differences in Self-Regulated Learning: Relating Grade, Sex, and Giftedness to Self-Efficacy and Strategy Use. *Journal of Educational Psychology*, vol.82(1), pp.51-59.

Zohar, A. (1999). Teacher's Metacognitive Knowledge and the Instruction of Higher-Order Thinking. *Teaching and Teacher Education*, vol.15(4), pp.413-429.

Zohar, A. & Barzilai, S. (2013). A review of research on metacognition in science education: current and future directions. *Studies in Science Education*. vol.49(2), pp.121-169.

Zohar, A. & Dori, Y. (2012) *Metacognition in Science Education: Trends in Current Research.* Springer.

CHAPTER 8

Critical and Creative Thinking in Science Education

Fatima Abazar

ABSTRACT

This chapter provides a critical analysis of creative and critical thinking research and how students can acquire it. It is widely believed by the psychologists and the educators internationally, and as presented through many of the research works, that the critical and creative thinking can be taught like other subjects. These thinking skills are either included in the teaching of a subject or they are learned separately. Creative thinking in the other side was identified by many researchers and its component of divergent thinking, capacity to produce numerous choices or solutions, attracted attention in the late 1950s, and since the time it nourished creativity has been fostered in convergent thinking, critical thinking, associational and analogical thinking. So, when we speak about critical and creative thinking in this chapter, we aim to discuss the status and the possibility of imparting students with the creative - not the ordinary- critical thinking skills.

INTRODUCTION

It is widely believed by psychologists and the educators internationally and educators in the UAE context, that thinking skills are to be included in teaching due to its essentiality in everyday life of the contemporary societies (Assaf, 2009; Fisher, 2003; Forawi & Mitchell, 2012; Kaddoura, 2011). As proven through many of the research works; the critical and creative thinking can be taught like any other learning stuffs. These thinking skills can either be included in the teaching of a subject, or they may be learned separately (Fisher, 2011; Weil & Kincheloe, 2004). Therefore, Instructors and curriculum developers need to understand, believe and value critical and creative thinking to better teach it (Moore, 2013).

A start-up definition of critical thinking is that it "is the art of thinking about thinking in order to make thinking better. It involves three interwoven phases: it analyzes thinking, it evaluates thinking, it improves thinking" (Paul & Elder, 2019, p.xvii)

As a concept, Paul and Elder (2010) considered several definitions together to form the following definition, "Thinking explicitly aimed at well-founded judgment, utilizing appropriate evaluative standards in an attempt to determine the true worth, merit, or value of something" p. xxiv. Siegel (1988) proposed in his book that the critical thinker is the one who is moved appropriately by the reasons and who can assess the situations in many contexts depending on the available evidences. A study by Moore, (2013) reported the investigated ideas about critical thinking as held by faculties working in three academic fields: history, philosophy and cultural studies. No less than seven definitional strands were identified in the informants' interviews, critical thinking: (I) as skepticism; (ii) judgment; (iii) as sensitive reading; (iv) as simple originality; (v) as rationality; (vi) as self-reflexivity; and (vii) as an extremist engagement with knowledge. This assortment of meanings is thought to have important ramifications for college instructing and learning. The outline of the investigation and the decisions drawn from it draw vigorously on Wittgenstein (1968) concept of meaning as use. In this manner, in the demonstration of endeavoring to comprehend and welcome these 'varieties of tradition and experience', what might emerge, suggested by Williams (1979) is an 'additional edge of awareness'. This evocative defining, which proposes an essentially empathic perspective of knowledge and of its makers and purveyors, might be as great a definition as any for the difficult term people have been thinking about. Undoubtedly, in endeavoring to understand 'critical thinking', and in working out how it may be best educated, it might be that it is mainly quality 'an additional edge of awareness' that we should plan to energize in students, in ourselves as educators, and furthermore on the world generally, regardless of the numerous difficulties that we face (Moore, 2013).

Creative thinking in the other side, was identified by many researchers and its component of divergent thinking attracted attention in the late 1950s (J. Park, 2011; W. Park, 2017). Divergent thinking as per W. Park (2017) is the capacity to produce numerous choices or solutions. And since the time it nourished creativity has been fostered in convergent thinking, critical thinking, associational and analogical thinking (J. Park, 2011). So when we speak about critical and creative thinking in this chapter we aim to discuss the status and the possibility of imparting students with creative not ordinary critical thinking skills. Laycock (as cited in Nadjafikhah et al., 2012)portrayed mathematical creativity as a capacity to break down a given issue from an alternate point of view, see patterns, contrasts and similarities, create various ideas and pick an appropriate technique to manage new mathematical circumstances. This type of thinking (as cited in Kandemir & Gür, 2009) is possible to be created as indicated by multiple researchers like Silver for example, who claims that creative thinking can be produced by practicing open-ended questions, nonetheless Meissner puts emphasis on the possibility that creative thinking can be produced through routine solving of challenging questions, and Fisher attracts attention to the need of solving questions relevant to daily life(Kandemir & Gür, 2009). Daud et al. (2012) showed that creative model indicates an arranged creative process that includes a critical and objective analysis, innovative idea and produce a critical evaluation. The procedure includes as per Daud et al. a balance between creativity, innovation and analysis. In the old approach, innovative ideas were thought to result from subconscious thoughts that are often cannot be controlled by the person. Thusly, considering Daud et al.' creativity model, strategies, and ways to deal with producing inventive thoughts are controlled by the person's thinking. This procedure requires the execution of actions and thoughts. It stresses on the quality of the imagination to produce new thoughts; however, we should likewise make this imagination a reality of nature (Daud et al., 2012).

United Arab Emirates society is a muslim society. If we are discussing critical an creative thinking in its context we are to mention that some Muslim researchers such as Al Karasneh and Mohammad Jubran Saleh (2010); have seen creative thinking as the mindfulness, usage or broadening of thoughts. Some muslim researshers even accentuated appropriate implementation of creativeness in all parts of life (Daud et al., 2012). In the meantime, Daud et al. think that the person must have an awareness of other's expectations to maintain the Muslim societies to the up next level, ready to confront future difficulties with effective and creative solutions. In this way, Al-Mazeidy (as cited in Al Karasneh & Mohammad Jubran Saleh, 2010) states another measurement in the meaning of creativeness from an Islamic point of view; which is "the capacity to make plans that can profit to mankind that depends on Shariah and Islamic standards." (p. 306).

Numerous specialists concede to the area-reliance of creativity, backing their believes with Gardner's Multiple Intelligence Theory (Park, 2011). That area-reliance implies that creativity in science may have different perspectives compared to creativity in literature or in Art. In this way, Prof. J. Park proposed a 3-dimensional model of scientific creativity consist of three dimensions: creative thinking, scientific knowledge, and scientific inquiry skills. On the off chance that somebody recommends another test technique while leading a "test" (scientific inquiry skills), identified with "Faraday's law" (scientific knowledge) by "thinking divergently" (creative thinking), it is said as pointed by Park, that his/her new experimental procedure has been developed via the feature of scientific creativity. Despite the prevalent view, J. Park pointed that, Weisberg, who broke down Watson and Crick's procedure of finding the DNA structure, inferred that there was no specific mindset in their procedure of the study. Besides, Prof. J. Park reported that researchers stress that creativity is more of a thinking custom, instead of an intellectual capacity. Following that rationale, creativity can and ought to be instructed to the ordinary, as well as the talented students, and people need to put more weight on outlining a science curriculum that includes scientific creativity for all students. Obviously, some people may rase the point of additional time. If we propose that an incubation stage is important for students to think inventively, this additional time could be seen by many as an obstruct for teaching scientific subjects in normal science classes where the time plan for running science curriculum is generally pre-decided. Nonetheless, as indicated by the view that creativity is a habit of thinking, short however iterative experiences of imaginative reasoning in science learning can be a conceivable solution for this snag (Park, 2011).

This chapter discusses the topic of critical and creative thinking in science education at the UAE in a thematic way after an intensive review of available literature. Following this introduction , there is a review of literatures dealing with the historical evolution of the critical and creative thinking in the world, the importance of this type of thinking, possible ways to produce and improve critical and creative thinking, assessment and possibilities of CT evaluation and the last section in the reviewed literature explores the critical and creative thinking issues particular to the United Arab Emirates (UAE) context. A discussion followed by summary and recommendations upon the reviewed material ended the chapter.

REVIEW OF THE LITERATURE

Evolution of Critical and Creative Thinking in Science Education.

Scientists have been linked to an image of isolated researchers working in laboratories with chemicals and instruments., wearing white coat, having untidy look. Such view was formed due to a science education system that has been taught as facts to be memorized and recalled in tests. It also required students to read thick textbooks and follow steps look like recipes to achieve experiments in the laboratories and reach certain outcomes. Such approach in teaching science didn't allow students to enjoy the potential and richness offered by the study of science. In such system also, teachers will not be able to be creative. The traditional approach of science teaching produced a meritocracy where two categories of students formed. A smaller group of students interested in science and was able to excel, and a bigger group slowly quit science study. Till 1983 the traditional approach was dominant in U.S. In 1983; the document of a nation at risk was published. The document invited for educational reform in the U.S. (Weil & Kincheloe, 2004). The creative thinking attracted attention as per J. Park (2011) in the late 1950s after the "sputnik shock" (J. Park, 2011). From that point forward, various special projects for encouraging creativity in critical, convergent, analogical or associational thinking have been created. Accordingly, an article in 1993 detailed that there were around 250 projects for enhancing creativity, and in 1998, a scientist characterized different teaching techniques utilized as a part of the projects into 170 techniques. Even though there have been wrangles about whether creativity is area-related or area-free, creativity is no ifs ands or buts an essential factor that characterizes the idea of science. Accordingly, numerous science teachers have focused on that learners need to recognize the inventive work required by nature in scientific fields. Project 2061 of the American Association for the Advancement of Science (AAAS) expressed that inquiry exercises, for example, creating hypothesis are creative work like compositing poetries; National Science Education Standards of the National Science Teachers Association (NSTA) in U.S.

noted additionally that understanding science and the procedures of science can add to the creative thinking. The recently reviewed Korean national science curriculum had also focused on the significance of creative thinking in instructing and learning science in schools. This flag sustaining creativeness has turned out to be one of the fundamental objectives of science learning in schools. Despite the long history and much exertion for enhancing instructing of creativeness, be that as it may, little studies for teaching creativity in the zone of science education have been accomplished. The picture of the critical thinker scientists has also become clear in the contemporary society, and according to Weil and Kincheloe (2004); that scientist recognizes the science he/she does and the decisions he/she makes in everyday life. He/she knows that the development in technology gave him/her access to information that needs evaluation with much more thinking than in the past. but at the same time, he/she is aware that science is tentative, and scientists need to question their assumptions continuously and they know that science is human field that is affected by politics, inconsistencies and errors like any other social activity (Weil & Kincheloe, 2004).That was supposedly the contemporary scientist, what about the contemporary educators scientists? They are the ones mostly responsible for graduating contemporary scientists. A study by Nicholas and Raider-Roth (2016) reveals the status of those educators in United States' two universities which are part of the voluntary system of accountability for reporting general education outcomes, such as CT. Nicholas and Raider-Roth study showed a misalignment between the educator's ideas about CT and their assessment of the CT abilities of the students. The misalignment was also found between the institutional approach and educators approaches to CT.

The Importance of Critical and Creative Thinking

Is it essential that our thinking become critical? What about non-critical thinking? What about being skilled in knowledge acquisition? In fact, knowledge acquisition is a dangerous habit that forms an obstacle in the way of any discovery. Because the illusion of knowledge as de Bono call it, will trap people in what they think they know and will not be open to new ideas (Fisher, 2003). Developing our minds is part of being educated, it is an essential to the person development, and every human have the right to develop his/her intellect. To develop as an educated human; our mind should think creatively and critically. Another reason for the need to think critically is to feel the pleasure of solving puzzles, Professor John Rawls expressed the link between problem solving and pleasure as the enjoyment of exercising the rational capacities. The more these capacities are the greater is the pleasure. (Fisher, 2003)

Teaching thinking is not a human right and a means for enjoyment only, thinking is essential to prepare the generation for unknown future issues in a rapidly changing world. Critical thinking is also essential to make sense of knowledge in any field. One of the major benefits also of the critical thinking is raising the moral quality of the person. Such moral qualities are like being open-minded, respect others, self-examination and perseverance (Fisher, 2003). As per GUNN et al. (2008) critical thinking makes the foundation for ethical and proficient consumers of scientific change.

Promoting Critical and Creative Thinking in Science Education

Two important characteristics of science have implications about linking it to philosophy and the pedagogical choice of teaching it. One characteristic is that science is generative; it is produced and formed out of knowledge. The other characteristic is, it needs understanding of the used concepts in describing knowledge and interpreting it into science. (Fisher, 2003). So accordingly, Fisher (2003) indicated that reasoning is one of the commonalities between science and philosophy. In philosophy it is the conceptual reasoning, and in science it's the empirical reasoning. To relate two or more facts to form science; analogical reasoning is needed. But as science is tentative, positive creative reasoning is also needed continuously to alter old ideas and old thinking. To teach scientific creativity, Park with his research colleagues suggested a learning model consisting of three steps: Spontaneous Activity, Guides for creative thinking, and Activity Again (AGA2 model). As indicated by this model, students are first given an issue to be illuminated, for example, to 'update the introduced common electroscope for a more assorted use', or to 'recommend new and fascinating scientific circumstances in which unordinary phenomena may be appeared'. Here, students try solving the assignments independently with no guides or aides. Along these lines, a few students can demonstrate a high number of creative thoughts, while some other may comprehend the assignments utilizing their common sense or in rather traditional ways. Or then again, a few students may experience issues in suggesting creative ideas. In this manner, in the second step, students are given directions or rules to enable them to think innovatively. 'Think contrarily', 'change the conditions', or 'utilize other similar circumstances' are great examples of such rules. In the wake of practicing how to think creatively to comprehend and solve the issues, students apply the given guides in new situations in the last step. Creative thoughts frequently can be produced by gathering, sharing, studying about different thoughts. Consequently, in different models, students are urged to share the underlying ideas with others to update or refine them through extra steps, examining the points of interest or disservices of their thoughts. Considering models, for example, above, different scientific creativity exercises have been as of late created and connected to students. However, there are yet numerous studies to be led identified with educating scientific creativity in schools. Prof. Park and his colleagues is recommending that more concrete and variety of bits of proof is first needed to demonstrate that scientific creativity can be supported by suitable training and instruction. Obviously, numerous specialists have revealed that their endeavors for enhancing creativity turned out to be viable. Be that as it may, huge numbers of those studies were about general creativity, not scientific creativity. For instance, Torrance demonstrated the viability of teaching creativity in different studies utilizing Torrance

Tests of Creative Thinking (TTCT), which was produced to test creativity. For this situation, the quantity of ideas is considered to test fluency, one of the characterizing components of divergent (creative) thinking. However, to be scientifically creative, more conditions should be available. That is, we must check which thoughts are quite conceivable or more helpful for scientific inquiry. Be that as it may, this sort of appraisal of scientific creativity is not so famous in the region of science instruction yet. Strikingly, numerous individuals believe that creative thinking does not generally apply to ordinary individuals or that it can't be created by commonplace assignments or in normal settings. Subsequently, some science instructors additionally feel that teaching creativity suits just for skilled, or talented students. Truth be told, the zone of skilled instruction has underscored educating and feeding creative thinking. Also, some science educators attempt to utilize irregular, peculiar, or in some cases, extremely hard to-solve situations when creating materials for teaching creativity, which can't be experienced in ordinary life setting. Along these lines, J. Park (2011) with his research team have built up another model called the 'small scale iterative experiences for teaching scientific creativity'. Their future studies will concentrate on discovering alternative approaches to imbed this model into the common science curriculum, and to gauge the viability of these techniques in enhancing creativity. Besides, Prof. Park and his team will strive for an extra evidence that these methods for creative thinking are still in connection with the nature of learning science in understanding scientific knowledge, applying scientific inquiry, and in supporting scientific thinking.

If the lack of creative CT in graduates pushed the responsible partners to approach organizations of higher education and staff to imbed CT in students; an accountable reaction is required from both academic staff and foundations. Such a reaction must be marked by the integrity and quality of the practice of the instructors and organizations of education. Having the capacity to answer inquiries concerning students learning is fundamental to the act of educating and learning. We must move from a "hopeful pedagogy" to supporting that hope with proof of student learning. Such act won't just fulfill the necessities for accountability however more imperatively, build up the integrity and quality of the service of instructors and foundations of higher education (Nicholas & Raider-Roth, 2016). In education in general, researchers worked hard to develop a theory and pedagogy to help students from kindergarten to college level to develop their thinking to become critical and creative. Researchers admitted that traditional teaching methods that make students passive while the instructor is the active element in the teaching process does not help in developing creative critical thinking in the students (Fisher, 2003).

As the Human Terrain System training and instruction directors in US. perceived how imperative critical thinking skills were to the graduates, and how ineffectively some were applying those aptitudes, they started to reconsider what was educated in their curriculum, how it was educated, and how the outcomes were evaluated (Griffin & McClary, 2015); for the design and administration of high-performing groups; the managers conducted research for five years. They reached and counseled experienced team members of human terrain and military administrators and staffs. They tried different things with various program plans and studies took place until, at long last, they recognized five keys to educating basic critical thinking aptitudes so human territory team members would utilize these aptitudes subsequent to leaving the classroom: 1. A compelling ability administration program to advise procuring and task rehearses and guarantee person work fit (with fit in this setting means compatibility) and person organization fit for team members and staff 2. An organizational atmosphere that esteems, expects, and remunerates critical thinking and development 3. A mutual comprehension of the particular critical thinking skills and practices most imperative for at work achievement 4. educators who incorporate basic critical thinking skills into all classes and viably show and model these skills 5. An exhaustive assessment program to assure organizational readiness. (Griffin, 2015).

In the health sciences education, huge shift in the Health Sciences Reasoning Test (HSRT) scores happened among first-year student pharmacists in the university of Mexico health sciences center following a course intended to educate foundational and explicit CT abilities. Incorporating dynamic learning while utilizing simulation, formative feedback, and clinical reasoning based on knowledge that students were acquiring in other courses and was put into the setting of patient-centered care. This study indicated that the HSRT could gauge the effect of an explicit critical thinking curriculum among pharmacy students. The outcomes speak to a noteworthy change and a potential clinically applicable progress for students in their critical thinking and clinical reasoning abilities. On the off chance that professional associations concur that critical thinking skills are essential for pharmacy students to master in their fields; these outcomes could be the impulse for the education system to coordinate foundational and explicit CT educational modules plus utilizing the HSRT or other approved test to evaluate its impact on students. (Cone et al., 2016)

Not only in military and pharmacy; but in the Nursing field as well, a study done by Kaddoura (2011) for a sample of the third-year nursing students enrolled in the nursing diploma program offered by the ministry of health Institutes of Nursing in the United Arab Emirates (UAE). The study revealed that it is basic that nurses have the capacity to think critically to confront the difficulties of the present quick paced innovatively propelled nursing practice. The didactic lecturing and case-based learning (CBL) techniques for instructing differ for the most part, in complexity and level of obligation set upon the undergraduates. the CBL students learn by cooperating in groups to achieve their assignments or exercises of shared learning objectives. undergraduates instructed in both procedures learned the content, but the CBL philosophy seemed, by all accounts, to be more successful in creating CT abilities for nursing undergraduates than the traditional lecturing (Kaddoura, 2011)

Creative and Critical Thinking Assessment.

Dr. E. Paul Torrance, "Father of Creativity," as per Kim (2006) is best known for creating the Torrance Tests of Creative Thinking (TTCT). The TTCT was produced by Torrance in 1966. It has been reformed in 1974, 1984, 1990, and 1998. There are 2 frames (A and B) of the TTCT-Verbal and 2 shapes (A and B) of the TTCT-Figural. As it were, even though the tests have been utilized for the most part for appraisal in the distinguishing proof of skilled kids, Torrance initially intended to utilize them as a mean for individualizing guideline for various students considering the test scores (Torrance, 1974). The test may yield a composite score (the Creativity Index[CI]), however Torrance disheartened translation of scores as a static measure of a man's capacity and, rather, contended for utilizing the profile of qualities to comprehend and support a man's creativity . Accordingly, the reasons for the TTCT are for experimentation and research, for instructional arranging, for general use, and for deciding strengths of the students (Kim, 2006).

In CT assessment, Verburgh et al, (2013) reported that the several concepts and dimensions of CT is reflected in a diversity of discipline- general and discipline-specific tests of CT. Even in the discipline-general tests, the developers place a different emphasis on specific aspects of CT. To demonstrate few examples, The Watson-Glaser Critical Thinking Appraisal, aims at measuring CT as it is defined by Glaser (as cite in Verburgh et al., 2013). the Reasoning about Current Issues Test (RCI) is grounded on the Reflective Judgement Model by King and Kitchener (as cite in Verburgh et al., 2013). The Cornell Critical Thinking Test (CCTT) is based on the Cornell/Illinois model of CT. The Halpern Critical Thinking Assessment (HCTA) is based on Halpern's definition as a last example, California Critical Thinking Disposition Inventory (CCTDI) claims to quantify the disposition or slant towards CT, as characterized by Facione (as cite in Verburgh et al. 2013)

In U.S., The Educational Testing Service (ETS) has composed a next generation evaluation-the HEIghtenTM critical thinking assessment- to gauge students' critical thinking aptitudes in synthetic and analytic measurements (Liu et al. 2016). The Health Sciences Reasoning Test (HSRT) as per Cone et al. (2014), is also a validated test to evaluate CT skills that would be superior to other educational activities or evaluations at foreseeing the measures of achievement.

IN THE UAE CONTEXT.

The Need for Critical and Creative Thinkers in the UAE Society

The education status in the UAE 2012 is set in the world economic forum as follow:

> Education and innovation: one of the UAE's key challenges is to ensure that its education system provides students with the skills demanded by its growing private sector, thereby helping to diversify the country's industries and redressing the demographic imbalance in its workforce. The scenarios demonstrate that ensuring highly-qualified Emirati workers with relevant skill sets are available in an innovative economy is crucial to the country,

> both in terms of capitalization on present oil wealth and achieving its own goals on the long-term economic stability" (World economic forum, 2012, p.8).

The teachers in the UAE then need to know and physically apply the ways of teaching thinking skills, which is difficult to be done if the teachers are not aware of the reasons behind teaching these skills, and if the teachers are not having continuous professional development program enabling them to teach in a style that promote critical thinking in the students. An example of a model for professional development of educators in the UAE, is the one developed by Prof. Forawi (Mansour & Al-Sharmani, 2015). Forawi's model is called the Science Teacher Professional Growth (STPG) model. Forawi had drawn three possible paths for planning and delivery of teachers' professional development; 'train the trainer' where a lead teacher is to attend a training by experts (in CT teaching for example) and then to deliver what he had learned to colleagues, 'school based' where a set of school's long goals to be set and professional development to take place accordingly, the third path is 'self-directed' where the professional development happen by initiative and activities done by the teacher himself. STPG involves the science educator in planning for the professional development and suggest rewarding that educator upon any evidenced achieved professional development considering the local context of the UAE educational system (Mansour & Al-Sharmani, 2015).

A study by Alzahmi and Imroz (2012) indicated that a discussion of the remarkable factors that impact the UAE's workforce education and improvement system will enable people to better understand the system's operation. Thusly, this analysis of the factors will empower important comparisons with systems in different countries and uncover fruitful systems that are in accordance with the requirements of the countries in which they are installed. These components were set as: 1) financial status, 2) social capital, 3) human capital, 4) physical environment, and 5) government climate. So, the new curriculum should concentrate on technical aptitudes and critical and rational thinking and the country's strategic plan ought to likewise be considered when growing new curriculum (Alzahmi & Imroz, 2012). As per Ashour and Fatima (2016) there is much potential on which higher education in the UAE can develop. This incorporates, firstly, the state's dedication towards building an aggressive competition education system, also, having settled, progressed advanced education foundations that are spread over the seven emirates, thirdly, having a very much organized quality confirmation framework set up, and, fourthly, having a focused situation that is pulling in lofty organizations to set up branches in the UAE. For the UAE to keep advancing in propelling its educational system and to better place itself in the district as an educational center point that pulls in students from abroad to study, in this way expanding enrolments and maintainability of its organizations, certain issues should be considered. These issues incorporate exceeding quality over amount of the graduates and in addition the establishments. The quick development of organizations should be controlled by putting more confinements on new foundations and appropriate statistical surveying led preceding opening any new establishment. The way that the private division is a noteworthy supplier of higher education in the UAE should make the government more careful and extremely strict as to the usage of the quality affirmation framework, the nature of projects, and the nature of graduates being produced (Ashour & Fatima 2016).

To reveal the status of CT in the UAE 2006 - 2016 the following studies may be of a benefit. The one by Dickson et al. (2015) demonstrated that since 2006, government funded schools in Abu Dhabi have been experiencing tremendous changes thus educational reforms. As a major aspect of a long haul strategic and operational plan to drastically adjust the educational system, public elementary schools were staffed by expatriate educators known as English Medium Teachers (EMTs). These educators were to teach science, math and English and were prepared in English medium colleges and selected from outside the UAE. The idea behind this recruitment strategy was that these instructors being taught, prepared and having worked in developed nations with since quite a while ago settled education system, would carry with them what is expected to be 'best practice' and execute it in schools of Abu Dhabi. Private schools in Abu Dhabi have a long history of assorted variety, yet a long history of selecting educators from the same background of the government funded schools' teachers, for the same reason of "best practice". Along these lines, science in both government funded schools and private schools is instructed mostly by expatriates, and nearly having a few years of involvement in their countries' educational systems. The U.A.E. had 483 private schools and 702 government schools in 2012 (UAE National Statistics Bureau, 2012 a), a figure which has been expanding even from that point forward as the proportion of expatriates to nationals keeps on expanding. The rates of nationals in the UAE remained at 11% at the 2010 evaluation (UAE National Statistics Bureau, 2012 b). The emirate of Abu Dhabi had 183 non-public schools in 2013 and is under expanding strain to make considerably more schools for the expanding populace (as cited in Dickson et al., 2015). Non-public schools in Abu Dhabi are enormous in assortment, tailing anything from British, American, Canadian to Indian and Pakistani Community Schools, however all need to take after parts of Abu Dhabi Education Council (ADEC) rules are liable to visit assessment by ADEC. Government funded schools are clearly substantially more intensely checked and entirely stick to the New School Model, ADEC's educational programs which is an adjustment of the Australian New South Wales educational modules executed at the beginning times of the changes in 2006.

What constitutes 'best' practice in the science classroom, precisely? ADEC's New School Model Teacher Guide underlines inquiry-based learning, comprehended as being experience-centered, with tactile experience assuming a huge part in the request (as cited in Dickson et al., 2015). ADEC characterizes inquiry-based learning in the classroom extensively as when students may be "cooperating, developing significance through coordinated effort with others, participating in basic reasoning and critical thinking" (ADEC New School Model Teacher Guide, 2013, p. 21). Inquiry-based learning is for the most part comprehended by science instructors, enables students to end up noticeably more self-coordinated and progressively free and self-ruling scholars. The study by Dickson and others (2015) demonstrates that there are exceedingly measurable huge contrasts in announced practices of science instructors in private and public schools, with private school educators showing practices, such as, collaborative, student-centered, and inquiry-based learning, in a more prominent arrangement with the acknowledged 'best practice' (Dickson et al., 2015).

In line with the previous article, a report titled ‘Ministry of Education introduces new world-class curriculum to enhance UAE's education system’. In 2016, by a news reporter-staff at “Education Business Weekly” reported that His Excellency Hussain bin Ibrahim Al Hammadi, Minister of Education for the United Arab Emirates declared another world-class educational program to upgrade the UAE's instruction framework. The seven-year contract has been signed with McGraw-Hill Education, a learning science organization, for all K-12 math and science instructional materials in digital and print books. The substance for each curricular program was chosen in view of the most exceptional and dynamic material created for U.S. principles and is lined up with the UAE National Standard Framework. All materials were made in Arabic and conveyed in August for the 2016/2017 academic year. Al Hammadi, Minister of Education for the UAE commented "The new instructional materials provided by McGraw-Hill Education will help us create the UAE of tomorrow,". "Our economic growth depends on investing in education to build a knowledge-based society, and we have made massive strides for the children, women and men of the UAE. These new instructional materials are one more step on our journey of providing world-class education for our citizens and residents." "It is the leadership's vision to provide students in the UAE with 21st century skills, education is a fundamental element for the development of a nation and the best investment in our youth. The new instructional materials clearly align school curricula with the country's vision of building a -robust knowledge and innovation economy." (McGraw Hill, 2016)

Critical and Creative Thinking Promotion in the UAE

According to several researchers (Forawi & Mitchell, 2012; Cone et al., 2016) institutions of the recent times must instigate the courses of thinking skills in their curriculum so that the students get the platform for developing their thinking skills. Examples of studies and initiatives made for the development of critical and creative thinking within the students are to be demonstrated here. A local example is (BEST) or Building and Enhancing Skillful Thinking, which was proposed by Assaf (2009) who is teaching in a school at Abu Dhabi. The program is an integration of two different techniques from two different thinking teaching approaches. It perceives writing as a thinking process that needs students to think creatively and critically. Another study conducted by Abuzaid and Elshami (2016) in Al Sharjah University-UAE, revealed that the application of scenario-based simulations in radiology instruction prompted a positive effect on learning results, formative interactive learning, and filling the hole amongst theory and practice. Also, it advanced critical thinking abilities and enabled radiology professionals (educators) to exhibit their knowledge of similar cases.

Those were few examples of initiatives and educators' efforts in enhancing CT in science education the UAE. Samples of local research papers are also following.

CT** Research Papers in the **UAE

Clarke and Otaky (2006) contend, as opposed to Richardson (2004) who sees the assumptions of reflective practice as incongruent with the values and beliefs of "Arab– Islamic culture", that culture can be conveniently comprehended as a never-finished site of competing verifiable and social talks, instead of as a got set of values and beliefs. Clarke and Otaky (2006) wanted to underline "the given and the conceivable" as opposed to only the "given" with a specific end goal to oppose what we see as another type of social imperialism (as cited in Clarke & Otaky, 2006).

According to Clarke and Otaky, we advocate a perspective of reflection as a "human" capacity like our capacities to make and utilize language and other "tools of the mind", even though the specific structures it takes will be formed by cultural, social and historical elements. In a comparable manner, we see figuring out how to instruct as the arranged apportionment and re-development of social and instructive talks, which shapes some portion of a progressing procedure of self-creating a way of life as an educator. Inside this system, we see reflective practice as an instructive talk accessible for student suitability as a component of their progressing personality development. This talk accentuates self-awareness, qualities of discourses, and a problem-solving orientation to the classroom as significant and suitable for instructors. Clarke and Otaky have introduced confirmation of Emirati instructors talking about, reflecting on, and taking part in reflective practices, both for themselves and with their own students. These cases of Emirati undergraduates -instructors exhibit according to Clarke and Otaky that they consider themselves to be both reflective, critical thinkers and as truly necessary specialists of progress in UAE education. And the researchers trust their discourse and specifically their students' words have offered something to think about to individuals who consider "culture" to be a frustrating requirement and a deterrent to Emirati undergraduates" engagement with reflective practices. All in all, they contend that if the Higher Colleges of Technology (HCT) is to fulfill the call for enhancing the nation's educational system, they can't preclude the reflective practice paradigm, i.e. a worry with deliberative ideas, reframing of issues, and the self-awareness (Roberts, 2016), for educator training in the UAE. We trust this is especially the case inside a system that perspectives students as full members in a progressing procedure of co-developing and re-framing their teaching knowledge inside their current experience and understandings.

One of the studies that aimed to assist in enhancing critical and analytical thinking skills at higher education level in the developing countries and took place in the UAE; is the one by Taleb and Chadwic(2016). These researchers took the case of the British University in Dubai (BUID) and studied it. The main findings include the suggestion that creating a single module for teaching critical thinking would be useful rather than formulating a new regulation or policy. Enhanced awareness of the concept among academics and encouraging them to apply critical thinking objectives to their module outcomes was also suggested. Provision of social learning environments such as those inherent in peer group works and small group activities to enable students to see different perspectives was another suggestion. Setting up a solid domain can help students to upgrade their critical thinking aptitudes and design suitable educational experiences. The researchers were also interested to find that completing this exploration has additionally had the startling effect of raising the consciousness of CT among the faculties at BUiD (Taleb & Chadwick, 2016).

DISCUSSION & CONCLUSIONS

Creative critical thinking is found to be a multi-dimensional concept that has many steps to be followed by the thinker, and the educational system in the country is fully responsible to physically train the citizens to practice creative critical thinking, and only by then the individuals will be able to solve open ended problems, analyze issues, imagine solutions, apply those solutions, evaluate those solutions and make necessary improvements. In my point of view, considering critical and creative thinking skills the most essential cognitive skills to be achieved; pushes the educational curriculum developers to give it the bigger percentage among the outcomes in the science curriculum. In the opposite side, content of knowledge in the science curriculum to be reduced and to be left for the students to acquire it by themselves. The assessment tools are accordingly to be selected and utilized to assess the same ratio of skills to content as the outcomes. So, the scientific content in the reformed curriculum is to be considered as a mean to master the aimed creative critical thinking skills and not to be targeted in itself.

About the direction of the curriculum reform, learning outcomes-wise, I would say the reform should be gradual from bottom to top (KG higher education) in the educational level. But pedagogical-wise the reform should happen instantaneously in all levels by giving the educators more space of applying CT pedagogies without worrying of being evaluated or complaint against by students. Institutions should also urge the newly recruited instructors to apply the CT-enhancing pedagogies while continually training the previously existing instructors to master these pedagogies promoting critical and creative thinking. The plan for professional development should involve the instructors' ideas and views and should depend on models formed and tested locally such as the UAE Science Teacher Professional Growth Model (STPG) developed and proposed by Prof. Forawi as per Mansour and Al-Shamrani (2015). That is to align with the vision of the government toward investing in education to achieve knowledge-based-economy.

Besides considering it to be a mean utilized in developing scientific fields and accordingly develop societies, critical and creative thinking in science education should at the same time be a learning outcome to be achieved when teaching students. I would say it is like a plant; depending on the seed planted by the educational system, the fruit can be expected. A lemon seed cannot give an apple tree. And teaching strategies that are far from promoting CT cannot form a critical thinker into a future educator. To close the previous loop (teacher- learner – future teacher); a soil and a climate with certain specifications monitored by people and institutions of the society must maintain and support a sustainable CT educational environment.

When we speak about the institutional climate in the UAE, it's good to discuss the impact of that climate on the educator and the student as well. From the reviewed literature it's clear that the UAE represented by its government is fully realizing the urgent need for equipping citizens with the skill of creative critical thinking. But it's also clear and as reflected by the perspectives of most educators in the public and private educational institutes; the institutional climate hinders the educators from taking the step of changing their traditional styles of teaching, especially as we go up the educational level. As the expatriate or the EMT educators who are according to Dickson et al. (2015) the qualified human resources to apply the new reformed curriculum- are predominant, and at the same time as they are not ready to waive their jobs after relocating to live in the UAE and waiving jobs at their home countries. These educators find it easier to adapt to the institutional culture set by students coming from a non-critical and a non-creative thinking educational experiences, and above all, they contribute to the failure of the instructors' efforts to conduct a CT promoting session by writing a disappointing student-instructor evaluation. This type of students (mostly in high schools and higher education institutes) prefer not to be the centre of the learning process, they find it easier to listen passively to the lecture and leave the classroom without putting an effort to obtain the information. CT then will be missed in the learning process, with the educator's full awareness of that. The educator then may imbed few techniques during teaching to be added to the course file as a CT enhancer. That course file or the few expected classroom evaluation sessions will be the element that satisfy the institute and decision makers in the country. But here comes the role of the institutes of keeping a closer eye at the actual teaching happening in the classrooms and keep the students-instructor evaluation optional for the teacher to give it or not to the students, the teacher may benefit from it as a confidential feedback for him, to access it and reflect on his own practice. In my opinion as an educator, the two previously recommended actions (getting rid of students' evaluation and intensifying the classrooms inspections) should accompany each other. Having one without the other will make the undesirable seen gap between the real measurements of CT in graduates and the reported CT enhancing curricula in educational institutes.

Once educated, the creative and critical thinking need to be assessed. And an unmistakable evidence of the enhancement is expected to be shown. Several instruments are available to serve that, but evaluators should ensure the appropriate utilization of these instruments in a proper setting, because as per Swartz (1988) variations in the procedures of testing would affect the accountability of the result (as cited in Kim, 2006).

In concluding this chapter, we may confidently admit that creative critical thinking in science education is the magic wand to create a knowledge-based society. That knowledge-based society, in the United Arab Emirates or anywhere in the world, will be able to control its present while deciding about and planning for its future with lofty standards of ethics and moralities.

References

Abu Dhabi Education Council. (2013). *Private Schools Policy and Guidance Manual.* Abu Dhabi. Retrieved from https://adek.gov.ae/-/media/Project/TAMM/ADEK/Downloads/Private-schools/Private-Schools-Policy-and-Guidance-Manual.pdf

Abuzaid, M., & Elshami, W. (2016). Integrating of scenario-based simulation into radiology education to improve critical thinking skills. *Reports in Medical Imaging, Volume 9*, 17–22. https://doi.org/10.2147/RMI.S110343

Al Karasneh, S., & Mohammad Jubran Saleh, A. (2010). Islamic perspective of creativity: A model for teachers of social studies as leaders. *Procedia Social and Behavioral Sciences, 2*, 412–426. https://doi.org/10.1016/j.sbspro.2010.03.036

Alzahmi, R. A., & Imroz, S. M. (2012). A look at factors influencing the uae workforce education and development system. *Journal of Global Intelligence & Policy, 5*(8), 23.

Ashour, S., & Fatima, S. K. (2016). Factors favouring or impeding building a stronger higher education system in the United Arab Emirates. *Journal of Higher Education Policy and Management, 38*(5), 576–591. https://doi.org/10.1080/1360080X.2016.1196925

Assaf, M. A. (2009). Teaching and Thinking: A Literature Review of the Teaching of Thinking Skills. In *Online Submission.* https://eric.ed.gov/?id=ED505029

Clarke, M., & Otaky, D. (2006). Reflection 'on' and 'in' teacher education in the United Arab Emirates. *International Journal of Educational Development, 26*(1), 111–122. https://doi.org/10.1016/j.ijedudev.2005.07.018

Cone, C., Godwin, D., Salazar, K., Bond, R., Thompson, M., & Myers, O. (2016). Incorporation of an Explicit Critical-Thinking Curriculum to Improve Pharmacy Students' Critical-Thinking Skills. *American Journal Of Pharmaceutical Education, 80*(3), 41. https://doi.org/10.5688/ajpe80341

Daud, A., Omar, J., Turiman, P., & Osman, K. (2012). Creativity in Science Education. *Procedia - Social And Behavioral Sciences, 59*, 467-474. https://doi.org/10.1016/j.sbspro.2012.09.302

Dickson, M., Kadbey, H., & McMinn, M. (2015). Comparing Reported Classroom Practice in Public and Private Schools in the United Arab Emirates. *Procedia Social and Behavioral Sciences, 186*, 209–215. https://doi.org/10.1016/j.sbspro.2015.04.079

Fisher, R. (2003). *Teaching thinking* (2nd ed.). Continuum.

Fisher, A. (2011). *Critical Thinking: An Introduction* (2 edition). Cambridge University Press.

Forawi, S. A., & Mitchell, R. M. (2012.). Pre-service teachers' perceptions of critical thinking attributes of the ohio and new york states' math and science content, *Jornal of Teaching and Education*, *1*(5), 379 –388.

Griffin, M. B. and B. McClary (2015). A Way to Teach Critical Thinking Skills so Learners Will Continue Using Them in Operations. *MILITARY REVIEW*, 12.GUNN, T. M., GRIGG, L. M., & POMAHAC, G. A. (2008). Critical Thinking in Science Education: Can Bioethical Issues and Questioning Strategies Increase Scientific Understandings? *The Journal of Educational Thought (JET) / Revue de La Pensée Éducative*, *42*(2), 165–183. JSTOR.

Kaddoura, M. A. (2011). Critical Thinking Skills of Nursing Students in Lecture-Based Teaching and Case-Based Learning. *International Journal for the Scholarship of Teaching and Learning*, *5*(2). https://doi.org/10.20429/ijsotl.2011.050220

Kandemir, M. A., & Gür, H. (2009). The use of creative problem-solving scenarios in mathematics education: Views of some prospective teachers. *Procedia Social and Behavioral Sciences*, *1*(1), 1628–1635. https://doi.org/10.1016/j.sbspro.2009.01.286

Kim, K. H. (2006). Can we trust creativity tests: A review of the Torrance Tests of Creative Thinking (TTCT)? Creativity Res. *J*, 3–14.

Liu, O. L., Mao, L., Frankel, L., & Xu, J. (2016). Assessing critical thinking in higher education: The HEIghten^{TM} approach and preliminary validity evidence. *Assessment & Evaluation in Higher Education*, *41*(5), 677–694. https://doi.org/10.1080/02602938.2016.1168358

Mansour, N., & Al-Shamrani, S. (Eds.). (2015). *Science Education in the Arab Gulf States: Visions, Sociocultural Contexts and Challenges*. Sense Publishers. https://doi.org/10.1007/978-94-6300-049-9

McGraw Hill. (2016). *Ministry of Education introduces new world-class curriculum to enhance UAE's education system*. Retrieved from https://www.mheducation.com/news-media/press-releases/uae-ministry-introduces-new-education-system-curriculum.htmlMoore, T. (2013). Critical thinking: Seven definitions in search of a concept. *Studies in Higher Education*, *38*(4), 506–522. https://doi.org/10.1080/03075079.2011.586995

Nadjafikhah, M., Yaftian, N., & Bakhshalizadeh, S. (2012). Mathematical creativity: Some definitions and characteristics. *Procedia Social and Behavioral Sciences*, *31*, 285–291. https://doi.org/10.1016/j.sbspro.2011.12.056

Nicholas, M. C., & Raider-Roth, M. (2016). A Hopeful Pedagogy to Critical Thinking. *International Journal for the Scholarship of Teaching and Learning*, *10*(2). https://doi.org/10.20429/ijsotl.2016.100203

Park, J. (2011). SCIENTIFIC CREATIVITY IN SCIENCE EDUCATION. *Journal of Baltic Science Education, 10*(3).

Park, W. (2017). *Abduction in Context* (Vol. 32). Springer International Publishing. https://doi.org/10.1007/978-3-319-48956-8

Paul, R., & Elder, L. (2019). *The Miniature Guide to Critical Thinking Concepts and Tools*. Rowman & Littlefield.

Richardson, P. M. (2004). Possible influences of Arabic-Islamic culture on the reflective practices proposed for an education degree at the Higher Colleges of Technology in the United Arab Emirates. *International Journal of Educational Development, 24*(4), 429–436. https://doi.org/10.1016/j.ijedudev.2004.02.003

Roberts, J. (2016). *Language Teacher Education*. Routledge.

Siegel, H. (1988). Rationality and epistemic dependence. *Educational Philosophy and Theory, 20*(1), 1–6. https://doi.org/10.1111/j.1469-5812.1988.tb00487.x

Taleb, H., & Chadwick, C. (2016). Enhancing student critical and analytical thinking skills at higher education level in developing countries: Case study of the British University in Dubai. In *2nd International Congress on Education, Distance Education and Educational Technology ICDET*. Antalya: Journal of Educational and instructional studies in the World.

UAE National Statistics Bureau. (2012). *UAE National Statistics Bureau 2012a*

Verburgh, A., François, S., Elen, J., & Janssen, R. (2013). The Assessment of Critical Thinking Critically Assessed in Higher Education: A Validation Study of the CCTT and the HCTA. *Education Research International, 2013*, 1–13. https://doi.org/10.1155/2013/198920

Weil, D. K., & Kincheloe, J. L. (2004). *Critical Thinking and Learning: An Encyclopedia for Parents and Teachers*. Greenwood Publishing Group.

Williams, B. R. (1979). *Education, Training and Employment: Report of the Committee of Inquiry into Education and Training*. Committee of Inquiry into Education and Training.Special Collections. http://hdl.handle.net/11343/191150

Wittgenstein, L. (1968). *Philosophical investigations*. Basil Blackwell.

World economic forum. (2012). *The global competitiveness report 2012-2013*. Geneva, Switzerland. Retrieved from http://www3.weforum.org/docs/WEF_GlobalCompetitivenessReport_2012-13.pdf

CHAPTER 9

Critical Thinking Skills in BSc Nursing Curriculum

Hadya Abboud

ABSTRACT

The nurse's clinical decision-making skills can have a significant impact on the nursing profession. Wise clinical decisions can influence the patients' health either positively or negatively. For that, the literature had stressed on the appropriate educational strategies to prepare a critical thinker nurse. This research paper aimed to critically study the critical thinking and decision-making skills in the BSc nursing curriculum among nursing students through a review of related literature. The literature review was undertaken by searching CINHAL, Eric and Medline databases, as well as Cochrane databases and Google Scholar by using the search terms: 'critical thinking,' 'clinical judgments,' 'critical thinker,' 'problem solving,' 'nursing,' 'education,' 'nursing educational institutions,' and 'nursing educational strategies.' After applying specific inclusion-exclusion criteria, 12 peer-reviewed published research-based articles resulted in and were used for this review; in conclusion, providing an active non-threatening environment and updating the teaching strategies with advanced simulations and planed clinical training programs will improve the student's critical thinking skills positively. However, no need to put it as a recommendation; the nursing educational systems have to play a significant role in empowering the health system with professional nurses.

INTRODUCTION

The current challenge and complexity of the health care facilities made many nursing educational systems to be more concerned in developing critical thinking in their nursing curricula as a vital need to improve the quality of the nursing profession. Recently, meeting the patients' needs and providing safe, professional nurses to give a high-class holistic patient's care has increasingly become the main target of most of the health care management (LaMartina & Ward-Smith, 2014). Adding to that, refining the interpersonal nursing student's capabilities and introducing qualified, knowledgeable independent nurses has become the top priority of the nursing faculty in the twenty-first century (Roland, Johnson & Swain, 2011). This targeted independence in the nursing care profession is based on enhancing the decision-making and problem-solving skills among the nursing students throughout their tertiary education and before their graduation (Cruz, Pimenta & Lunney, 2009; Moattari et al., 2014; Rezaei, Saatsaz, Nia, Moulookzadeh & Behedhti, 2015). The applications of the clinical nursing decisions may have a positive or negative impact on the patient's health, and the nurse's clinical reasoning skills and the nurse's contentious guide this! Progress in clinical judgment experiences (Geist & Kahveci, 2012; LaMartina & Ward-Smith, 2014). Accordingly, the close relationship between the decision-making, clinical reasoning, and clinical judgment made many pieces of literature to use these terms interchangeably in their arguments on nursing educations while others used them purposefully with a clear understanding of each term individually (LaMartina & Ward-Smith, 2014; Huang, Lindell, Jaffe & Sullivan, 2016; Mahmoud & Mohamed, 2017).

Rezaei et al., (2015) had argued that working on improving the critical thinking skills among the nursing students in order to improve the quality of the nursing education in many universities was considered as one of the primary goals for the university's academic recognition. The need to reform the nursing curricula to adopt the critical thinking development and increase the student's awareness and practices of the clinical reasoning skills has emerged in the clinical settings where the complex chronic diseases have called for a complex clinical practice (Mahmoud & Mohamed, 2017). Today, most of the patients admitted to the hospitals with more than three different medical diagnoses would require the nurses to apply their critical care practices away from classical care. Additionally, the new technology, the complicated medical diagnoses of some new diseases as well as the other ethical–cultural factors have created a more challenging environment at the hospitals for the nurses to implement their critical thinking. The nursing students cannot acquire safe and appropriate nursing interventions unless nursing educational institutions develop their teaching strategies accurately (Lee, Lee, Bae & Seo, 2016).

In contrast, Ward and Morris (2016) have emphasized the role of the nursing faculty in facilitating the needed changes in the educational strategies used to enhance the students' critical thinking abilities and improve their techniques in solving the patient's problems. Active learning is one of the used strategies to fill the gap between the critical thinking and the students' success, by engaging them in real cases and asking them to think like a nurse to understand the fundamental and essential concepts of critical thinking. In the same way, other science teaching research studies have found that the teachers' instructions and guidance in the science and mathematics classrooms have a significant influence on the students' critical thinking development and cognitive growth. Besides, the classroom teacher can make the classroom-learning environment a meaningful experience to the students by integrating the critical thinking skills into their classroom instructions for targeting better students' preparations and outcomes (Forawi, 2012).

Studies have found that there is an excellent correlation between the students' critical thinking and other factors, such as the prescribed curriculum and standards, achievement level, and motivation (Forawi, 2012, 2016). The more critical thinking abilities the students gain, the less stress and anxiety they would experience at the clinical settings (Rezaei et al., 2015). While others found that having more critical thinking skills will positively influence the student's levels of the accuracy scale (Cruz et al.,2009; Pimenta & Lunney, 2009; Lee et al., 2016) and their clinical performance (Choi & Cho, 2011), Kaddoura (2011) has ascertained the student's improvement in problem-solving skills when using case-based learning and active learning environment. Furthermore, the student's self-confidence has become in most of the studies as a result of developing the student's critical thinking (Romeo, 2010; Le Roux & Khanyile, 2011; Chesser-Smyth & Long. 2012; Duque 2013; Moattari et al., 2014; Rezaei et al., 2015).

Consequently, the American Psychological Association (APA) (1999) and other recent researchers argued that a good critical thinking student has to recognize the required personal characteristics and work on developing them such as being habitually inquisitive, open-minded, flexible, reasonable, honest, confident of abilities, clear-minded, objective and away from personal biases. Besides, a critical thinking student should have a contextual perspective to understand the whole situation, ethical reflection, and good self-evaluation. (Secrest, Keatley & Norwood, 1999; Moattari et al., 2014; LaMartina & Ward-Smith, 2014).

In providing the best possible critical thinking nurses, improving the nursing curricula, and enabling and empowering the nursing educational institutions to deliver their best nursing educational strategies are highly demanded. For instant, the predictable students who practiced and developed their problem solving and critical thinking skills will influence their patient's holistic health positively. Consistently, the health care facilities workforce will significantly progress with these professional nurses who demonstrated their outstanding expected abilities, including combining sound clinical judgments with their competent hand skills. The purpose of this literature review is to examine the existing research on the critical thinking skills in the BSc nursing curriculum and the quality of clinical decision making among nursing students.

Conceptual Framework Discussing Key Related Studies

Critical thinking, problem-solving, decision-making, and clinical judgment were the key terms that emanated from the study topic search. There is a close relationship between these concepts, and the students' clinical experience would vitally contribute to having better clinical judgments and improving their clinical response quality whenever they were exposed to new clinical settings. Therefore, for a better understanding of the learning process and elaborate on its application on the CT skills, this study finds it necessary to consult and understand the nursing learning theories and their application in the nursing education in both the classrooms and the clinical setting. Always for many years, the learning theories were hardly working on explaining the learning process and its applications on putting a clear framework for the educators. All the educational psychologists were trying to support the educational institutions with their best practices. (LaMartina & Ward-Smith, 2014; Aliakbari et al., 2015).

The deep thinking and the internal understanding process were the main interest of the cognitive psychologists dissimilar to the behaviorists who believed that learning comes after observing the behavior of other in stimulation events that could change the observer's emotions accordingly. At the same time, Humanists were interested in studying the feelings and experiences of the learners. Besides, all beginners feel that learners should be prepared with the problem-solving and inquiry skills within a reinforcement environment (Skinner, 1974; Graham & George, 2017).

Piaget's cognitive development theory (1896) is corresponding with the current study topic as he concentrated on exploratory learning through the student's discovery and practical experience. In addition, Vygotsky's cognitive theory (1978) was adopted by the nursing educational leaders for its focus on the social interaction between the learner and his environment with particular respect to the socio-cultural issues (Aliakbari et al., 2015). Never the less, Dewey (1910) stressed the importance of the students' reflections on a planned simulation experience environment, which can help in developing their problem-solving skills and clinical judgment. The nursing students are facing many complex clinical situations that require them to make decisions and pass the best clinical judgment when trying to solve their patients' problems (Yildirim & Ozkahraman, 2011).

Literature Review / Historical Overview

This literature review study has carried out extensive search on CT skills in the nursing curriculum with a particular focus on clinical decision making for better problem solving and clinical judgment. The nursing education institutions for many years because of the contentious complexity of the health care systems had an interest in studying this issue (Cruz et al.,2009) also specifically attend to this topic. The American Association of Colleges of Nursing (AACN) (2008) to assure the nurse's professionalism and motivate the quality of patients' care put a new policy to test the graduate nursing students' critical reasoning competencies to be able to deal with the hospitals complex situations. The AACN have added a clear statement in their documents demanding the academic institutions to elaborate and prepare their students with the CT skills extensively as part of their academic preparation and capacity building. A definite highlight has been given to the importance of the student's CT abilities for assuring the patient's safety. '"This document emphasizes such concepts as patient-centered care, inter-professional teams, evidence-based practice, quality improvement, patient safety, informatics, clinical reasoning/critical thinking, genetics and genomics, cultural sensitivity, professionalism, and practice across the lifespan in an ever-changing and complex healthcare environment"' (AACN, 2008, p .3).

Thenceforth, the American Psychological Association (APA) sponsored a considerable project to prepare the 'ideal critical thinker.' This project was called the 'Delphi Project 'conducted in 1990, and it took them two years to come up with the ideal critical thinker criteria that the nursing students need to maintain to improve their clinical judgment skills. The Delphi's ideal critical thinker criteria "routinely inquisitive, well-informed, trustful of reason, open-minded, flexible, fair-minded in evaluation, honest in facing personal biases, practical in making judgments, willing to reassess, vibrant about issues, orderly in complex problems, hardworking in looking for relevant information, sensible in the selection of criteria, attentive in inquiry, and persistent in looking for consequences which are as precise as the subject and the circumstances of inquiry permit" (LaMartina & Ward-Smith 2014, p. 156). In contrast, Ennis (1985) was satisfied to minimize the Delphi's ideal critical thinker criteria's as a tool of inquiry and self-regulation into one focused criterion informing that it is imperative and essential for the person to believe in his decisions and believe that his decisions to be the ideal and the best decision ever. Critical thinking and self-regulation will empower the people as a powerful resource in one's civic life. (Yildirim & Ozkahraman, 2011).

Historical Overview

In the old days, the Greek philosophers were very interested in studying critical thinking. They argued that the truth of any dilemma could not be reached unless the people drive through an in-depth analysis process to the situation and its related aspects. During an argument, two persons will sit against each other and discuss the issue critically to encounter which of the ideas is pitted against each other. Criticism and critical discussions were the only way to get close to the truth (Ennis, 1989). The ability of the students to ask relevant questions for the seen phenomena and be able to analyze solutions without creating alternatives was another definition of the CT (Ennis, 1989). The sequence of using interpretation, explanation, and self-regulation has been argued to be central to any critical thinking process (Edwards, 2007).

John Dewey's idea of reflective thinking was the ground of the educational critical thinking theories as he discriminated between the thinking process and thinking product. Dewey (1910) argued that a 'perplexed' situation would be the best technique to stimulate reflective thinking within any educational experience. This used to happen by giving the student some prompts and hints to resolve the situation while the teacher kept silent until the solver developed his own hypothesis after he had realized the problem clearly. Modifications for the hypothesis would be followed by close observation and reason-guide tests as very important CT phases. Additionally, Dewey argued that judgment is a reflective thinking that is twisted to disagreement based on the presented facts and suggestions and how the critical thinker valued them. Dewey also disputed a comparison between the memories and the judgments of the persons. He referred our memories to previous old events that have been stored in the person's brain, while the peoples' judgments are based on selecting and adopting specific events to be used in any emergencies. The educational systems influence the student's problem solving and critical judgment either positively or negatively. The scientific methodologies, highlighting the students' interests, assimilating experiences, and reflection, are the bases of any successful educational process. Continuously, many philosophers have defined critical thinking in different ways for the reason of their believing in its importance in enhancing the educational process. (Dewey, 1910; Yildirim & Ozkahraman, 2011; Geist & Kahveci, 2012). For an instant, Fini, Hajibagheri, and Hajbaghery (2015) defined critical thinking as "a process of purposeful, interactive reasoning, criticism and judg-ment about what we believe and do" (P.1).

Research Method

Search Strategy

A literature review was undertaken by searching the EBSCO database: CINHAL, Eric and Medline, and Cochrane databases as well as the Google Scholar, using the search terms: 'critical thinking,' 'clinical judgments,' 'critical thinker,' 'problem solving,' and adding 'nursing,' 'education,' 'nursing educational institutions,' 'nursing educational strategies.' The initial search ended up with 948 different articles. These articles went through skimming and scanning of abstracts and conclusions of the articles for filtering the needed articles. That resulted in reducing the search result in 31 internationally published articles. Then the search was limited again to the articles that were published between the years 2010-2017 to make sure that the reviewed findings are up to date and assure the reader that the reviewed outcomes are very recent to the nursing education and are revealing to the nursing education field in general. Furthermore, two books have been hand searched to provide this review with the topic of background information.

Then, a final inclusion and exclusion criteria have been applied to reduce bias in selecting the articles and limit the search results to those that are directly related to this literature review. For that, the articles included in this paper met the criteria of those research articles published between 2010 and 2017, peer-reviewed articles written in the English language, and research-based literature. The selected research studies were also included if they were focusing on critical thinking and nursing students as well as nursing education and clinical judgment. On the other hand, the articles excluded have met the following exclusion criteria: studies investigating critical thinking of the postgraduate nursing students, articles and studies that dealt with the critical nursing thinking of the clinical setting after gaining the clinical experience away from an educational background, the non-peer reviewed articles, and finally the non-published studies.

As a result of the above pre-planned selection criteria, the search ended by using a 12 peer-reviewed published research-based articles in the above selected period and that presented in table.1..

However, two research articles were added to the study based on their significant findings that would contribute final findings of the current study, although they were excluded from the first search. The first one was testing the CT of the nurses at the hospital. Professional nurses were excluded from the search criteria, but this particular article used the 'California Critical Thinking Skills Test' that commonly used to test the student's CT by others. While the second one used the same test but for the Nursing Diploma programs, students in UAE and the study was significant since it related to the country of interest.

Findings and Discussion

This literature review discussed the three common themes that have been emerged from the used articles. To be precise, they are professional identity and CT student characteristics, accuracy in clinical decision-making effects on the patients' outcomes, and instructional teaching strategies will be discussed deeply in the following sections below.

Alharbi (2019) conducted an empirical study titled 'Using Simulation in Nursing Didactic Classes to Enhance Students' Critical Thinking and Knowledge' on 39 nursing students at the Indiana University Kokomo (IUK) campus. The data was collected using the Watson Glaser Critical Thinking Appraisal II (WGCTA) measured the contributors' critical thinking pre and post simulations. The National League of Nursing Student Satisfaction and Self-Confidence in Learning Survey used to measure students' satisfaction and self-confidence after the simulation sessions. The study claimed that 'No significant increase in the nursing students critical thinking skills after their exposure to the simulation experience. No increase on the grades of nursing students after simulation. Recommendations for larger further research studies to validate the effectiveness of simulation in improving the critical thinking skills and decreasing the attrition rates. The study limitation was the lack of a control group due to the small sample size that can give a comparison data to enable analysis of the real influence of simulation experience on students' critical thinking'. In the same year, a descriptive qualitative study aimed to explore the perspectives of nursing students on critical thinking on 65 nursing students from one school of nursing "Nursing students' view of critical thinking as 'Own thinking, searching for truth, and cultural influences'. A focus group study utilized eleven focus group interviews were conducted in Chinese and translated into English and thematic analysis has been adopted. The study found that Nursing education and students should employ critical thinking to assure quality and safety of patient care. If nurses have more time to think about what is going on, they could think critically about what they are actually doing. This could assure safety of patients and reduce medical accidents. Future studies recommended to be conducted on how these factors might affect critical thinking' (Chan, 2019).

Teaching critical thinking needs the cooperation of students, teaching staff and the education institutions. Graduating the students with respectable CT skills will label them to be highly recognized and more professional than other nurses in the field. The teaching faculty recommended enhancing the students 'cognitive and metacognitive' thinking levels by asking them more questions and direct them through hints to link their theoretical knowledge to their clinical experiences (Huang et al. 2016). However, the theoretical definition of the terms: 'critical thinking, clinical reasoning and clinical decision making in nursing' can be used interchangeably to evaluate the students' thinking process in nursing education. The qualitative study 'Addressing the Challenge of Developing a Conceptual Definition for Clinical Judgment' from a 23 articles conclude that the main goal of the nursing educational leaders were directed to utilizing theses terms to improve the student's clinical judgment skills (Jacobs et al. 2016). For better nursing practices and judgments, preparing professional nurses who are utilizing their 'cognitive, metacognitive, psychomotor, and affective processes' efficiently through clinical thinking and clinical reasoning at their clinical settings. Proper educational strategies are needed to apply the change in using the critical thinking of the students and improving the assessment tools used in measuring that change (Victor-Chmil, 2013).

Quasi-experimental study, 'Concept mapping – an effective tool to promote critical thinking skills among nurses' on 45 students found that although the used assessment tool was very advanced and the students scored poorly, the study results illustrated great improvements among the nursing student's critical thinking. 'First a tool of '32 multiple-choice questions', then followed with a 'concept mapping lecture for the experimental group pre-assessment, then the 'Experimental group' 8 groups/ 5 members. Both 'the experimental group and control group' critical thinking were evaluated for the second time by the same tool'. Revision for the used tool and more studies on concept mapping were recommended. Promoting the students' CT should be the concern of the nursing educational institutions. Revolution in that field should be the major concern for the educational leaders and 'Concept mapping' was one tool for that reason (Nirmala & Shakuntala, 2011). Another quasi-experimental study titled 'Clinical concept mapping: Does it improve discipline based critical thinking of nursing students?' that examined 32-year four nursing students using the post-test only design'. The study recommended the use of the 'clinical concept mapping based on the nursing process ' to stimulate the brain to think and analyze critically till it becomes a habit of the nursing students at their early stages of learning in the clinical settings. Also, they recommended to use it as a tool to measure the student's critical thinking at the clinical (Moattari et al., 2014). This is in contrast to another quantitative study 'A Comparative Study on Critical Thinking Skills of Bachelor and Master's Degree Students in Critical Care Nursing'. This cross-sectional study sample was 123 nursing students, 79 BSc nursing students and 44 MSc in three different Universities. The analysis and the critical thinking skills were very limited within the nursing students at all levels although it was more advanced at the BSc level. Recommendations on improving the educational strategies at the nursing educational systems by emphasizing the classroom 'conceptual learning', critical reasoning and critical thinking skills'(Babamohamadi, Fakhr-Movahedi, Soleimani, & Emadi, 2016).

The 'Effects of Multi-mode Simulation Learning on Nursing Students' Critical Thinking Disposition, Problem Solving Process, and Clinical Competence' was a quasi-experimental study on 65 nursing students. The study focused on the nursing students studying 'Emergency and critical nursing course at N university'. Nonequivalent control group with a pre-test-post-test design collected the data from a treatment group of 33 students in 2010 and the control group of 32 students in 2011'. The study results stated the 'Multi-mode simulation' influenced the nursing students' clinical nursing skills and enhanced their problem-solving abilities but it had a minor effect on their clinical thinking nature. Recommendations on improving the teaching strategies in classrooms rather than using the 'Multi-mode simulation' for its economical and physical necessities have been elaborated by the study' (Ko & Kim, 2014).

The nursing students displayed a very low capacity on their critical thinking skills and that was correlated with the student's anxiety level in a descriptive analytical study of 245 nursing students utilizing the 'Watson-Glaser Critical Thinking Appraisal and Spielberger's State Anxiety Inventory'. The authors claimed that the more the nursing students were stressed the fewer abilities' of critical thinking appears. Great recommendations were given to work on the applied educational strategies to enhance critical thinking and reduce the student's anxiety to be ready for the current hospital's complexity' (Rezaei et al., 2015).

Mahmoud and Mohamed (2017) had a descriptive research design study 'Critical Thinking Disposition among Nurses Working in Public Hospitals at Port-Said Governorate' on 196 nurses using the 'California Critical Thinking Disposition Inventory (CCTDI)'. The study showed that ¾ of the nurses were not truth- seekers besides they were not sure about their critical thinking skills. The study recommended raising up the CT and problem-solving awareness in the clinical setting using updated teaching strategies for better patient's health outcomes. The California Critical Thinking Skills Test, form B questionnaire was used in another quantitative comparative study to examine the 'Critical Thinking Skills in Nursing Students: a Comparison Between Freshmen and Senior Students' on 150 undergraduate freshmen and senior nursing students. 'The nursing students critical thinking skills' were obviously very law from the start of their nursing study at the University. They did not show any improvement or progress over the studying years at the nursing program. The study recommended assessing the students' progress over a long period through longitudinal research at the University to assess the student's critical thinking development deeply. Intensive curriculum reforming to adopt active learning and provide a professional development training sessions were required for the teaching staff to train them to utilize the CT in their teaching strategies (Azizi-Fini, Hajibagheri & Adib-Hajbaghery, 2015).

Kaddoura (2011) quantitative research 'Critical Thinking Skills of Nursing Students in Lecture-Based Teaching and Case-Based Learning' inspected 103 nursing students from the diploma programs in the UAE who went through the two ways of learning strategies the classical lecture based way beside the CBL. The group-shared activities were divided as 65 CBL and 38 traditional CBL using 'a comparative descriptive survey and California Critical Thinking Skills Test (CCTST) form B. The study presented that the CT skills of the nursing students were improved effectively through the CBL. The study results have great implications on both nursing education and health care professional research. The author concludes, "CBL should be encouraged in the nursing curricula to develop the learners' CT, which might impact nursing care to improve patient outcomes" (Kaddoura 2011, p.15).

A positive influence on the student's critical thinking came as a result of using the Essentials of Critical Care Orientation ECCO program by improving and increasing the theoretical knowledge among most of the nursing fresh graduates at their last year training. Few nursing students felt that it has limitation for their critical thinking because of not having their teachers or peers around to elaborate on the theory part and have further discussions. The main aim of any nursing educational institution is to 'provide safe, critically thinking nurses to deal with the critical cases at the hospital like ICU patients. Such orientation programs can help in improving the CT of the nurses that will be the reason to maximize the rate of retention and staff burnout and increase their job satisfaction. Developing the nursing students CT will help in controlling the global nursing shortage in the hospitals through better recruitment and orientation training nursing programs' (Kaddoura, 2010).

In the same year the author had another exploratory qualitative descriptive study 'New Graduate Nurses' Perceptions of the Effects of Clinical Simulation on Their Critical Thinking, Learning, and Confidence' on BCs nursing graduates' at their last year of nursing utilizing a 'demographic questionnaires and Semi-structured interviews and were analyzed using content analysis'. The sample of the students were selected to be trained at the biggest free educational hospital in USA. The hospital setting provides an advanced 'clinical simulation centers' for the students that are training them in exceptionally high quality and advanced nursing care skills programs. All of the student's critical thinking skills were improved at the end of the training program. The simulation lab influenced the student's critical thinking skills and improved their relations with the health team staff after knowing how to communicate effectively and professionally. Moreover, the students' stress level became less after this simulation-training program with clear progress and improvements in their leadership management skills with evidence of supporting and non-threatening teaching environments. Interactive training with other nurses at the simulation program in one specialty improved the student's clinical competencies within a teamwork manner. There was a significant improvement in the patient's health outcomes in critical areas. The study recommended the educational leaders to reform the nursing educational curricula by adopting the 'simulation as a teaching-learning strategy' (Kaddoura, 2010).

Practice-Based Simulation Model: a curriculum innovation to enhance the critical thinking skills of nursing students discussed in a qualitative research paper describing the (PBSM) Model fine elements as a pedagogical framework for the nursing education'. The Practice-Based Simulation Model (PBSM) used as an educational framework for integrating the simulation as a tool to assure developing the nursing students' critical thinking skills as a main aim of the educational learning strategies'. Constructivists supported the use of the 'Practice-Based Simulation Model' to reform the nursing teaching strategies. They found this model to be a good way to assure the development of the CT of the nursing students. They perceived that the more they implemented the CT within the nursing curriculum the better learning outcomes they gain and innovate better nursing competences. The five essential elements of the model were 'practice situation, simulation, structured learning, inquiry process, and assessment'. These elements are designing a clear educational framework that can be implemented in the nursing curricula to assess the students' knowledge in a systematic and constructive way. The study elaborated that the PBSM is an active integration of simulation and a key factor that can improve any educational curriculum'. The authors declared "simulated learning experiences need to be integrated into a curriculum underpinned by sound pedagogy, such as the PBSM, in order to ensure that learning facilitates the development of the critical thinking abilities deemed essential for nursing" (Park et al. 2013, p. 49). Since 2010 the PBSM is considered the most common clinical educational framework at the nursing educational institutions that have been used in various nursing posts and undergraduates' educational systems. Its reputation came from its flexibility and efficacy for the 'simulation integrated teaching' and 'learning practice' and enhancement of the nursing students' critical thinking. The authors recommended a longitudinal study to link the simulation and the CT correlation in a way to evaluate the PBSM (Park et al. 2013).

This literature review revealed that most of the students demonstrated low critical thinking abilities in the clinical field (Kaddoura, 2010; Kaddoura, 2011). It also showed that there is no clear strategy in building critical thinking skills during their educational studies at the university (Rezaei et al., 2015; Mahmoud & Mohamed, 2017). Besides, most of the studies discussed the nursing students' CT during their clinical practice when they get exposed to interacting with real patients in clinical settings. Most of college educational settings ignored students' CT preparation during the pre-clinical preparation period as an essential aspect of their college curriculum and education.

1. *Instructional Teaching Strategies of CT*

Huang et al. (2016) highlighted that the teaching faculty suggested a relevant educational strategy that would enhance the students' 'cognitive and metacognitive' thinking levels by asking them more questions and direct them through hints. Then, linking their theoretical knowledge to their clinical experiences. Teaching critical thinking needs the cooperation of the students, teaching staff and the institutions. Graduating the students with respectable CT skills will label them to be highly recognized and more professional than other nurses in the field. For that, one of the main goals of the nursing educational leaders is directed to utilizing these terms to improve the student's clinical judgment skills (Jacobs et al., 2016).

Promoting the students' critical thinking should be a significant concern for the nursing educational leaders and institutions, and 'Concept mapping' was one of the tools introduced for that reason. In 2011, Nirmala & Shakuntala presented the results of their empirical study which they carried out to scrutinize the effects of concept mapping on CT. The 45 students who participated in the study were divided into eight groups with five members in each group, and the researchers found that although the used assessment tool was very advanced and the students scored poorly, the study results illustrated significant improvements in the nursing students' critical thinking. Revision for the used tool and more studies on concept mapping were recommended. Proper instructional strategies are needed to apply the change in using the critical thinking of the students and improving the used assessment tools in measuring that change (Victor-Chmil, 2013).

Ko and Kim's (2014) empirical study of 65 nursing students in (2010-2011) found that the 'Multi-mode Simulation' strategy influenced the nursing student's clinical nursing skills. 'Multi-mode Simulation' can enhance their problem-solving abilities, but it had a minor effect on their clinical thinking nature. This quasi-experimental study put some recommendations on improving the teaching strategies in classrooms rather than facing the challenges of the 'Multi-mode Simulation.' The economic and physical necessities of using critical thinking skills in emergency cases have been probed in the comparative study performed by Moattari et al. (2014) on 123 bachelor and master's nursing students testing their CT in 'critical care nursing.' The analysis and the critical thinking skills were very limited within the nursing students at all levels of this cross-sectional study although it was more advanced at the BSc level. Recommendations highlighted the importance of improving the educational strategies in the nursing educational systems by having more emphasis on the classroom's conceptual learning,' critical reasoning, and critical thinking skills.

In another study ran by Kaddoura (2010), a positive influence on the level of 'student's critical thinking' was shown as a result of using the ECCO program, which focused on improving and increasing theoretical knowledge among most of the participants. Few students felt that it is imitating their critical thinking because of not having their teachers or peers around to elaborate on the theory part and have further discussions.

In the same year, another study for Kaddoura (2010) claimed that the hospital setting is providing an advanced 'clinical simulation center' for the students where they received training with exceptional high quality and advanced nursing care skills programs. All of the student's critical thinking skills improved by the end of the training program. The simulation lab positively influenced the student's critical thinking skills and improved their relations with the health team staff after knowing how to communicate effectively and professionally. Interactive training with other nurses at the simulation program in one specialty improved the student's clinical competencies within a teamwork manner. There was a significant improvement in the patient's health outcomes in critical areas. This study demonstrated that the educational leaders should consider reforming the nursing educational curricula by adopting the 'simulation as a teaching-learning strategy,' and preparing nursing educators to employ better and implement "innovative and active teaching strategies (Azizi-Fini, Hajibagheri & Adib-Hajbaghery, 2015).

Again, the same author conducted another study in 2011 on the diploma program students in the UAE examining 103 nursing students by using quantitative research using the 'California Critical Thinking' Skills Test' (CCTST) Form B survey. The nursing students went through the two learning strategies, the classical lecture-based learning and the CBL in the group shared activities. The study presented that the CT skills of the nursing students have improved effectively through the CBL. The study results have enormous widespread implications on both nursing education and health care professional research. The author concluded, "that CBL should be encouraged in the nursing curricula to develop the needed learners' CT" (Kaddoura, 2011, p.15), which might have a significant impact on the nursing care and the patient's outcomes".

In contrast, Azizi-Fini, Hajibagheri, and Adib-Hajbaghery (2015) and Mahmoud and Mohamed (2017) have used the same assessment tool, but they both found that 'The nursing students critical thinking skills' were very law from the start of their nursing education at the University. They did not show any improvement or progress over their studying years at the nursing program.

2. Accuracy in Clinical Decision-Making Effects on the Patient's Outcomes

Jacobs et al. (2016) argued that the theoretical definition of the terms: 'critical thinking, clinical reasoning and clinical decision making in nursing could be used interchangeably to evaluate the students' thought process in nursing education. The main goal of the nursing educational leaders is directed to utilizing these terms to improve the student's clinical judgment skills. The importance of critical reasoning and clinical decision – making has been tested in many studies. In addition, the author of one study concluded that CBL "should be encouraged in the nursing curricula to develop the needed learners' CT, which might impact the nursing care to improve patient outcomes" (Kaddoura, 2011, p.15).

For better nursing practices and wise clinical judgments, preparing professional nurses who are utilizing their' cognitive, metacognitive, psychomotor, and affective processes' efficiently through clinical reasoning in their clinical settings is highly desirable. Proper educational strategies are needed to apply the change in using the critical thinking of the students and improving the used assessment tools in measuring that change (Victor-Chmil, 2013). Mahmoud and Mohamed (2017) conducted a descriptive research study in which they tested 196 nurses on their CT skills and decision- making abilities, and results demonstrated the students' weaknesses in these areas. They recommended to raise students' awareness of the CT and to problem-solve in the clinical settings using updated teaching strategies for better patients' health outcomes. Moreover, There was a significant improvement in the patients' health outcomes in the critical areas at the health care facilities after implementing the CT training programs through providing an advanced 'clinical simulation center' for the students where they were able to have adequate training with high quality in advanced nursing care skills programs (Kaddoura, 2010). Others perceived that the more they embed the CT within the nursing curriculum, the better the learning outcomes they gain and innovate better nursing competencies. (Park et al., 2013).

In contrast, Azizi-Fini, Hajibagheri, and Adib-Hajbaghery (2015), in their comparative study of 150 undergraduate and senior nursing students, we are not sure of the link between the CT of the students and its influence on the patients' health. They recommended further studies assessing the students' progress over a long period through longitudinal research at the University to assess the student's critical thinking development deeply. Since 2010 the PBSM is considered as the most common clinical, educational framework in the nursing educational institutions that have been used in various nursing post and under graduate's educational systems. Its reputation came from its flexibility and efficacy for the 'simulation integrated teaching' and 'learning practice' and enhancement of the nursing students' critical thinkers. The authors recommended a longitudinal study to examine the link between the simulation and the CT in a way to evaluate the PBSM effectively. The authors declared, "That simulated learning experiences need to be integrated into a curriculum underpinned by sound pedagogy, such as the PBSM, in order to ensure that learning facilitates the development of the critical thinking abilities deemed essential for nursing" (Park et al., 2013, p. 49).

On the other hand, the students' CT and stress levels were commonly investigated together in many studies to explain the effect of the accuracy of the nurse decision on the patients' health. A 245 nursing students' level of stress was tested in a cross-sectional study, and the findings presented that the nursing students displayed an insufficient capacity on their critical thinking skills, and that was correlated with the student's anxiety level. The more they are stressed the fewer abilities of critical thinking appear. Great recommendations were given to work on the applied educational strategies to enhance critical thinking and reduce the student's anxiety to be ready for the current hospital's complexity (Rezaei et al., 2015). Moreover, the students' stress level became less after the simulation-training program with clear progress and improvements in their leadership management skills with evidence of supporting and non-threatening teaching environments (Kaddoura, 2010).

3. Professional Nurses' Identity and CT

Most of the pieces of literature were very concerned with the nursing profession and the ways in building the nursing students' professional identity. The critical thinking of the nursing students has been linked to the student's professional identity in many articles for its direct effect on how the students are going to present themselves in clinical settings. Graduating the students with respectable CT skills will label them to be highly recognized and more professional than other nurses in the field (Huang et al., 2016). Others found that 'the nursing students critical thinking skills' were obviously very law from the start of their nursing study at the university (Azizi-Fini et al., 2015).

Some studies linked the nursing professional identity to safety. They argued that the main aim of any nursing educational institution should be directed to provide safe critical thinking nurses to deal with the critical cases and patients at the hospital. Improving the CT of the nurses would contribute to maximizing their job satisfaction and the rate of retention and minimizing staff burnout. Developing nursing students, CT, would possibly help in controlling the global nursing shortage in hospitals through better recruitment and effective nursing training programs (Kaddoura, 2010). Babamohamadi et al. (2016) conducted a cross-sectional comparative study on critical thinking skills among bachelor and master's students in evaluating critical care nursing testing. This study was applied to 123 nursing students, 79 BSc nursing students, and 44 MSc in 3 Universities. The data analysis showed that the critical thinking skills were minimal within the nursing students at all levels, and the researchers recommended improving the educational strategies in the nursing educational systems by giving more emphases on the classroom' conceptual learning', critical reasoning and critical thinking skills for nurturing better nursing graduates.

Constructivist supported the use of the CT to improve the nursing profession in general and argued that it advances the nurses' confidence in specific. They perceived that the more they embedded the CT learning within the nursing curriculum, the better learning outcomes they gain and better nursing competencies (Park et al., 2013), which would, in turn, improve their self-confidence in critical clinical situations. In Mahmoud and Mohamed's (2017) study, they tested the nurse's professionalism among those who were working in the field and graduated from nursing schools. The results showed that 75% of the nurses were not truth- seekers, and they were not sure about their critical thinking skills. Others had supported the idea of having low critical thinking abilities among most of the nurse's graduates after many previous studies, while they had emphasized a lot over the importance of evaluating the student's critical thinking competencies to assure the patient's safety at the health care facilities (Azizi-Fini, Hajibagheri & Adib-Hajbaghery, 2015; Rezaei et al., 2015; Huang et al., 2016).

Conclusions and Recommendations

Pieces of evidence from the works of literature of critical thinking in nursing education suggested that nursing students need to improve their CT skills during their years of studying. The need for critical thinking nurses increases as a result of the contentious complexity in the health care systems to meet the patients' health needs. However, the better patient's health outcomes were proven to be achieved through having excellent critical thinking professional nurses in the field. Professionalism in the career will lead to the professional nurse's identity that has been linked to the improvement of the nurses CT and problem-solving skills. Anxiety and stress are seen to be less whenever the students were prepared to practice their CT in a non-threatening environment. For that, the nursing educational systems were recommended to utilize the CT skills of the nursing students at a very early stage through massive innovative educational strategies, such as using Practice-Based Simulation Model (PBSM) and advanced 'clinical simulation center,' CBL, conceptual mapping frameworks and other orientation programs for the new students in the classrooms and the clinical setting. Never the less, many pieces of literature were looking to have future longitudinal studies to follow the student's critical thinking and clinical judgment improvements over a long period for a more in-depth evaluation.

The study review found that the students' CT is significant for improving and assuring the patients' outcomes. However, then again, more researches are needed to follow up on the student's preparations in their universities on routine bases. It is essential to test the students' CT by implementing some simulation lab and case-based tests to evaluate the student's abilities to have wise and safe decisions to their patients before they have direct contact with them. Never the less, the pre-clinical CT and the decision making competencies should be tested to guarantee safe nursing practices at the college like testing the student's clinical hand skills. Finally, the components of the CT assessment tools need to be prepared and tested based on the theory of the nursing curriculum. In the end, it is worth mentioning that international collaborations are needed to reform the nursing pre-clinical competencies and tests to include the CT questions to assure the patient's safety.

REFERENCES

Aliakbari, F., Parvin, N., Heidari, M. & Haghani, F. (2015). Learning theories application in nursing education. *Journal of Education and Health Promotion 4*(2).

The American Association of Colleges of Nursing (AACN) Homepage. Retrieved 29 June 2017, from http://www.aacn.nche.edu/education-resources/baccessentials08.pdf

Azizi-Fini, I., Hajibagheri, A., & Adib-Hajbaghery, M. (2015). Critical Thinking Skills in Nursing Students: A Comparison Between Freshmen and Senior Students. *Nursing and Midwifery Studies, 4*(1).

Babamohamadi, H., Fakhr-Movahedi, A., Soleimani, M., & Emadi, A. (2016). A Comparative Study on Critical Thinking Skills of Bachelor and Master's Degree Students in Critical Care Nursing. *Nursing and Midwifery Studies, 6*(2).

Burkhardt, M., & Nathaniel, A. (2014). *Ethics & issues in contemporary nursing* (4th ed., pp. 100-102). Cengage Learning.

Chesser-Smyth, P., & Long, T. (2012). Understanding the influences on self-confidence among first-year undergraduate nursing students in Ireland. *Journal of Advanced Nursing, 69*(1), 145-157.

Choi, H., & Cho, D. (2011). Influence of Nurses' Performance with Critical Thinking and Problem-Solving Process. *Korean Journal of Women Health Nursing, 17*(3), 265.

Cruz, D., Pimenta, C., & Lunney, M. (2009). Improving Critical Thinking and Clinical Reasoning with a Continuing Education Course. *The Journal of Continuing Education in Nursing, 40*(3), 121-127.

Dewey, J. (1910). How We Think. John Dewey. *The School Review, 18*(9), 642-645.

Duque, L. (2013). A framework for analysing higher education performance: students' satisfaction, perceived learning outcomes, and dropout intentions. *Total Quality Management & Business Excellence, 25*(1-2), 1-21.

Edwards, S. (2007). Critical thinking: A two-phase framework. *Nurse Education in Practice, 7*(5), 303-314.

Ennis, R. (1989). Critical Thinking and Subject Specificity: Clarification and Needed Research. *Educational Researcher, 18*(3), 4-10.

Forawi, S. (2012). Pre-Service teachers' perceptions of critical thinking attributes of the Ohio and New York states' science and math content standards. *Journal of Teaching and Education, 1*(5), 379–388.

Geist, M. & Kahveci, K. (2012). Engaging Students in Clinical Reasoning When Caring for Older Adults. *Nursing Education Perspectives, 33*(3), 190-192.

Huang, G., Lindell, D., Jaffe, L., & Sullivan, A. (2016). A multi-site study of strategies to teach critical thinking: 'why do you think that?'. *Medical Education, 50*(2), 236-249.

Jacobs, S., Wilkes, L., Taylor, C., & Dixon, K. (2016). Addressing the Challenge of Developing a Conceptual Definition for Clinical Judgment. *Nursing and Health, 4*(1), 1-8.

Kaddoura, M. (2010). Effect of the Essentials of Critical Care Orientation (ECCO) Program on the Development of Nurses' Critical Thinking Skills. *The Journal of Continuing Education in Nursing, 41*(9), 424-432.

Kaddoura, M. (2010). New Graduate Nurses' Perceptions of the Effects of Clinical Simulation on Their Critical Thinking, Learning, and Confidence. *The Journal of Continuing Education in Nursing, 41*(11), 506-516.

Kaddoura, M. (2011). Critical Thinking Skills of Nursing Students in Lecture-Based Teaching and Case-Based Learning. *International Journal for the Scholarship of Teaching and Learning, 5*(2), 1-18.

Ko, E., & Kim, H. (2014). Effects of Multi-mode Simulation Learning on Nursing Students' Critical Thinking Disposition, Problem Solving Process, and Clinical Competence. *Korean Journal of Adult Nursing, 26*(1), 107-116.

LaMartina, K. & Ward-Smith, P. (2014). Developing critical thinking skills in undergraduate nursing students: The potential for strategic management simulations. *Journal of Nursing Education and Practice, 4*(9), 155-162.

Le Roux, L., & Khanyile, T. (2011). A cross-sectional survey to compare the competence of learners registered for the Baccalaureus Curationis programme using different learning approaches at the University of the Western Cape. *Curationis, 34*(1).

Lee, J., Lee, Y., Bae, J., & Seo, M. (2016). Registered nurses' clinical reasoning skills and reasoning process: A think-aloud study. *Nurse Education Today, 46*, 75-80.

Mahmoud, A., & Mohamed, H. (2017). Critical Thinking Disposition among Nurses Working in Puplic Hospitals at Port-Said Governorate. *International Journal of Nursing Sciences, 4*(2), 128-134.

Moattari, M., Soleimani, S., Moghaddam, N. & Mehbodi, F. (2014). Clinical concept mapping: Does it improve discipline based critical thinking of nursing students? *Iranian Journal of Nursing and Midwifery Research, 19*(1), 70-76.

Nirmala, T. & Shakuntala, B. (2011). Concept mapping an effective tool to promote critical thinking skills among nurses. *Nitte University Journal of Health Science, 1*(4), 21-26.

Park, M., McMillan, M., Conway, J., Cleary, S., Murphy, L. & Griffiths, S. (2013). Practice-based simulation model: a curriculum innovation to enhance the critical thinking skills of nursing students. *The Australian Journal of Advanced Nursing*, *30*(3), 41-51.

Rezaei, R., Saatsaz, S., Nia, H., Moulookzadeh, S., & Behedhti, Z. (2015). Anxiety and Critical Thinking Skills in Nursing Students. *British Journal of Education, Society & Behavioural Science*, *10*(2), 1-7.

Roland, E., Johnson, C., & Swain, D. (2011). "Blogging" As an Educational Enhancement Tool for Improved Student Performance: A Pilot Study in Undergraduate Nursing Education. *New Review of Information Networking*, *16*(2), 151-166.

Romeo, E. (2010). Quantitative Research on Critical Thinking and Predicting Nursing Students' NCLEX-RN Performance. *Journal of Nursing Education*, *49*(7), 378-386.

Salzinger, K. (1982). Methodological behaviorism is not radical behaviorism. *Contemporary Psychology: A Journal of Reviews*, *27*(1), 70-70.

Secrest, J., Keatley, V., & Norwood, B. (1999). Integrating the AACN "Essentials Baccalaureate Education for Professional Nursing Practice": A Teaching Project. *Nurse Educator*, *24*(6), 37-44.

Victor-Chmil, J. (2013). Critical Thinking versus Clinical Reasoning versus Clinical Judgment. *Nurse Educator*, *38*(1), 34-36.

Ward, T. & Morris, T. (2016). Think Like a Nurse: A Critical Thinking Initiative. *The ABNF Journal*, *27*(3), 64-66.

Yildirim, B. & Özkahraman, Ş. (2011). Critical Thinking Theory and Nursing Education. *International Journal of Humanities and Social Science*, *1*(17), 176-185.

CHAPTER 10

Concept Mapping in Chemistry

Nimmy Thomas

ABSTRACT

The present study aimed to determine the effects of concept mapping; a constructivist-based learning strategy, on middle school students' academic achievement in chemistry in a private school in UAE. The study employed a quasi-experimental, pre-test, post-test design using an achievement test called CMAT (Chemistry Matter Achievement Test). Results showed that concept mapping is an effective strategy for teaching and learning chemistry concepts. The data analysis also revealed that the concept mapping strategy is capable of enhancing learners' mastery of content at higher order levels of cognition. The findings of this study recommended that chemistry educators should use concept mapping strategy as a pedagogical and evaluation tool as it can enhance students' meaningful understanding of chemistry concepts, longer retention of information and academic performance in the subject content.

INTRODUCTION

Science education has continued to be a relevant discipline of studies in academia and there has been several reform efforts in the methodology and pedagogy of K12 science education globally. Several reform efforts have been made to develop effective teaching and learning strategies to increase the pace of meaningful science learning globally and concept maps are considered to be one of the effective teaching strategies in science education. Novak (1998) justify that concept mapping help students learn how to learn, as it is an effective pedagogical tool that is metacognitive in nature.

CONCEPT MAPS

Concept maps are graphical tools used to link concepts hierarchically, starting from more general to specific, by using propositional statements and connected label lines dictate interrelationships among concepts (Novak, 1987). Several studies have indicated that meaningful learning of scientific concepts can be enhanced by concept mapping, an effective constructivist teaching and learning strategy which has its orientation under David Ausubel's assimilation theory of cognitive learning (Novak & Gowin, 1984; Novak, 1998; Novak, 1990; Novak, 2010; Correia, 2012; Correia & Cicuto, 2014; Jack, 2013; Novak & Cañas, 2008). Concurrently, the literature review on the effectiveness of concept mapping on learning indicates a plethora of studies (Novak,1990; Novak ,1998; Nesbit & Adesope, 2006; Villalon & Calvo, 2011; Novak & Cañas, 2008; Correia, 2012; Novak, 2010; Moon *et.al.,* 2011; Karakuyu, 2010; Jack, 2013; Kinchin, 2014).

According to Novak and Gowin (1984), concept maps are effective because they let students, structure and organize their understanding of knowledge. Concepts maps are designed to enhance learners' understanding of concepts by expressing concepts and propositions known to the student and making them visually ostensible to enable their relationship with newly learnt concepts (Jack, 2013). Concept maps are based on visual imagery and are easily comprehended than the words that represent an abstract information (Terry, 2003). One significance of concept map is that the prior knowledge which has qualitatively, and quantitatively different knowledge structures is visible in it (Hay, Kinchin & Lygo-Baker, 2008; Popuva-Gonci & Lamb, 2012). According to Guastello, Beasley and Sinatra (2000), concept mapping can be used to translate ideas from texts into graphic representations that exhibit whole relationships, by creating a network of content ideas. Davies (2011) pointed out that the main benefit of concept mapping is its relational aim as it allows relational linking between the concepts.

Construction of Concept Maps

Novak and Gowin (1984) pioneered the first development of concept maps based on Ausubel's assimilation theory of learning in which a focal concept is used along with several key concepts (Chularut & Oklahoma, 2001). According to Novak and Gowin (1984), the steps of constructing a concept map are as follows:

- Read the material thoroughly and identify the concepts to be mapped.
- Select the focal concept and other key concepts and group them based on their interrelationships. A good focal question helps to keep our thoughts focused on the key concepts to be mapped and also a relevant parameter to select important concepts and linking phrases to make the network (Correia, 2012).
- Arrange the focal concepts and other clustered key concepts and link those using lines or drawing arrows from the most abstract to the most specific and include relevant examples at the lowest point of the hierarchical representation.

- Label the connecting lines between the concepts to represent meaningful relationships between the concepts. The use of linking phrases to elucidate relationships between concepts makes concept maps powerful than other graphical organizers (Davies, 2011).
- The maps can be redesigned or modified as more knowledge is gained by research.

Kiliç and Çakmak (2013) further support the steps of concept map construction in their study of using concept maps as a tool for meaningful learning and teaching in chemistry education.

Figure 1, shows a list of concepts to address a focal question, what is a Plant? A simple Novakian concept map, which incorporates the concepts into a network of meaningful learning material by representing the relationships between the concepts with connecting lines and linking phrases (Novak, 1998).

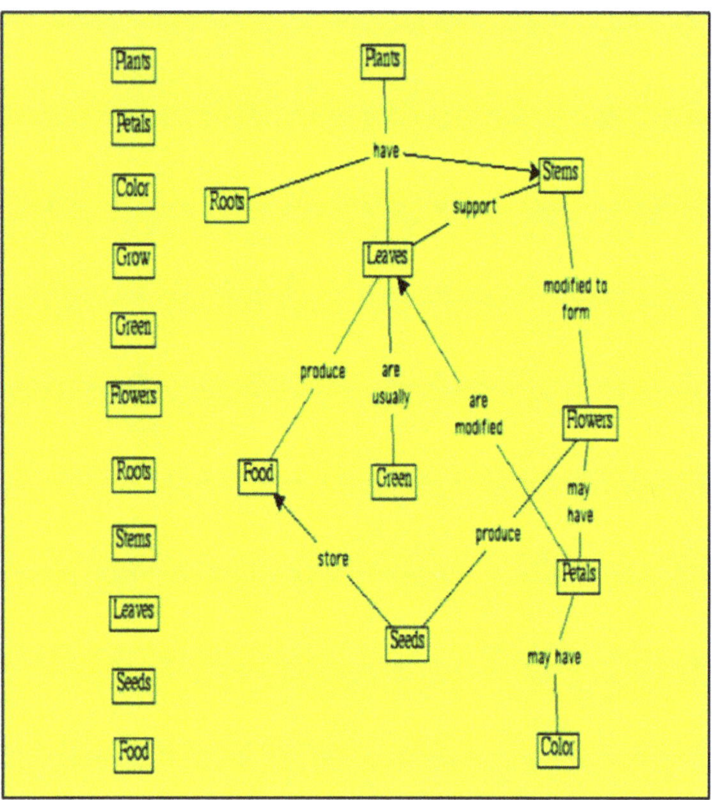

Figure 1: A Simple Novakian Concept Map

RATIONALE AND SIGNIFICANCE OF THE STUDY

According to Vanides *et.al.* (2005), it is challenging for a science educator to create an exciting, inspiring and inquisitive learning environment that engages all students at the middle school level. They also indicate that monitoring the progress of each student in science and ensuring that they understand all the taught scientific concepts is significant. These challenges make the science teaching at the middle school level demanding and rewarding for science educators. Teaching and learning chemistry is difficult as it involves highly complex conceptual relations, which includes concepts, which are unfamiliar and abstract in nature (Ghassan, 2007; Aderogba & Olarundare, 2009; Emmanuel, 2013; Meerah *et.al.* 2013; Burrows & Mooring, 2015). The best way of improving students' performance in chemistry is to exploit effective teaching and learning strategies which are constructivist or metacognitive in nature that prevents rote learning and enhance meaningful learning. Several researchers point out that concept maps are effective instructional tools that enhance meaningful learning by inculcating thinking skills, creative thinking, motivating learning, enhances knowledge transfer performance etc. (Krajcik & Czerniak, 2007; Meerah *et.al.* 2013; Tseng *et. al.*2012; Kostova & Radoynovska, 2010; Novak, 2010; Novak & Gowin, 1984; Burrows & Mooring, 2015). A comprehensive review of research proved that concept mapping as an instructional tool, is an effective strategy in enhancing meaningful learning in science, though very few recent studies in chemistry (Karakuyu, 2010; Meerah *et.al.* 2013; Jack, 2013; Burrows & Mooring, 2015), especially very few (Buldu & Buldu, 2010) within UAE or Dubai at the middle school level with students who are native Arabic speakers.

The current study is quite relevant and appropriate in Dubai considering the significant proportion of students' persistently scoring low in international standardized assessments like TIMSS, PIRLS or PISA (KHDA, 2012, 2013), which indicates that students in Dubai are performing below the international benchmarks especially in the cognitive domains such as application and reasoning domains in science. The achievement pattern in students in private and public MoE (Ministry of Education) curriculum schools shows that achievement was lowest in chemistry compared to other content domains in science (KHDA, 2013). The report has raised concerns about the lack of application and reasoning skills in students in science. The reason for poor performance has been blamed upon the ineffective and improper teaching strategies and lack of cognitive skills or understanding of concepts (KHDA, 2012). KHDA's (2012) report on TIMSS and PIRLS achievement and KHDA (2013) school inspection handbook states the importance of equipping students with a range of cognitive skills, which is directly linked to students' attainment and progress in science. The UAE Vision 2021 National agenda emphasizes on the need of a complete transformation of the current education system and teaching methods and has set as a target that students in UAE rank among the top 15 in TIMSS and top 20 in PISA exams by 2021. According to PISA 2015 draft science framework (OECD, 2013), scientific literacy is defined by the three scientific competencies such as ability to "explain phenomenon scientifically, evaluate and design scientific enquiry and interpret data and evidence scientifically" which, needs content, procedural and epistemic knowledge. A key new feature of the PISA science draft framework for the year 2015 was defining different levels of cognition within the assessment of scientific literacy, across the three competencies mentioned above (OECD, 2013).

The demand for development in the UAE education system calls for a paradigm shift in learning and teaching science. Several studies emphasis on the use of metacognitive strategies in learning and assessment in the higher education institutions in UAE (Forawi, Almekhlafi & Al-Mekhlafy, 2011; Tubaishat, Lansari & Al-Rawi, 2009). Metacognition can empower meaningful learning (Collins, 2011; Anderson & Krathwohl, 2001) and the metacognitive skills can be enhanced by cognitive visualizations (Villalon & Calvo, 2011). A review of studies (Nesbit and Adesope, 2006; Karakuyu, 2010; Jack 2013; Villalon & Calvo, 2011; Novak & Gowin, 1984; Novak, 1990; Novak, 1998; Meerah *et.al.* 2013; Tseng *et.al.* 2012; Kostova & Radoynovska, 2010; Emmanuel, 2013; Burrows & Mooring, 2015) maintain the effectiveness of concept mapping strategy with extensive evidence and justify that concept maps are highly effective cognitive visualization technique and engage students in higher cognitive functions, creates a concrete understanding of concepts and their relationships (Novak, 1998; Novak & Cañas, 2008; Jack, 2013), promote active and meaningful learning rather than memorization (Correia,2012; Erasmus, 2013; Burrows & Mooring, 2015). The concept mapping strategy can be used to determine the nature of existing ideas of students, make evident the new concepts to be learned, and suggest links between the two, which, can make learning, and teaching effective. The current study is significant as it has confirmed the importance of concept mapping in enhancing meaningful learning by metacognition and constructivism, increasing student achievement by improving cognitive skills, mainly at the higher order levels of cognition and above all develop a genuine interest for science and related disciplines like chemistry right from the elementary and middle school level. The current study's findings can provide relevant contributions to the field of education to understand the effectiveness of concept mapping when used as an instructional strategy to enhance meaningful learning of difficult concepts and therefore increase the academic achievement of students and develop a positive attitude towards chemistry, especially the native Arabic speakers who receive education in the medium of English. It is hoped that the current study will fill the gaps between the theories of constructivism and metacognition and its application on learning and academic achievement.

Purpose of the Study

The purpose of this study is to investigate the use of concept mapping as an instructional strategy in enhancing students' achievement and meaningful understanding of difficult chemistry concepts at the middle school level. The study also attempts to determine whether the use of concept mapping strategy would affect students' higher order levels of cognition such as analysing, evaluating and creating levels of the domain. The study was mainly built on several empirical studies on effectiveness of concept mapping approach on wide range of disciplines within diverse contexts and a range of variables and was guided to some extent by the revised bloom's taxonomy of educational objectives in designing the instrument (Bloom, 1969; Anderson & Krathwohl, 2001). It has been assumed that to enhance meaningful learning of difficult concepts, effective teaching and learning strategies based on constructivist and metacognitive theories should be used. A quasi-experimental method to determine the cause and effect relationship between variables assigned for an experimental group and control group using a pre-test post-test design on a selected topic of chemistry, considered to be the appropriate method for this investigation. The researcher being the participants' general science teacher and her knowledge about the learners' learning needs supported in designing lesson transcripts, learning resources and the CMAT (Chemistry Matter Achievement Test).

Research Questions and Hypotheses

The current study addresses the following questions,

1. To what extent does the concept mapping strategy affect students' academic achievement in chemistry?
2. To what extent does concept-mapping strategy affects the students' achievement scores in higher order cognitive objectives such as analyzing, evaluating and creating levels of chemistry instruction?

Hypotheses formulated for the study are as follows:

1. There would be significant difference in the achievement scores of students taught by concept mapping approach than that of those taught by the lecture method.
2. There would be significant differences in the students' achievement scores in analyzing, evaluating and creating levels of cognition after being taught by concept mapping strategy than by conventional method.

THEORETICAL FOUNDATIONS

Literature review reveals psychological and epistemological foundations for the development of concept maps (Novak, 1990; Novak & Gowin, 1984; Novak & Cañas, 2007; Novak, 2010)

Ausubel's Assimilation Theory of Meaningful Learning: A Psychological Foundation

David Ausubel is one of the educational psychologists who made relevant contributions to the field of education which include his assimilation theory of meaningful learning. Ausubel's assimilation theory of meaningful learning was published in 1963 and mainly provides psychological foundations of concept maps. The pivotal concept of his learning theory is the concept of meaningful learning (Ausubel *et.al.* 1978). The main idea of his theory is the distinction between the surface learning and meaningful learning which deals with three concerns such as curriculum content, learning and instructions. According to Ausubel, "meaningful learning requires three conditions: (1) the material to be learned must be conceptually clear and presented with language and examples relatable to the learner's previous knowledge. (2) The learner must possess relevant prior knowledge. (3) The learner must choose to learn meaningfully" (Novak & Cañas, 2007, p.30). In his view students must relate new knowledge to their existing knowledge to learn meaningfully. Ausubel (1963) and Ausubel *et.al.* (1978) claim that new concepts to be learned can be incorporated into graphic representations of concepts. He further emphasizes on teacher's role for organizing and presenting concepts to be learned to enhance meaningful learning and for motivating students to choose to learn meaningfully by incorporating new ideas into existing ones rather than rote learning the concepts (Novak & Gowin, 1984; Novak, 1990; Novak, 2010; Otor, 2011).

Based on Ausubel's ideas of assimilation theory of meaningful learning, Novak and Gowin (1984) have developed an instructional theory incorporating concept maps that graphically represents new concepts linked to previous knowledge. Novak (2010, p.23) claimed that if thinking, feeling and acting is constructively incorporated that enhances meaningful learning, can empower learners to be committed and responsible. The theory of meaningful learning emphasis on transforming information to long term memory from short term memory when students learn complex ideas from their own experience and linking new information to prior knowledge.

Epistemological Foundations Constructivist and Metacognitive Nature of Concept Maps

Novak (1987; 1993) as cited in (Novak, 2010) explains a close relationship between constructivist epistemology (Kuhn, 1962) and Ausubel's theory of human learning. Novak describes this as human constructivism in his refined theory of education (Novak, 2010). Learners learn through concept maps create new knowledge and create new meanings (Novak & Cañas, 2007; Novak, 2010). Novak & Cañas (2007) assert that concept maps facilitate meaningful learning and new knowledge creation, which has constructivist epistemology, has its foundation in which knowledge is a human construction from which new ideas are evolved (Novak, 1987).

The constructivist approaches of learning emphasize on interaction between learner and learning environment and during this interaction prior knowledge is used to interpret and construct new ideas. Concept map is a metacognitive strategy that uses affective as well as cognitive skills to enhance pattern recognition and stimulate meaningful learning (Irvine, 1995; Novak & Gowin, 1984) and to promote critical thinking skills (Ku & Ho, 2010).

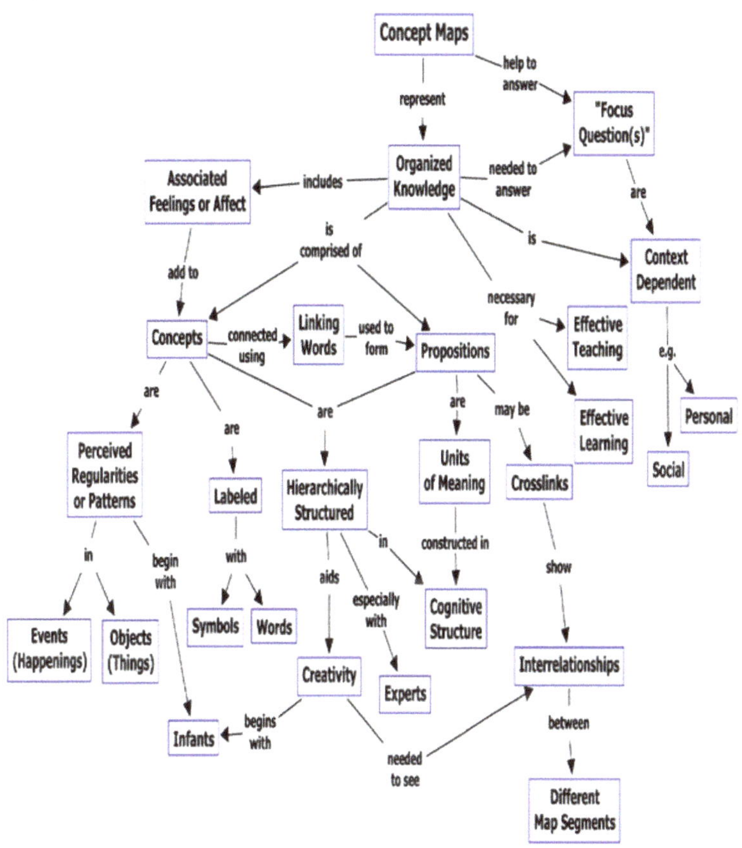

Figure 2: A Conceptual Overview of Concept Maps (Novak & Cañas, 2006).

Villalon and Calvo's (2011) study further extend evidences that drawing a concept map requires students to engage in higher cognitive functions. Correia (2012) emphasized on the use of concept maps to promote meaningful learning rather than surface learning. Correia (2012) also points out the relevance of concept maps in supporting collaborative construction of knowledge, which improves communication and thus meet the demands of our contemporary society for strategic knowledge management. According to Novak and Cañas (2008) the hierarchical structure and the ability to search for and characterize new cross-links are the important features of concept maps that facilitate creative thinking.

Concept Mapping as an Instructional Strategy Related Studies

Joseph Novak and the members of his research group introduced concept maps as a graphical tool to evaluate children's understanding of science by organizing information (Novak & Gowin, 1984; Novak, 2010; Novak & Cañas, 2007). Concept maps have been extensively used and verified as pedagogical tools since its inception (Novak & Cañas, 2007).

Concept Mapping as a Pedagogical and Evaluation Tool

The concept mapping strategy has been adequately advocated in literature as effective tools to facilitate meaningful learning of abstract concepts (Novak, 1990), as a way of graphical representation of frameworks for the relationship between concepts (Novak & Gowin, 1984; Stewart *et.al.*1979) as a teaching and assessment tool to foster purposeful learning (Liu & Hinchey, 1996). To sum up, concept maps are used as means for meaningful learning and knowledge construction, instructional planning and curriculum development, assessment and evaluation, identifying understanding and misconceptions of key ideas and helping them achieving correct conceptual learning (Novak, 1990; Mintzes *et.al.*, 1997; Nesbit & Adesope, 2006; Novak & Cañas ,2007; Ikeobi, 2010) and for assessing learning processes and as an alternative to science assessment in classrooms (Fleener & Marek, 1992; Meerah *et.al.*, 2013; Soika & Reiska, 2014).

Soika and Reiska (2014) reported in their study that concept mapping, as an evaluation tool is indispensable even in large-scale studies as it provides a visualization of the student's knowledge structure. Correia (2014) investigated neighborhood analysis to foster meaningful learning using concept maps and concluded that they are applicable for assessing science lessons at secondary and higher level if the instructor deals with the cognitive conflicts of students and providing scaffolding and timely feedback. Kinchin (2014) in his study recommended the use of concept mapping in compatible curriculum settings so that the constructivist foundations of the strategy are reflected and should be utilized as a research tool in learning.

Concept Mapping in Various Discipline of Studies

Several studies have found the use of concept mapping strategy very effective in teaching and learning difficult concepts across different subjects, in mathematics

(Awofala & Awofala , 2011; Awofala, 2011), in biology (Udeani & Okafor, 2012; Adlaon, 2012) in physics (Karakuyu, 2010; Martínez et. al., 2012), in chemistry (Emmanuel, 2013; Jack, 2013; Burrows &Mooring, 2015; Meerah *et.al.*, 2013) in English literature (Leahy, 1989; Villalon & Calvo, 2011). According to Leahy (1989), concept-mapping strategy can be adopted effectively in learning literature in which students can link characters, actions, and symbols, which are the concepts in literature and make a network of content ideas to exhibit what they have learned. They can also use concept mapping to condense their experience through reading and give reflections. These research studies concluded that the concept mapping is a valid tool in teaching and learning. Several research studies revealed that concept mapping is a valid and reliable teaching and learning tool in higher education (Hay, 2007; Hay, Kinchin & Lygo-Baker, 2008; Grice, 2016), especially in nursing education (Lee *et. al.,* 2013; Hunter Revell, 2012; Gerdeman, Lux & Jacko, 2013) and in enhancing critical thinking in nursing students (Wheeler & Collins, 2003; Nirmala & Shakuntala, 2011; Kaddoura, Van-Dyke & Yang, 2016). However, a very few have addressed the effectiveness of concept mapping strategy at the elementary level. The practice of using concept maps as an instructional tool is becoming more extensive in the areas of science and mathematics education.

Effects of Concept Mapping on Students' Academic Achievement in Science/Chemistry

Emmanuel (2013) used a quasi-experimental study to investigate the effects of concept mapping strategy on the achievement in chemistry of 1357 secondary school students in Nigeria. The study reported that the students taught using concept-mapping strategy achieved higher scores than those taught by traditional teaching strategy. The study recommended to provide adequate training to teachers on the use of concept maps to teach difficult concepts in secondary level. These findings are in agreement with findings of several research studies that investigated the effects of concept mapping strategy on the academic achievement of students in chemistry (Jack, 2013; Meerah *et. Al,* 2013; Burrows & Mooring, 2015). Meerah *et.al.*, (2013) in their study have provided empirical evidence of the effectiveness of concept mapping as reliable techniques to foster thinking in chemistry lessons, especially on topics such as atomic structure, the periodic table and chemical bonding. Jack (2013) in his comparative study concluded that concept mapping approach was more effective in facilitating meaningful understanding of chemistry concepts which favored long time retention of information and was superior to even guided inquiry approach, which is considered as one of the effective methods in science instruction. This may be due to lack of background information and comprehension of concepts. This can be solved by encouraging students' active participation in constructing concept maps of the learning material which can help students create a cognitive schema to integrate and relate new information (Guastello, Beasley & Sinatra, 2000). Concept mapping strategy was recommended as the most effective instructional strategy over expository and guided inquiry methods of instructions by pointing out that strong conceptual understanding is relevant for inquiry lessons to be meaningful and effective (Jack, 2013).

METHODOLOGY

The study was a four-week investigation, which was carried out in a K-12 British IGCSE curriculum School in Dubai, the UAE. The participants were from two classes of grade seven (7), consisted of sixty-four (64) native Arabic speaking students, which includes thirty-one (31) males and thirty-three (33) females within an age range of 11 to 12. A quasi-experimental research design is employed in the study. A non-equivalent group pretest-posttest experimental design with maximum naturally occurring comparisons as possible is considered. A pre-test consists of questions to examine the cognitive skills like remembering, understanding, applying, analysing, evaluating and creating was done on the topic taught. Thirty-one (31) students (15 males and 16 females) were assigned as the experimental group, who were intervened with concept mapping strategy and thirty-three (33) students (16 males and 17 females) were assigned as the control group who were exposed to the conventional method of teaching. The researcher being a science teacher of the participants had taught a topic of chemistry to both the experimental and control groups for four weeks, meeting five times a week. The same questions of pre-test was given as a post-test to measure and compare the effectiveness of concept mapping approach on the academic achievement of students. A stratified random sampling was employed in the present study in which the researcher did an equivalent distribution of students from intact groups into experimental and control groups. The instrument was piloted in a different British curriculum School with thirty-six (36) grade 7 students to ensure its reliability and validity.

Data Analysis and Results

In the present study data collected were analysed by utilizing descriptive and inferential statistics (independent *t* test and paired *t* test) for the analysis of different data to test the two hypotheses of this study. Descriptive statistics were used to compare the mean pre-test and post-test scores of the experimental group and control group in the study and to report the statistical significant difference between them. Inferential statistical methods such as a paired *t* test was employed to compare the independent variable within each group to determine whether there is any statistical significant differences between the mean scores of pretest and posttest within each group. The *t* test compares the means of two groups and compares the differences in variables between two groups.

Difference Between the Pre and Post-Test Achievement Scores Within Experimental and Control Groups

The pre-test and post-test scores of participants within the same group are compared and explained using descriptive and inferential statistical methods. A paired t test was applied to find out whether there is any significant difference in the CMAT pre and posttest scores of students within the same group. Table 1 represents the descriptive and inferential statistics of CMAT scores for experimental and control groups.

Table 1: Descriptive Statistics and Paired T-Test Results

Groups	**Tests**	**N**	**M**	**SD**	**df**	**Paired *t* test**	**P**
Experimental E1	Pre (O1)	31	14.12	3.01	30	-19.88	0.000
	Post (O3)	31	21.38	3.49			
Control	Pre (O2)	33	13.33	2.52	32	-22.22	0.000
E2	Post (O4)	33	17.87	2.52			

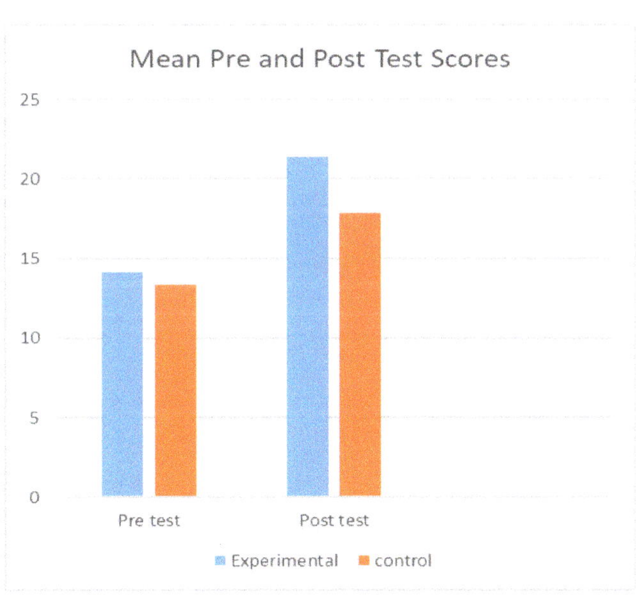

Figure 3: Difference between Mean CMAT Pre and Post Test Scores of Experimental and Control Groups

As shown in the table 1, the paired *t* test results for experimental group, t(30) = -19.88 and value of $p\angle 0.05$ shows that there is a significant difference in the pre and post test scores of students with in the experimental group. It is found that there is also a significant difference in the pre and post test scores of students within the control group as determined by the paired *t* test, t (32) = -22.22 and $p\angle 0.05$.The descriptive and inferential statistic results explain that the experimental group and control group have achieved better scores in the CMAT post-test than the pre-test irrespective of the teaching intervention they have received during the period of this study. Graph 1 represents the mean pre and posttest scores of experimental and control group in the CMAT test.

Effect of Concept Mapping Approach on Achievement Scores at Higher Levels of Cognition

Descriptive statistics and independent t tests were used to compare the mean CMAT post test scores on the analysing, evaluating and creating components of the test for the experimental and control groups. The results are represented in table 2.

Table 2: Comparison of Students' Post-test Analysing, Evaluating and Creating scores

Taxonomy	Group	N	M	SD	Df	Independent t test	P
Analysing	Experimental Control	31 33	3.61 2.52	0.80 0.71	62	5.792	0.000
Evaluating	Experimental Control	31 33	3.26 2.39	0.68 0.56	62	5.574	0.000
Creating	Experimental Control	31 33	2.48 1.06	1.00 0.70	62	6.633	0.000

As shown in table 2, students' mean post test scores in the analysing, evaluating and creating levels of cognition of the experimental and control groups indicate that the experimental group had greater gains in the higher order levels of cognition in the post-test than the control group. The results indicated significant outcomes in the students' scores at analysing level (t (62) = 5.792, $p \angle 0.05$), evaluating level (t (62) = 5.574, $p \angle 0.05$) and creating level (t (62) = 6.633, $p \angle 0.05$) of cognition. These differences showed that the students exposed to concept mapping approach achieved better than those exposed to traditional teaching method at their higher levels of cognition in the CMAT test as shown in graph 2 given below.

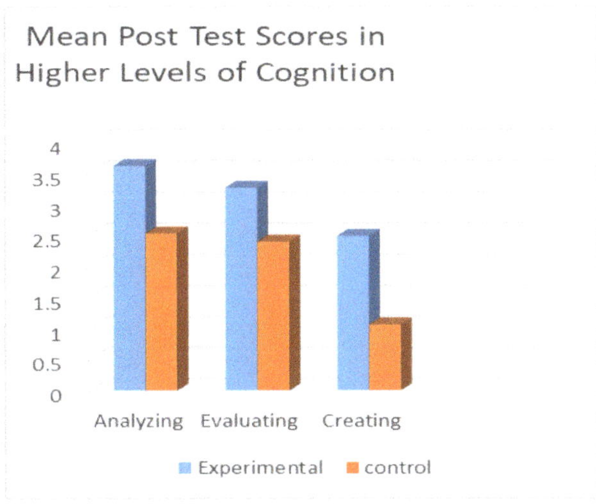

Figure 4: Mean Post Test for Higher Levels of Cognition

These outcomes corroborate the findings that using concept mapping as an instructional strategy has positively affected the students' academic achievement in chemistry as well as their higher levels of cognition. Thus, based on the above findings, it can be concluded that there has been an increase in the academic achievement of students in the experimental group than the students in the control group.

Conclusions

The findings of the study indicated that the concept mapping approach was more effective and superior to the lecture method in teaching and learning science/ chemistry concepts which, are aligned with several previous studies (Meerah *et.al.*, 2013; Jack, 2013; Burrows & Mooring, 2015; Emmanuel, 2013). It can be concluded that concept mapping has the potential to foster meaningful learning, which, leads to retention of information for a long time. In the current study, all the students who were exposed to concept mapping performed better in their post-test and also attained higher scores in higher levels of cognition such as analysing, evaluating and creating compared to the students who were not exposed to concept mapping. This is because instruction-using concept mapping approach has helped students to identify the relationship among concepts, link their previous knowledge and create new knowledge, which has enhanced their cognitive skills due to its metacognitive and constructivist nature. The UAE Vision 2021 National agenda accentuates the need of a complete makeover of the present education system and teaching methods and has set a target for students in UAE to rank among the top 15 in TIMSS and top 20 in

PISA exams by 2021. This can be achieved by using effective strategies like concept mapping in science instruction that will foster meaningful learning of concepts and promote longer retention of information which will help students to perform better in science/chemistry achievement tests and also develop interest and positive attitude towards the subject and motivated to pursue higher studies in the subjects which they considered difficult to study.

The findings of this study can provide relevant contributions in the field of education, to understand the need for using strategies such as concept mapping effectively for meaningful instruction, effective curriculum designing, effective evaluation etc. to provide enhanced learning outcomes. Chemistry educators must acquaint with concept mapping in classrooms in the absence of well-equipped laboratories and lack of appropriate instructional aids for hands- on learning as it can be a best alternative to laboratory experience which provides relationships between concepts that will enhance students' performance in the subject. Text books and other learning resources should include concept maps at the end of each lesson as a summary of the learning material. It is hoped that the present study will add to the body of knowledge regarding constructivist and metacognitive learning and will aid in linking the gap that currently exists between these learning theories and its applications in science teaching and learning.

REFERENCES

Aderogba, G.A & Olorundare, A.S. (2009). Comparative effects of concept mapping, analogy and expository startegies on secondary school students' performance in chemsitry in Ilesa, Nigeria. *Journal of Curriculum and Instruction*, 7(1,2), 112-126.

Adlaon, B.R. (2012). *Assessing Effectiveness of Concept Map as Instructional Tool in High School Biology.* M.Sc. Thesis. Louisiana State University.

Anderson, L.W & Krathwohl, D.R. (2001). *A taxonomy of learning, teaching and assessing: A review of Bloom's taxonomy of educational objectives.* Longman Publication.

Ausubel D.P. (1963). *The Psychology of Meaningful Verbal Learning.* Grune and Stratton.

Ausubel D.P., Novak, J.D & Hanesian, H. (1978). *Educational Psychology: A Cognitive View.* 2nd edn. Holt, Rinehart, and Winston.

Awofala, A.O.A & Awolola, S.A. (2011). Curriculum Value Orientation and Reform in the 9-year Basic Education Mathematics Curriculum. In O.S. Abonyi (Ed.) Reforms in STEM Education. 52nd Annual Conference of Science Teachers Association of Nigeria, 297-304.

Awofala, A.O.A. (2011). Effect of Concept Mapping strategy on Studnets' Achievement in Junior Secondary School Mathematics. *Intenational Journal of Mathematics Trends and Technology*, [Online] vol. 2(3), pp. 11-16. Retived on 12th January 2016. http://www.internationaljournalssrg.org

Bloom, B. (1969). *Taxonomy of educational objectives: The classification of educational goals.* McKay.

Buldu, M. & Buldu, N. (2010). Concept Mapping as a formative assessment in college classrooms: Measuring usefulness and student satisfaction. Procedia Social and Behavioural Sciences. http://www.sciencedirect.com

Burrows, N & Mooring, S. (2015). Using concept mapping to uncover students' knowledge structures of chemical bonding concepts. *Chem. Educ. Res. Pract*, vol.16(1), pp.53-66.

Chularut, P & Oklahoma, N. (2001). The influence of concept mapping on achievement, self regulation, and self-efficacy in students of English as a second language. *Contemporary Educational Psychology*, 29(3), 248-254.

Collins, T. (2011). *Science Inquiry as Knowledge Transformation: Investigating Metacognitive and Self-regulation Strategies to Assist Students in Writing about Scientific Inquiry Tasks.* Ph.D. Dissertation. Oregon State University.

Correia, P.R.M. (2012). The use of concept maps for knowledge management: from classrooms to research labs. *ABCS of Teaching Analytical Science*, 402, 1979-1986.

Correia, P.R.M. & Cicuto, C.A.T. (2014). Neighbourhood analysis to foster meaningful learning using concept mapping in science education. *Science Education Inernational, col.*, 24(3), 259-282.

Davies, M. (2011). Mind mapping, concept mapping, argument mapping: what are the differences, and do they matter? *High Educ.*, 62, 279-30.

Emmanuel, O. E. (2013). Effects of concept mapping strategy on students' achievement in difficult chemistry concepts. *International Research Journals*, 4(2), 182-189.

Erasmus, J. E (2013). Concept mapping as a strategy to enhance learning and engage students in the classroom. *Journal of Family and Consumer Science Education*, vol. 31(1), pp. 27-35. [online]. Retrieved on 11th March 2016 http://www.natefacs.org/JFCSE/v31no1/v31no1Erasmus.pdf

Fleener, M & Marek, E. (1992). Testing in the Learning Cycle. *Science Scope*, 15, 48-49.

Forawi, S., Almekhlafi, A & Al-Mekhlafy, M. (2011). Development and validation of electronic portfolios: The UAE pre-service teachers' experiences. *Computer Research and Development (ICCRD): 3rd International Conference.* United Arab Emirates University, Al-Ain. 11-13 March. ICCRD.

Gerdeman, J.L., Lux, K & Jacko, J. (2013). Using cocnept mapping to build clinical judgement skills. *Nurse Education in Practice*, vol. (13), pp. 11-17.

Ghassan, S. (2007). Learning difficulties in chemistry: An overview. *Jourrnal of Turkish Science Education,* vol 4(2), pp. 2-20.

Grice, K. (2016). Concept Mapping as a Learning Tool in Occupational Therapy Education. *Occupational Therapy in Health Care.* 25th February 2016. [online] [retrived on 14th March 2016], pp.1-10. Available at: http://www.ncbi.nlm.nih.gov/pubmed/26914229

Guastello, F.E., Beasley, M. T & Sinatra, C.R. (2000). Concept Mapping Effects on Science Content Comprehension of Low- Achieving Inner-City Seventh Graders. (online)., vol. 21(6). [retrived 17th February 2016]. Available at: www.researchgate.net/publications/249835034

Hay, D.B. (2007). Using concept maps to measure deep, surface and non-learning outcomes. *Studies in Higher Education,* (32), 39-58.

Hay, D.B., Kinchin, I.M & Lygo-Baker, S. (2008). Making learning visible: The role of concept mapping in higher education. *Studies in Higher Education*, 33, 295-311.

Hunter Revell, S.M. (2012). Concept maps and nursing theory: A pedagogical approach. *Nurse Educator*, 37, 131-135.

Ikeobi, I.O. (2010). *Beyond the Sterotype: Thoughts and Reflections on Education.* Lagos Nigeria. The CIBN Press Limited.

Irvine, L.M.C. (1995). Can concept mapping be used to promote meaningful learning in nurse education? *Journal of Advanced Nursing*, 21, 1175-1179.

Jack, G.U. (2013). Concept mapping and guided inquiry as effective techniques for teaching difficult concepts in chemistry: Effect on students' academic Achievement. *Journal of Education and Practice*, (4(5), 9-15.

Kaddoura, M., Van-Dyke, O & Yang, Q. (2016). Impact of a concept map teaching approach on nursing students' critical thinking skills. *Nursing & Health Sciences*. [online]. [Accessed 23rd April 2016]. Available at: http://onlinelibrary.wiley.com/doi/10.1111/nhs.12277/full

Karakuyu, Y. (2010). The effect of concept mapping on attitude and achievement in a physics course. *International Journal of Physical Sciences*, vol. 5(6), pp. 724-737.

Kiliç, M & Çakmak, M. (2013). Concept maps as a tool for meaningful learning and teaching in chemistry education. *International Journal on New Trends in Eductaion and Their Implications*, vol. 4(4), pp. 152-164.

Kinchin, I. M. (2014). Concept mapping as a learning tool in higher education: A critical analysis of recent reviews. *The Journal of Continuing Higher Education*, vol. 62, pp. 39-49.

Knowledge and Human Development Authority (2012). 'Dubai: TIMS and PIRLS 2011 Report' [online]. Dubai: KHDA. [Accessed 21 January 2016] Available at: http://www.khda.gov.ae/CMS/WebParts/TextEditor/Documents/TIMSS_2011_Report_EN.pdf

Knowledge and Human Development Authority (2013). 'PISA 2012 Report' [online]. Dubai: KHDA. [Accessed 18 January 2016]. Available at: https://www.khda.gov.ae/CMS/WebParts/TextEditor/Documents/PISA2013EnglishReport.pdf

Knowledge and Human Development Authority (2013). 'Inspection Handbook 2013-2014: Dubai Schools Inspection Bureau' [online]. Dubai: KHDA. [Accessed 21 January 2016]. Available at: http://www.khda.gov.ae/CMS/WebParts/TextEditor/Documents/handbook%202013_4-7-13_English.pdf

Kostova, Z & Radoynovska, B. (2010). Motivating students' learning using word association test and concept maps. *Bulgarian Journal of science and Education Policy*, vol. 4(1), pp. 62-98.

Krajcik, J & Czerniak, C. (2007). *Teaching science in elementary and middle School: A project-based approach.* 4th edn. New York, NY: Routledge.

Ku, K & Ho, I. (2010). Metacognitive strategies that enhance critical thinking. *Metacognition Learning*. vol.5, pp. 251-267.

Kuhn, T.S. (1962). *The structure of scientifi revolutions*. Chicago, IL: University of Chicago Press.

Leahy, R. (1989). Concept mapping: Developing guides to literature. *College Teaching*, vol. 37(3), pp.62-69.

Lee, W., Chiang, C.H., Liao, I.C., Lee, M.L., Chen, S.L & Liang, T. (2013). The longitudinal effect of concept mapping teaching on critical thinking. *Nurse Education Today*, vol 33, pp. 1219-1223.

Liu, X. & Hinchey, M. (1996). The internal consistency of a concept mapping scoring scheme and its effect on prediction validity. *International Journal of Science Education*, vol. 18 (8), pp. 21-937.

Martínez, G., Pérez, Á., Suero, M & Pardo, P (2012). The Effectiveness of concept maps in teaching physics concepts applied to engineering education: Experimental comparison of the amount of learning achieved with and without concept maps. *Journal of Science Education and Technology*, vol. 22(2), pp.204-214.

Meerah, S.M., Osman, K & Wahidin. (2013). Concept mapping in chemistry lessons: Tools for inculcating thinking skills in chemistry learning. *Journal of Baltic Science Education*, vol. 12(5), pp. 666-681.

Mintzes, J.J., Wandersee, J.H. & Novak, J.D. (1997). *Teaching Science for Understanding*. San Diego. Academic Press.

Moon, B.M., Hoffman, R.R., Novak, J.D & Cañas, A. J. (2011). *Applied concept mapping: capturing, analysing, and organizing knowledge*. CRC Press, Boca Raton.

Nesbit, J & Adesope, O. (2006). Learning with Concept and Knowledge Maps: A meta-analysis. *Review of Educational Research*, vol. 76(3), pp.413-448.

Nirmala, T & Shakuntala, B.S. (2011). Concept mapping: An effective tool to promote critical thinking skills among nurses. Nitte University Journal of Health Science, vol. 1(4), pp. 21-26.

Novak, J.D. (1990). Concept mapping: A useful tool for science education. *J. Res. Sci. Teach.*, vol. 27(10), pp.937-949.

Novak, J.D & Cañas A. (2006). The origins of the concept mapping tool and the continuing evolution of the tools. *Inf Vis*, vol.5 (3), pp.175-184.

Novak, J.D & Cañas A. (2007). Theoretical origins of concept maps, how to construct them, and uses in education, vol. 3(1), pp. 29-42 [online]. [Accessed on 23rd February 2016] Available at: http://www.researchgaet.net/publication/228/61562.

Novak, J & Gowin, D. (1984). *Learning how to learn*. Cambridge [Cambridgeshire]: Cambridge University Press.

Novak, J. D. (1993). Human constructivism: A unification of psychological and epistemological phenomena in meaning making. *International Journal of Personal Construct Psychology*, vol. (6), pp. 167-193.

Novak, J.D & Cañas, A.J. (2008). The theory underlying concept maps and how to construct and use them. *Pensacola: Florida Institute for Human and Machine Cognition*. [online].[Accessed 17th

February 2016], Available at: https://www.uibk.ac.at/tuxtrans/docs/TheoryUnderlyingConceptMaps-1.pdf

Novak, J.D. (1987). Human Constructivism: Human Constructivism: A unification of psychological and epistemological phenomena in meaning making. *Proceedings of the Second International Misconceptions and Educational Strategies in Science and Mathematics Conference*, June 1987. Ithaca, NY: Department of Education, Cornell University.

Novak, J.D. (1998). Learning, creating, and using knowledge: concept maps as facilitative tools in schools and corporations. *Choice Reviews Online*, vol. 36(02), pp.36-1103-36-1103.

Novak, J.D. (2010). *Learning, creating and using knowledge: concept maps as facilitative tools in schools and corporations*, 2nd Edition. Routledge, New York.

OECD (2013). PISA 2015, Science Draft Framework (online) [Accessed 15 November 2015]. Available at: https://www.oecd.org/pisa/pisaproducts/Draft%20PISA%202015%20Science%20Framework%20.pdf

Otor, E.E. (2011). *Effects of concept mapping strategy on students' attitude and achievement in difficult chemistry concepts.* A PhD Thesis Submitted to the Postgraduate School, Benue State University, Makurdi.

Popova-Gonci, V & Lamb, M.C. (2012). Assessment of integrated learning: Suggested application of concept mapping to prior learning assessment practices. *The Journal of Continuing Higher Education,* vol. 60, pp. 186-191.

Soika, K & Reiska, P. (2014). Using Concept mapping for assessment in science Education. *Journal of Baltic Science Education*, vol. 13(5), pp. 662-673.

Stewart, J., Van Kirk & Rowell, R. (1979). Concept maps: A Tool for use in biology Teaching. *The American Biology Teacher*, vol. 41, pp. 171-175.

Terry, W. S. (2003). *Learning and memory basic principles, processes, and procedures.* 2nd Edition. Boston: New York.

Tseng, K.H., Chang, C, -C., Lou, S.-J & Chiu, C.-J. (2012). How concept-mapping perception navigates student knowledge transfer performance. *Educational Technology & Society*, vol. 15(1), pp. 102-115.

Tubaishat, A., Lansari, A & Al-Rawi, A. (2009). E-portfolio assessment system for an outcome-based information technology curriculum. *Journal of Information Technology Education: Innovations in Practice,* vol. 8, pp. 43-54

Udeani, U & Okafor, P. N. (2012). The effect of concept mapping instructional strategy on the biology achievement of senior secondary school slow learners. *Journal of Emerging Trends in Educational Research and Policy Studies*, vol. 3(2), pp. 137-142.

Vanides, J., Ruiz-Primo, M. A, Tomita, M & Yin, Y. (2005). Using concept maps in science classrooms. *National Science Teachers Association*, vol 28 (8).

Villalon, J. & Calvo, R.A. (2011). Concept maps as cognitive visualizations of writing assignments. *Educational Technology & Society*, vol. 14(3), pp. 16-27.

Wheeler, L. & Collins, S. (2003). The influence of concept mapping on critical thinking in baccalaureate nursing students. *Journal of Professional Nursing*, vol.196), pp.339-346.

PART 4

CHAPTER 11

Innovative Science Education Trends

Hind Abou Nasr Kassir

ABSTRACT

The United Arab Emirates (UAE) within the last twenty years has been facing economic growth and global challenges. Forces that challenge policy implementation are specific to each country's economic, political authority and social demands and affect education policies. Another concern is the level of student's enrollment in public schools and universities. Therefore, the analysis would suggest how Emirati leaders within the Ministry of Education (MoE) respond to these forces and, furthermore, how society understands such new regimes of policy change. That means that students need to be working on new indicators and their teaching strategies need to reflect the market need and the future labor force and can be placed at the center of the institutional decision-making. The current quantitative study shows with evidence why the science inquiry as a teaching strategy is the innovative tool to be used in teaching science and how it includes entrepreneurship, critical thinking and can be the vehicle for the science reform and is the first step towards sustainability and long-term learning.

INTRODUCTION

The Middle East in the late ten years has shown through surveys an increase of interest in the scientific professions in high schools, while the top course preference from around the world was Bachelor of Business Administration (BBA). Yet, around 70 per cent of the students indicated engineering as their preferred choice and more than 55 per cent of students from the Middle East are likely to go abroad to the United States, United Kingdom, Canada and Spain for higher studies. These numbers published by the gulf news in September 2013, showing the pressing need to more hands-on and inquiry-based curricula in our schools and colleges (Chaudhary, 2013).

Since 1962, Education in the United Arab Emirates is witnessing significant growth and evolvement on both quantitative and qualitative levels. The progress touches public and private education institutions towards the highest quality standards that could contribute to a better qualitative performance of the existing institutions and raise the components of their efficiency on the applied sciences, by maximizing modern technologies benefit and by introducing new disciplines aligned with the labour market needs (Ministry of Education and Higher education vision in the United Arab Emirates 2013).

Boosting science inquiry and scientific research and innovation are on the top priority of the United Arab Emirates government and notably the Ministry of Education and Higher Education levels in the United Arab Emirates (UAE). Under the guidance of His Highness Sheikh Mohammed bin Rashid Al Maktoum, UAE Vice President and Prime Minister and Ruler of Dubai, and his Excellency Mr. Hussein Ibrahim al Hammadi Minister of the Ministry of Education- An educational reform has been designed to support the country's vision, and to build a knowledge-based society through effective school management and academic curriculum reform (The National January 2014). In consequence, the United Arab Emirates (UAE), will offer with the new reform to all her citizens a wider range of educational watercourses especially in the advanced stream at the high school levels and building new educational streams that develop the vocational skills as a part of the applied experimental inquiry-based approach.

The project-based learning as an open- ended science-inquiry approach that builds both growth mind and entrepreneurship, is an endeavor for scientific courses and the tool for reform implementation. The English as a medium of instruction is an effective tool to help students' literacy and to bridge the gap between the public sector and the freshmen year at university levels (The National July 2013).

The UAE government, took part in the new educational reform, Moreover, the UAE adapted the United States of America educational policy values "no child is left behind" (2006) and science is for all (AAAS), and released the federal law No. 29 in the UAE:" School for all" (2006). Science and math were both on the top agenda of the stakeholders notably how to increase the students' outcome in the international tests such as TIMSS and PISA. An experimental project was launched in 2007 to restructure 56 schools out of 465 schools of the public sector; the project began with grade 1 from cycle 1, grade 6 from cycle 2 and grade 10 from cycle 3 schools. The main purpose of the project is first to integrate the standard-based curriculum and second to improve the literacy of the students in the English language and later on in the scientific courses after five years of the beginning of the program (Education council Abu-Dhabi, Boujaoude & Dagher, 2009 p.1).

In order to monitor the change in the new reform, all science educators in the program, should integrate the science- inquiry teaching approach as the main endeavour and an integral part of the curriculum taught from pre-K to grade 12 (Donnelly, 2009), acknowledging the difficulty to determine its effectiveness on the short-term basis (Hudson, 2007).

The purpose of the present study is to evaluate the effectiveness of the science-inquiry teaching approach on the improvement of students' achievement. This study is considered as pioneer in the UAE region, as it highlights "the effectiveness of the science-inquiry teaching approach and its effect on improving the students' achievement".

The paper is a case study that uses an evaluative quantitative approach to answer the following question:

> How effective is the science-inquiry teaching approach on the improvement of students' achievement in the UAE public schools following a standard-based curriculum?

LITERATURE REVIEW

The rapid growth of the online teaching and learning has been well noted in the United States of America from early 2008 (Shea & Bidjerano, 2009). With this rapid pace of technology that reached an estimation of around 3.5 million students of full online courses and a positive exponential of 20% each single year of students enrolled fully in online courses in the US, teachers were facing the challenge of the traditional setting of classroom-based strategies and what could retain students in schools (Morueta et al.,2016), while online teaching and learning numbers are six times higher than the face to face learning numbers (US Department of Education and National Center for Education Statistics, 2016).

Within the conceptual perspective of what science is and what science entails the inquiry-based approach is the tool that equals "learning is doing" (Shea & Bidjerano, 2009, p.544). The science inquiry is a solid epistemological approach that translated Kuhn cycle and can reach into the paradigm shift. It starts with the normal science where you define a hypothesis from real life, ask a question to initiate the model drift where capable instructors develop within the students' cognition the critical thinking as well as the reflection to answer a question from a different perspective. I see the Model crisis in Kuhn cycle as a classroom discussion where students are not all at the same page which will lead to a thoughts' revolution that will build the new concept constructed by students' themselves. As a consequence, new synapses are formed, and new ideas are built while breaking misconceptions. This is where the paradigm shift happens from a focused teaching and learning into an individualized learning (Warner, 2016). Afterwards norms are established, and learning is deepened where new domains in students' mind are built but this time with more knowledge built in our mind figure 1.

In the new science pedagogy, the science inquiry follows the 5 E model where teachers need to engage the student, elaborate the concept, explain the knowledge, evaluate the understanding which is the learning and extent where the teacher can see if a shift has occurred or not.

The current study is epistemological quantitative study that assesses the impact of the use of the effective science inquiry on the scientific reasoning. We present our theoretical framework by contrasting the effective use of scientific inquiry in a standard based curriculum facing the teacher centered instruction and textbook-based science curricula figure 2.

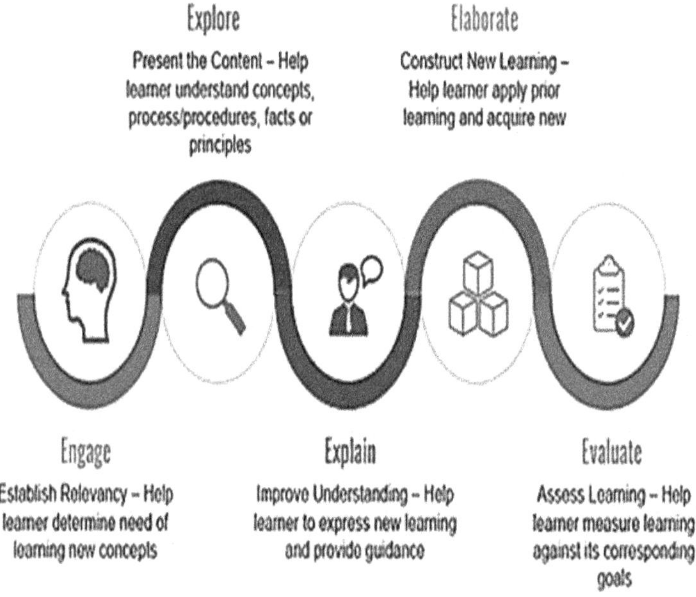
Explore
Present the Content – Help learner understand concepts, process/procedures, facts or principles
Elaborate
Construct New Learning – Help learner apply prior learning and acquire new
Engage
Establish Relevancy – Help learner determine need of learning new concepts
Explain
Improve Understanding – Help learner to express new learning and provide guidance
Evaluate
Assess Learning – Help learner measure learning against its corresponding goals

Figure 2: Sophie Isabell 5 E's

Whenever an effective instruction occurs, an effective teaching is taking place figure 3. Effective instructional strategies and effective teaching both are derived from a psychological perspective on thinking about teaching (Kyriacou, 2009). The UAE government requires teachers' commitment and demonstrated support by providing training opportunities to improve teaching skills and to help them develop the maximum potential in the students' mind based on challenging scientific hands-on activities as a part of the UAE School for all framework (Dahmashi, 2007).

The emphasis is placed on identifying observable behaviour in the classroom that can be linked to observable outcomes (Kyriacou 2009, pp. 11); it is all about making a difference in the students' lives. If the effective instruction in teaching could be defined as a teaching that could fruitfully help students to attend the required learning outcome by the teacher himself, this statement infers that the teacher should have a clear idea in mind about what he is going to teach and how he should be able to create a learning experience that could involve the students in their own learning to achieve specific learning outcome (Marzano, 2001); Therefore effective teaching evolves with the teacher's experiences and his own beliefs about teaching (Wideen, 1998 in Hudson 2007).

Scientific Inquiry/Learning Cycle

evaluate...engage...explore...

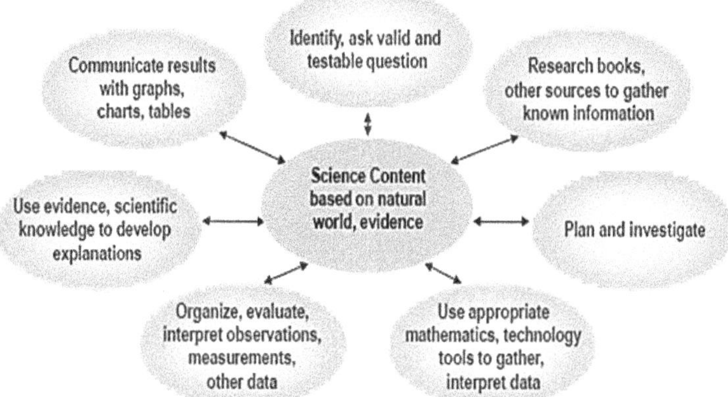
Identify, ask valid and testable question
Research books, other sources to gather known information
Plan and investigate
Use appropriate mathematics, technology tools to gather, interpret data
Organize, evaluate, interpret observations, measurements, other data
Use evidence, scientific knowledge to develop explanations
Communicate results with graphs, charts, tables
Science Content based on natural world, evidence

explain...extend...evaluate

Figure 3: Nasa Model for Scientific inquiry learning cycle

Some specific instructional strategies could affect the students' achievement such as summarizing and taking notes, identifying similarities and differences, questions, cues and organizers, and cooperative learning. Further research on the topic shows that students' concentration begins to decline 15 to 20 minutes after the beginning of the instruction (Bligh, 2000), therefore, active learning strategies are required. Active learning strategies include students taking substantive decisions together using their knowledge. Critical and creative thinking will progress if students learn how to plan and process some shared ideas with peers. Communication and how presentation is shaped to a particular audience is one of the 21st century skills required from students. Receiving feedback from peers could help students to explore their own personal attitudes and ethical values, giving and receiving feedback and reflecting on their own learning process (Bonwell & Eison, 1991). These active learning procedures will diminish risks of science misconceptions in the students' mind.

Assessing the effectiveness of an instruction in teaching, is evaluating what has been taught according to the observable and/or measurable students' outcomes. This evaluation is not done by expressing positive feelings about what is happening within the classroom that does not infer a learning process- yet it is an issue that should be concerned with the learning experiences' aspects that contribute to the efficiency of the instruction and reciprocally, which means, how these aspects have the effect they do on the effectiveness of the instructional teaching strategy (Kyriacou, 2009).In consequence, the study will not evaluate the instructional technique itself, but how this specific instruction is being effective on the overall learning process and specifically the students' achievement level. Research has not proven yet, the effectiveness of some specific instructional strategies in certain subject areas or per grade level than others. Research has not recognized if some instructions are better than others for students from different backgrounds and demographics (Marzano, 2007).

An article published by Lee & Luykx (2006) discusses factors that could influence on the students' achievement following the science-inquiry teaching approach as a hands-on teaching approach. The three main concerns that could have a direct influence on the student' achievement is the choice of the appropriate curricula, the teaching quality i.e. the pedagogical knowledge of the teachers using appropriate pedagogical materials, finally the students' attitude towards science and their level of engagement.

The choice of the curriculum taught in schools is a crucial factor that could increase or decrease the science achievement. Most of the curricula taught in the United Arab Emirates are imported from western culture, and all textbooks before the reform plan adopted by the ministry of education in 2014-2015 were not-relevant to the students' culture and choices (UAE vision 20121). The mainstream textbooks and content do not take in consideration students with low income families, second language learners, students with low background experience in science, students with special needs. Moreover, all these factors could influence on the students' achievement level their scientific skills and competencies.

The teaching' quality is the other part of the equilibrium. Quality teachers' influence positively on the variation in student achievement. The teachers' attributes and characteristics are a dependent variable in the students' performance. In reform-oriented practices effective science instruction is always comes from teachers having the same students' background, because they understand the students' culture and understand some of the local codes which could establish mutual respect and easiness in the teaching and learning process. Moreover, the major purpose of schooling is to provide for students' flexible means for adaptation to new problems and settings from every day's knowledge and practices (Brandsford, 2000); (Cuevas et al., 2005). Effective teachers need pedagogical content knowledge literacy. Pedagogical content knowledge is how to structure the cognitive road maps they master in their content knowledge. This type of knowledge can help teachers to design effective assessments for students' learning. Educational assessment goals in the twenty first century is different from the goals of earlier times; Science assessment may become more valid and equitable if its examples reflect real world problems and towards innovation. Teaching and learning should be relevant to students' environment. Frequent feedback is mandatory so teachers can monitor learning and evaluate actively their strategies and the students' understanding (Brandsford, 2000).

Research has shown that students' attitudes toward science were significantly related to their achievement. A case study done at school district in different American states discovered that boys' achievement test scores were more positively related to their attitudes toward science than were girls' attitudes. Moreover, the case study showed that teachers who followed a standardized –based instructional strategy and who participated in professional development programs focused on construction of science understanding and adaptation of science materials within science-inquiry practices using software, modular science curriculum and information search tools had better students' scores (Kahle et al.,2000 & Marx et al.,2004).The effect of scientific inquiry instruction on students' achievement, as well as the students' engagement, is a great interest of policy makers. This is particularly true as the No child left behind Act (2002) required testing students' achievement and engagement level. However, despite these investments and heightened emphasis on science achievement, few studies were made on findings across individual studies investigating aspects of science- inquiry instruction and students' achievement (Minner, 2009).

Setting

The purpose of the study is to measure the students' achievement through a quantitative analysis of the students' progress scores using a pre-test and a post-test data collection.

This case study is pioneer in the United Arabs Emirates (UAE) as it reflects on the practices of the English as a medium of instruction (EMI) in science teaching and learning in government schools. The sample of the study is fifty-two students from grade 6 girl's public schools, located in the eastern educational zone of the United Arabs Emirates (UAE), in four months and a half period time. The sample of students is a part of a population of eight hundred fifty students' boys and girls of 12 years old, following the standard-based program and implementing the EMI in science. The students' parents agreed to have their children take part of the four months and a half study as a part of the study project in which the science reform curriculum is the Next Generation of Science standards and the teaching and learning strategies will focus on the science-inquiry as the main teaching approach in the science classes. Students' parents signed a letter of agreement (appendices) about the case study and the safety procedures taken.

The present study could use concurrent procedures, in which both quantitative and qualitative data converges and go together to provide a comprehensive analysis of both questions asked in the introduction section. In this type of design, data will be collected from both tests and class observations were done at the same time during the study timeline and the interpretation was integrated with the overall results. Within this study paradigm, the quantitative method is the best way to assess students' achievement from data collection and analysis.

The quantitative method using students' scores collection emphasizes data based on measurements, which need to be collected and analysed. This procedure should follow a scientific model based on the 21st century skills: elaborate the hypothesis, ask the questions, analyse, discuss, conclude and recommend. Tools are pre-test and post-test for objective knowledge, significant for the case study (Common Wealth, 2004).

The content knowledge for the case study is grade 6 science content aligned with the next generation of science standards. The experimental teacher was advised to use the same science resources provided by the school along with some extra resources provided by the researcher to reinforce the science-inquiry teaching and to design extra science-inquiry activities.

The timeline of the project is the first term of the academic year and the content respected the science department scope and sequence.

Methodology Procedure

Some bureaucratic procedures were necessary before entering the schools. First getting the approval of the research department in the Ministry of Education Dubai for clearance by presenting the researcher's personal papers, the main objective of the study, and a reassurance that all personal information will be classified and school name will not be revealed. All science tests had to go through the curriculum department for review and reassurance of skills and learning outcome alignment, later on tests should be sent to the assessment department as they are the official body to administer the test.

Second step is the ethical approvals, as mentioned earlier in the setting, first the parents, later the teachers'. The researcher should highlight that the students' scores analysed and evaluated in this case study will not put them in any kind of accountability facing the administration office of the school, and will not affect negatively on their yearly appraisal (MoE appraisal document 2013).

Third step is to design a clear timeline for classroom observation, field work and students' assessment that should take into account several factors such as the teacher's timetable, the public holidays, the local events in the school such as martyr Day and National day, UAE clean-up campaign as well as the international celebrations the water day; the earth day. The timeline should consider the teachers' professional development and trainings required by the ministry of education and for the case study as well. Finally, begin the field work and the data collection. The sample of fifty-two students of grade 6 will be divided in two groups.

The first sample of the twenty-six students received the science content with a classical instruction by their science teacher This is the controlled sample classroom observations were made in addition students had to do the same assessments designed for the current study for analysis purpose. On the other hand, the second sample of the twenty-six students was exposed to the science-inquiry teaching approach for four months and Third step is to design a clear timeline that includes classroom observation, field work and students' assessment and takes into account several factors such as the teacher's timetable, the public holidays, the local events in the school such as the UAE clean-up campaign, the international celebrations such as the water day; the earth day, the teachers' professional development and trainings required by the ministry of education, finally the administrative requirements of the school's principal.

The first sample is the controlled group. The population had twenty-six students who received the instruction by a science teacher who had not received any new instructional strategies on science teaching and learning from 4 years. As a controlled group, classroom observation and written report were designed for comparison purpose as well as assessment tests. On the other hand, the second sample is the experimental group. The population had twenty-six students receiving science via the scientific inquiry teaching approach for four months and fifteen days, equal to eighty-four hours of effective science-inquiry teaching. The big amount of teaching hours could affect the students' outcome either positively or negatively, therefore, the teacher of the experimental sample, should guarantee the effectiveness of each science inquiry to improve students' learning process. The role of the teachers is shifting from teller to coaching, guidance and learner; a continuous professional development program has been tailored so the teacher could easily provide students a rich environment involving high level thinking: problem solving and decision making. Training needs follow up for immediate feedback and reorganisation of the timeline along with classroom observation during the whole case study timeline. The trained teacher (TT) was expected to design new activities to help students link their prior knowledge with the novice, just learnt.

On the other hand, the lecturer teacher (LT) teacher did not follow any continuous professional development sessions and did not receive any of the tutoring sessions. The (LT) teacher designed, instructed the seven sessions and was observed, and the number of times students showed engagement in his sessions was noted but no feedback was given for the controlled group of the science teacher.

The Study Tool

According to the NSES (2009), the scientific inquiry as a learning process engages the students mentally in a constructivist environment and the scientific inquiry in her multiple stages such as writing, communicating and reflecting within a range of specific activities involve the students in peers and cooperative learning (Anderson 2002). The present study delineates and assesses quantitatively the students' achievement improvement after a science-inquiry teaching approach and a classic teaching session for both experimental and controlled group using a pre-test that students did in the first beginning of the year and another post-test done at the end of the four months and a half science-inquiry teaching approach.

The Design of the Quantitative Tool

The study assessed the students' achievement based on the 6th grade standardized tests from California, Tennessee, and the University of New York as well as Virginia, writers and inspirers of the NGSS; and the ministry of Education in the United Arabs Emirates (2009).

The assessment of the students' knowledge had a two phase's process: phase I is the phase to administer the pre- test in September 2013, to place all students' and check their level regarding the science content as well as their attainment in inquiry-skills, phase II is the final test to assess both students' samples achievement using their scores after four months and a half of effective instruction received by the experimental group, and the non-effective instruction received by the controlled group.

The case study administrated the final test in two rounds: December 2013 (end of term I) the first round using the final test part I that covers the Life science content, and in March 2013 (end of term II) the second round using the final test part II, that covers the Earth science and Geology content. The reason behind this type of administration is mainly the students' age and their limited cognitive ability to understand and verbalize a certain amount of information in a short period especially if they are Arabic native speakers with English language barriers (Piaget, 2009).

Taking into account all these listed factors above, the standards, the language barrier, the Ministry of education requirements, and the students' cognitive ability, the quantitative tools which are the pre-test as well as the post-test 1 and post-test 2, were designed as per the TIMSS layout and the local National Assessment Program (NAP) and aligned with the NGSS standards. The pre-test had three sections: the first section constituted of multiple-choice exercises, the second section fill in the blank and the third section matching, as for the post-test 1and post-test 2 they had both the multiple-choice section that proved to be the best tool to assess content as well as skills for second language speakers and the open-ended questions to assess their ability to communicate and verbalize the science concept understood and the modelling which is transforming a text into a diagram. Moreover, because of the students' language barriers some Arabic translation was added to the main question in order to assess the skill in adequate manner regardless the language competency of the student moreover the language will not be a negative influencing factor on the students' achievement (table 7). The questions assessed the science-inquiry skills such as infer, predict, compare and contrast as per the NGSS (2013) requirements for the 6th grade. On the other hand the test scoring was distributed according to the weight of the standards taught during the period of the study: the multiple choice questions each one had one point and a total of 25 marks to assess the content knowledge. The Fill in the blank part where students should conclude what could be the correct word had 2 marks for each correct answer. As per the experimental questions and the questions that delineate some abstract representations, they were the lowest questions graded as to differentiate between students who had higher cognitive abilities more than others.

Table 1. Sample of Science Post-test 1 Questions

Type	Standard	Sample items
Multiple-choice	6.3.2. Observe and infer an experimental protocol identify and label the experiment and analyse its usage	1. **Which statement is correct about the experiment below?**  **a.** Evaporation & condensation **b.** water cycle **c.** infiltrat
Fill in the blank	6.5 draw conclusions and apply in a text format	**Non-living groundwater disappear Water cycle Scientific method Popu** **a.** The hypothesis and the experiments are **two parts** from the .... b. **Classification** is grouping organisms with similar .... **c.** The temperature and the water **are two** .... factors d. If one living thing **dies** from the food web all the **food web** will .... **e.** A **community** in an ecosystem **is made of different** .... living together. **f.** When the **water** turns around the ecosystem it **is called** .... g. The **treatment** of the water **keeps the** .... Safe to use
Matching	6.5.2. Correlate a conclusion to an earlier experience and try to extend into a new hypothesis	**2- Match the correct number from column 2 to the correct num** Column 1 / Co (..........) **a.** *Felis domesticus* / **1.** Linnaean S (..........) **b.** classification / **2.** scientific me (..........) **c.** boiled water has a .... temperature / **3.** 100 °C (..........) **d.** give conclusions / **4.** the cat

Table 2: Sample of Science Post-test 2 Questions

Type	Standard	Sample items
Multi ple-choic e	6.2.2 Compare and contrast between phenomen a such as mitosis, meiosis	2- Picture □a or □b, below is Meiosis. (2pts) Plant cell division Animal cell division a. b.
Open-ended	6.5 draw conclusion s and apply in a text format	26- Answer the following questions أجب عن الأسئلة التالية مستعينا بالرسم (4p 1- What type of boundary has formed the mountains? (2pts) 2- What could happen to the mountain after this heavy rain ? (1pt) 3- Whv is runoff important to nature? (1pt) Lake
Model ling	6.4.3. Transform a table into a graph, chart, or a concept map using computer application s, or the traditional ways	Parents → X Offspring → Key G Gray (dominant) g White (recessive) g g

The results of post-test 1 and post-test 2 were gathered and added and the mean was calculated in order to have one result of the post-test. The percentages results are shown in chart 1 and chart 15 for both group samples: the controlled group taught by the teacher who didn't receive any of the new science inquiry teaching approach training and was teaching using the traditional ways of teaching and the experimental group where students were receiving the science instruction as an inquiry-teaching strategy.

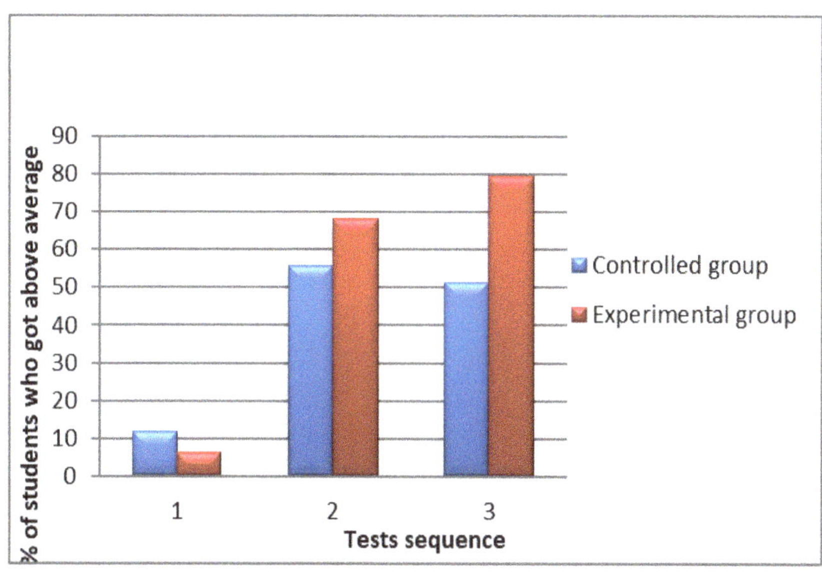

Figure 1: Percentage of experimental and controlled groups

After doing the data collection and the data analysis using the percentage methodology, the findings of the students' scores were different for both groups. Both groups were at a very low starting point with 11.8 % of success for the controlled group and 6.45 % of success for the experimental group, in addition the controlled group students' were having a higher level of success in the pre-test compared to the experimental group 11.8%> 6.45%. Moreover, both groups improved in the second assessment, with a 55.6% for the controlled group and 68.1% for the experimental group. The results infer that both teaching methods were fine, and the difference between the controlled group and the experimental group was not quite high; yet the experimental group showed a progress and increase of the average higher in the last post-test designated for both students' samples. The final post-test designated as number 3 in chart 14, shows the difference between the experimental group and the controlled group with 79.6% versus 51.2% for the controlled group that indicates that students decreased in their engagement as well as their commitment to the science discipline showing a demotivation for taking the test and a non-willingness for the improvement (Zhu & Leung, 2011).

These results and findings are aligned with earlier research on science-inquiry teaching approach and inquiry science learning that reports positive relationship between inquiry approach and students' science achievement (Hakan, 2012). The chart M11 above shows explicitly the scores' improvement of the students following the effective science-inquiry teaching approach from 6.4% to 79.6 % the level of students' 'achievement. Traditional teaching methods are no longer effective to create long-term learning in students which could be implied with the decrease in the students' achievement scores for the controlled group from 55.65 to 51.2%, because the twenty first century students are evolving and demand immediate feedback and answers, by their own investigation abilities, not through lecturing Healy (2000). The science inquiry teaching approach is not an easy task that aims to make students move in the classroom, yet it is an active learning that totally engage the student mentally and physically (Enger & Yager, 2001), the student in that case will be engaged in his own learning and this improvement will show in his achievement scores as per the experimental group scores that are not considered as high compared to other private schools where students usually begin their academic year with 65%, however show an improvement in the students' achievement an issue that was challenging by the beginning of the academic year.

Table 1: Pre and post test mean and standard deviation results

			Bootstrap[a]			
					95% Confidence Interval	
		Statistic	Bias	Std. Error	Lower	Upper
Controlled group Pretest	N	26				
	Mean	10.58	.01	.54	9.54	11.65
	Std. Deviation	2.730	-.085	.385	1.990	3.426
	Std. Error Mean	.535				
Controlled group Postest	N	26				
	Mean	53.69	-.05	3.05	47.96	59.81
	Std. Deviation	15.252	-.394	1.855	11.061	18.459
	Std. Error Mean	2.991				
experimental group Pretest	N	26				
	Mean	7.00	.02	.49	6.12	8.00
	Std. Deviation	2.561	-.074	.406	1.705	3.283
	Std. Error Mean	.502				
experimental group Postest	N	26				
	Mean	79.15	.22	3.45	72.38	86.04
	Std. Deviation	17.301	-.496	1.865	12.380	19.829
	Std. Error Mean	3.393				

a. Unless otherwise noted, bootstrap results are based on 1000 bootstrap samples

To check the level of significance of the students' achievement scores, the t-test presents a 95% of confidence level the opportunity to compare the mean, the standard deviation for both samples of students with N equal 26 which is the number of students per class. The t-test as per the percentage shows that the mean of the pre-test is higher for the controlled group with 10.58 > 7.00 which shows some insignificance in the start of the experiment; on the other hand the standard deviation of the post-test for the experimental group (17.301) is higher than the standard deviation of the controlled group (15.252) which shows that the experimental group showed a significant improvement compared to the controlled group. The comparison of our two samples shows that the t value in the controlled group from the pre-test towards the post-test with $t1 = 19.754 > t2 = 17.950$ and the difference in the mean between the pre-test and post-test is equal to 43.115. On the other hand, the t value in the experimental group is increasing from the pre-test towards the post-test with $t1 = 13.936 < t2 = 23.328$ and the difference in the mean between the pre-test and post-test is equal to 72.154, greater than the mean difference of the controlled group. All of these comparisons show that the experimental group performed better than the controlled group which is significant with a p value $0.000 < 0.05$. So the results infer by the data analysis that the null-hypothesis is not true, therefore rejected in favour of the study with a level of confidence of 95% in the mean, It is therefore concluded, that there is an actual difference in the students' achievement scores between the ones receiving the science-inquiry teaching approach (that had an increase in their scores), and the ones receiving the science instruction as a lecturing (and had a decrease in the achievement scores, in favour of the experimental group receiving the science-inquiry teaching approach.

Table 2: T-test result for experimental **and controlled groups**

One-Sample Test

	Test Value = 0			
	t	df	Sig. (2-tailed)	Mean Difference
Controlled group Pretest	19.754	25	.000	10.577
Controlled group Postest	17.950	25	.000	53.692
experimental group Pretest	13.936	25	.000	7.000
experimental group Postest	23.328	25	.000	79.154

Conclusion and Study Recommendations

In conclusion the science-inquiry was effective in improving both the students' achievement as well as their engagement level because students reached their comfort level. First, they explored, defined and tested their hypothesis, second, they tried more than one probable answer through the cooperative learning work strategy Therefore their attitudes were positive regarding the science concept as well as the learning process.

The results of this study were encouraging and demonstrated that science-inquiry teaching approach promoted effectively both research items. Students were able to ask effective questions and to plan and begin a scientific investigation.

Future studies could also shed more light on the relationship between language development in the areas of reading, language arts and writing and science- inquiry teaching instruction. Studies already examined the relationship between teacher's professional development, students' assessment, science notebooks writing and students' improvement. Further studies could be on the correlation between science-inquiry teaching approach and the science achievement in the boys and the girls as well as the difference between both scores or to what extent science-inquiry teaching approach could decrease the level of drop-outs in the boys' government schools.

REFERENCES

ADEC footprint recruiting hiring (2013). http://www.footprintsrecruiting.com/forum/adec/7245-adec-now-hiring-up-to-1200-teachers-for-august-2013.

Al Muhairi, J. & Al Karam, A. (2010). Dubai school of inspection report bureau. DSIB

American Association for the Advancement of Science (1993). Benchmarks for science literacy. New York: Oxford University Press.

American Association for the Advancement of Science (AAAS) (1990). Science for all Americans.

American Association for the Advancement of Science (AAAS) (1965). The science technology engineering and Mathematics (STEM) Education and Skills in Queensland. http://education.qld.gov.au/projects/stemplan/docs/stem-discussion-paper.pdf

Anderson, R.D. (2002). Reforming Science Teaching: What Research says about Inquiry. *Journal of Science Teacher Education*, Vol 13(1). pp.1-12.

Atar, H. Y.& Atar, B. (2012). Examining the Effects of Turkish Education Reform on Students' TIMSS 2007 Science Achievements. *Educational Sciences: Theory & Practice* 12(4), 2632-2636

https://www.academia.edu/2422230/Examining_the_Effects_of_Turkish_Education_Reform_on_Students_TIMSS_2007_Science_Achievements

Bailey, K., & Jacobsen, M. (2019). Connecting theory to practice: Using the Community of Inquiry theoretical framework to examine library instruction. *Journal of Information Literacy*, *13*(2).

Bligh, D. A. (2000). What's the use of lectures? Jossey-Bass publishers.

BouJaoude, S. & Dagher, Z. (2009). Status of Science Education in Arab Countries. Paper presented at the International Science Education Conference 2009, Singapore, from Nov. 24-26

Bonwell, C.& Eison, J. (1991). Active learning: Creating excitement in the classroom. ERIC Higher Education Report Vol 1. [Accessed on July 25th, 2013]. http://www.ed.gov/databases/ERIC_Digests/ed340272.html.

Chaudhary S.B. (2013). higher education trends in the middle east dubai counsellor examines why the preferred undergraduate degree among students in this region is engineering rather than bba. http://gulfnews.com/culture/education/higher-education-trends-in-the-middle-east-1.1234827

Cleveland-Innes, M. (2019). The Community of Inquiry Theoretical Framework. *Rethinking Pedagogy for a Digital Age: Principles and Practices of Design*, 43.

Cohen, J.W. (1988). Statistical Power Analysis for the Behavioural Sciences. Hillsdale NJ: Lawrence Erlbaum. *Department of Education, Training and the Arts* (2007).

Enger, S. K. & Yager, R. E. (2001). Assessing student understanding in science: A standards-based K-12 handbook. Corwin Press, Inc. Thousand Oaks, CA. English learners, grades 4–8. NCBE Program Information Guide Series.

Gautreau B. T, Binns I. C. (2012): Investigating student attitudes and achievements in an environmental place-based inquiry in secondary classrooms. *International Journal of Environmental & Science Education, 7*(2), 167-195. http://www.ijese.com/

Hudson, R. (2007). Language networks: The new Word Grammar. Oxford: Oxford University Press. Improving Head Start Teacher Qualifications Requires Increased Investment. Center for law and social policy: head start series. 1, 1-16. http://www.leg.state.vt.us/PreKEducationStudyCommittee/Documents/hs_policy_paper_1.pdf.

Isabell, S. (2018). 5 E's instructional model and propose in E-Learning industry Scooped by onto **Error! Hyperlink reference not valid.** . From www.swiftelearningservices.com.

Johnson, C.C & Kahle, B. (2007): A Study of the Effect of Sustained, Whole-School Professional Development on Student Achievement in Science. *Journal of Research in Science Teaching. 00 (0), 1–12.*

Kyriacou Ch. (2009). Effective teaching in schools: Theory and practice 3rd ed. Nelson Thornes publication.

Minner, D.D & Levy, A.J.& Century, J. (2009). Inquiry-Based Science Instruction—What Is It and Does It Matter? Research synthesis Years 1984 to 2002. *Journal of research in science teaching*. 20, 1-24.

Nasa Model. http://www.nasa.gov/audience/foreducators/nasaeclips/5eteachingmodels/index.html

No Child Left Behind Act (2001), P.L.107-110.

Piaget, J. (1952). The Child's Conception of Number. London: Routledge & Paul Press: psychological bases of Science—A Process Approach (SAPA).

Settlage, J. (2007). Demythologizing science teacher education: Conquering the false ideal of open inquiry. *Journal of Science Teacher Education, 18*, 461-467.

Shea, P., & Bidjerano, T. (n.d.). (2009). Community of inquiry as a theoretical framework to foster "epistemic engagement" and "cognitive presence" in online education. *Computer & Education 52*(3), 543–553. https://doi- org.liverpool.idm.oclc.org/10.1016/j.compedu.2008.10.007

The National (2014). UAE ministry of higher education outlines strategic plan for 2014\ the national staff January 2, 2014.

Wilson, E. O. (1998). Consilience: The unity of knowledge.

Wilson, S. M. (2013). Teachers' professional development. 340 (6130), 310-313. http://www.sciencemag.org/content/340/6130/310.abstract

Yager, R.E. & Akcay, H. (2009). A Comparison of Student Learning in STS vs Those in Directed Inquiry Classes. 13(2),186-204.

CHAPTER 12

Formative Assessment in Science Inquiry

Sura Sabri

ABSTRACT

The use of inquiry-based learning (IBL) in science education is one of the effective strategies used to develop students' critical thinking skills. IBL trains students to become problem solvers and leaders ready for future challenges in their careers. Little research has been conducted in the Arab world, specifically in the United Arab Emirates (UAE), to cover the topic of science education and its teaching and learning practices. The following case-study investigates students and teachers' perceptions on formative assessment FA and IBL in science classrooms and will provide recommendations for its best practices in the educational field. The paper uses the mixed method approach; the defined tool consists of two versions of FA: an IBL questionnaire for teachers and students, and a lesson observation form to identify the best practices in a classroom environment and support data analysis. Three groups of participants contributed to the study: 535 students, 51 teachers answered the questionnaire, and 10 of the teachers have been observed during an active classroom session. The findings of the case study found a positive relationship between FA and IBL. The case study results recommend the use of FA and are in favor of implementing IBL in classroom practices in the UAE.

INTRODUCTION

One of the major challenges of education in the UAE is coping with the growing pace of internalization and supporting the coming generations by enabling them to contribute effectively to the evolving global market. The UAE needs successful leaders, able to take an active part in the future knowledge economy (UAE ministry of education mission, 2015). To cope with the growth, individuals need to develop their skills to become effective leaders, who can approach future problems using critical thinking and problem-solving skills. These skills are nurtured through science education, which plays a major part in empowering students' cognitive abilities. (Hanauer. D. I and Bauerle. C., 2012) Therefore, improving science education by adopting the IBL approach would equip students with the necessary skills to solve future challenges, as it encourages them to practice critical thinking and problem solving (Harrison, C. 2014). In addition, assessment is a tool teacher use to collect solid evidence of the level of mastery of a scientific concept. This evidence will define the type of teaching and learning practices teachers can use in future (Clark, I. 2012).

BACKGROUND AND SIGNIFICANCE OF STUDY

Education is meant to provide students with the essential skills to solve future challenges (Clark, 2012). Students' proficiency in those skills is measured through FA, which also enhances students' self-regulated learning skills (SRL), according to Torrance (2012) and Clark (2012). They also recommend various teaching strategies that would allow students to generate feedback, reflect and transform from being passive to active learners; If students were responsible for their own learning, they will develop metacognitive skills, which improve their ability to reflect upon their understanding, and plan ways to gain the required knowledge and skills to achieve a certain target (Clark, 2012). According to Tabari (2014), students in the UAE need more practice to become self-regulated learners and take responsibility of their own learning. Teachers in his study pointed out that students' current attitude towards education is still negative, and the UAE's education system needs to be less formal.

Research on science education indicates a great impact associated with incorporating inquiry-based learning activities using a student-centered approach on students' level of understanding and performance (Forawi & Liang 2011; Hanauer & Bauerle, 2012; Harrison, 2014). On another note, Nagle (2013) recommended teaching for the sake of understanding, where students should experience challenging tasks that require the application of their understanding, rather than evaluate their knowledge solely based on their ability to memorize. Similarly, introducing a challenging task would enhance students' understanding of the concept, and improve their skills (Clark, 2015). As reported by science education literature, IBL would be an ideal method that allows students to experience challenging tasks that require higher order thinking skills (Asay & Orgill, 2009; Llewellyn, 2010; Nehring, Tiemann & Belzen, 2013; Stone, 2014).

In expounding FA and IBL implementations, researchers have highlighted several challenges that could hinder the smooth application of both strategies. These challenges include teachers' ability to utilize the IBL process using various FA strategies, and students' ability to understand and successfully engage in inquiry-based activities (IBA) (DiBiase & McDonald, 2015) which are the same challenges that were also identified in the UAE.

Science education researchers recommend further studies to identify the best practices and most efficient methods in FA, in order to assess and evaluate students' daily performance, and accordingly improve their conceptual understanding and achievement (Clark, 2015). Furthermore, additional studies suggest putting IBL into action in science classrooms, by proposing new procedures and providing the required professional development to support science teachers (DiBiase & McDonald, 2015).

To investigate the current situation of science education quality in the UAE, various inspection reports from the Knowledge and Human Development Authority (KHDA) and Abu Dhabi Education Council (ADEC) were reviewed. Results revealed that in ADEC, none of the private schools were categorized as outstanding in the academic year 2014-2015.Areas of improvement included promoting continuous assessment of students' progress and integrating scientific activities that promote higher order thinking skills. On the other hand, the KHDA report included only two schools categorized as outstanding, with their strengths including the presence of an innovative curriculum paired with teaching and learning approaches that were novel and led to high levels of engagement. Inspection reports issued by the KHDA and ADEC have indicated that schools should promote challenging activities to enhance students' critical thinking and problem-solving skills (Irtiqa'a, 2015; KHDA, 2015). This situation demands the necessity of undertaking research that investigates the current situation of educational practices in science classrooms in the UAE.

A limited number of previous studies have linked the use of FA to evaluate and assess students' progress in science classrooms. Therefore, the intention of this study is to investigate students' perceptions towards FA and identify best methodologies that can be applied in the UAE. In addition, this study aims to investigate students' and teachers' perceptions of inquiry-based learning, and to identify the most efficient strategies that effectively use IBL as a central teaching approach in science classrooms. This study also aims to link between FA instructional strategies and the phases of IBL, which involve collaboration, discussion and questioning, implemented within class groups. Hence, this study will elaborate on how to properly assess IBA in science classrooms and is expected to formalize the reciprocal instructions that hinder the FA of IBL activities to develop students' conceptual understanding and achievement in science subjects accordingly.

The purpose of the current study is to investigate teachers' and students' perceptions and practices of FA of IBL in the UAE. This will be achieved through utilizing various comprehensive studies, which identified attributes of FA and IBL that can enhance students' SRL skills and make them independent learners, able to contribute to future innovation (Clark, 2015; Pedaste, 2015).

The following research questions drive this study:

1. What are the teachers' and students' conceptions of FA in science classrooms?
2. What are the teachers' and students' conceptions of inquiry-based learning in science classrooms?
3. How well do teachers use formative assessment of inquiry-based activities in science classrooms?

Study Design

The hypothetical basis behind this study is deemed to be practical, as it aims to identify best practices of using FA of IBL activities in science classrooms, located within a group of high schools in the UAE. The study will apply a simultaneous mixed-method approach, which means collecting and analyzing data using quantitative and qualitative tools concurrently, then merging the data to come up with answers and explanations to the questions driving the study(Creswell, 2011; Laban, 2012). Applying quantitative methods in this study follows the positivistic paradigm, which requires collecting data through closed-ended questions regarding the FA of IBA in science classrooms and using the statistical data to obtain generalizations and recommendations (Creswell, 2002; Laban, 2012).

Study Methods

One of the best tools that can be used to track the effectiveness of teaching strategies to improve a classroom's environment is obtaining information on students' and teachers' perceptions and daily practices. Questionnaires are suitable to assess the degree to which the classroom environment is consistent with new insights in teaching science and provide the required support for teachers to reshape their teaching practices (Abell & Lederman, 2010). Particularly, in the current study, questionnaires were used to investigate the implementation of FA of IBL in the science classroom. Two types of questionnaires were developed to detect teachers' and students' perceptions and practices, both included two main scale terminologies: IBL activities and FA practices. The quantitative part of this study was complemented by qualitative lesson observations, which was used to identify best practices of formative assessment of inquiry-based learning activities.

The study design and methods are described in Figure 1 below. (Bell, 2010; Cresswell, 2011). The steps of data collection will be as follows:

1. Quantitative data from teachers' questionnaire "Formative Assessment of Inquiry-Based Activities Questionnaire" to identify actual teaching practices in the classroom, regarding formative assessment and the inquiry-based learning approach.
2. Quantitative data from students' questionnaire "Formative Assessment of Inquiry-Based Instructions Questionnaire" to identify actual practices in the classroom regarding formative assessment and students' conceptual understanding of the inquiry-based learning.

Qualitative lesson observations to record data regarding best practices of using formative assessment of inquiry-based learning.

To investigate the use of FA of IBL activities in the science classroom, it is important to explore the implementation from different perspectives; any educational situation can be observed from the teachers' view, students' view and the actual daily practices in the classroom. Therefore, questionnaire items were designed to identify students' perceptions of FA and conceptions of IBL. Additionally, they were meant to detect teachers' perceptions and actual practices in their instructional methodologies. Hence, different items were included to detect the actual implementation of FA and IBL processes in the class. More details about the tools are described in the study instruments section.

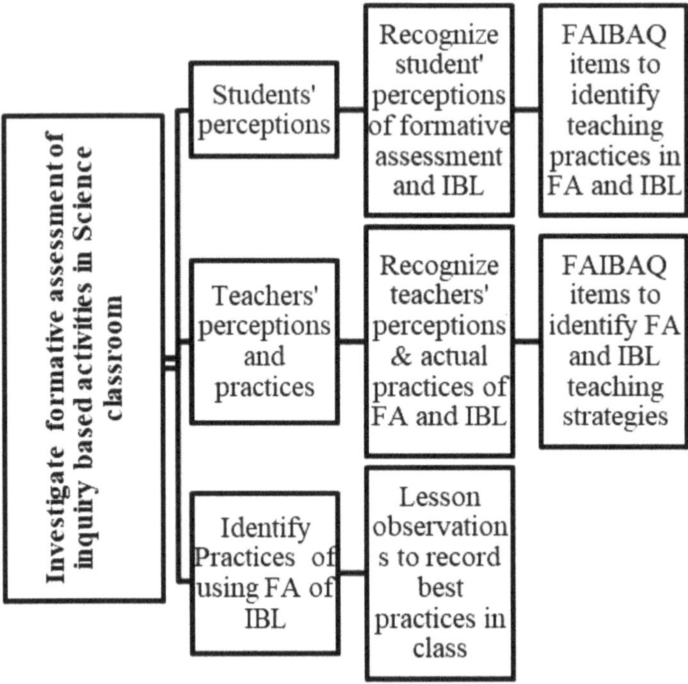

Figure 1: Design of the study and data collection instruments

Study Instruments

The nature of this study requires a mixed-method approach, in which quantitative and qualitative methods are applied with three different instruments. The quantitative data was collected via teachers' and students' responses to the Formative Assessment of Inquiry-Based Activities Questionnaire (FAIBAQ), which is used to identify teachers' and students' perceptions and practices regarding the FA of IBL. The qualitative part was completed by observing ten various science lessons to investigate teaching and learning practices related to the FA of IBL activities in science classrooms.

Teacher Questionnaire

A teacher quantitative questionnaire was developed to investigate teachers' perceptions and current practices regarding IBL and FA (Abell & Lederman, 2010; Bell, 2010; Cohen, Manion & Morrison, 2000; Creswell, 2002; Creswell, 2011). It consists of three main parts: First, demographic items to identify teachers' gender, age, and teaching experience. Second, items that focus on the main criteria of IBL activities in a classroom which were adopted from Harlen, Nowak, Tiemann and Belzen (2013) who identified the requirements of IBL. These criteria include observing, procedure of collecting data or experimenting, and proofs to construct new conclusions, and concepts to provide explanations to natural phenomena. The third part consists of 20 questions to examine daily formative assessment practices. Responses in the second and third sections were collected though a semantic differential scale (Cohen et al., 2000).

Validity was checked by an expert panel including a university professor and three educators in the field who advised that options be changed to make them more convenient to context conditions.

Students' Questionnaire

The students' questionnaire was used to measure students' perceptions regarding FA and IBL activities implemented in science classrooms. The tool consists of three main parts: First, information asking about students' grade, campus location and the subject in question. The second part included 8 questions focusing on students' understanding of IBL, and the extent to which students are involved in classroom activities. The questions were modified from the principles of inquiry questionnaire developed by Campbell, Abd-Hamid, and Chapman (2010) for students.

The third part regarding formative assessment was adopted from questionnaire (WIHIC) developed by Dorman, Aldridge and Fraser (2006). Responses to all items were in the form of Semantic Differential Scale to indicate how frequently each item was implemented in the classroom (Cohen et al., 2000). The questionnaire was available in both Arabic and English to avoid misunderstandings in students' responses. It was sent to a professional translator, then sent to the supervisor for approval.

Lesson Observation Tool

For the lesson observations, a qualitative observation tool (Bell, 2010; Creswell, 2002; Cresswell, 2011; Richards, 2003) was used as the instrument to investigate teaching situations comprehensively and identify best teaching practices in class. The lesson observation tool used in the current study consists of two main parts. First, the inquiry cycle looked at steps of inquiry including orientation, conceptualization, investigation, data collection and organization, conclusion, and discussion. The second part focused on classroom interactions; including cooperative learning, promotion of critical thinking and problem solving and allowing assessment of individual effort. In addition, teacher- student interactions were also observed, which included both ongoing constructive feedback and the availability of scaffolding techniques. The researcher in this study remained solely as an observer of the process and did not participate during the instructional process (Creswell, 2002; Symon & Cassell, 2004;). However, teachers were given feedback to acknowledge best practices, and identify how to improve the missing items. Volunteer teachers from three different schools participated in lesson observations; classes were conducted in the laboratories or in normal classrooms.

Data collected from the lesson observations provided the necessary information to answer the third research question: How well do teachers use FA of IBA in science classrooms? According to previous research, the existence of inquiry cycles is the key factor that influences the use of FA strategies to assess students' progress in IBL activities. Various formative assessment strategies are used to evaluate each cycle (Harrison, 2014; Pedaste et al., 2015).

The researcher ensured that the data collected regarding IBA and FA practices in the observed lessons is used to describe the best practices of interactions that may aide students' learning. In addition, diverse sampling in lesson observations (e.g. male and female teachers, subjects observed were biology, physics and chemistry) also indicated the degree to which the recommendations from this research could be generalized.

Questionnaire, Results

Demographic Information:

Students' FAIBAQ questionnaire was sent to 2800 students, 535 responded 55% were from grade 9 and 45% from grade 10. They were distributed across female and male campuses, 57.7% of them were males and 43.3% were female students.

The teachers' questionnaire was sent to 80 science teachers, 51 of them responded: 39.6% being male and 60.4% being female. 50% of those who answered were physics teachers, while about 24% of them were chemistry teachers, and 26% were biology teachers. 24% had more than 20 years of experience, 14% had 16-20 years, 20% had between 11-15 years, 22% had 6-10 years' experience and the last 20% had 1-5 years of teaching experience.

For lesson observations, 10 teachers participated, 4 males and 6 females. 50% of the participants were physics teachers, 24% were chemistry teachers and 26% were biology teachers. 20% had more than 20 years of experience, 30% had between 11-15 years, 20% had 6-10 years of experience and 30% had 1-5 years of teaching experience.

Quantitative Results

The reliability test SPSS analysis was used to find Cronbach's Alpha presented in Table 1 below (Frankel, Wallen, Hyun, 2015).

Table 1: Reliability test results

Questionnaire Items	Cronbach's Alpha
Students' Q items related to IBL	.943
Students' Q items related to FA	.966
Teachers' Q	.905

Students' Perceptions of Formative Assessment and Inquiry-Based Learning Strategies.

Students' FAIBLA questionnaire was designed to recognize students' perceptions of FA and IBL learning strategies implemented in science classrooms.

The results demonstrated that about half of the students had experienced various formative assessment strategies during physics (50%) and chemistry (48%) lessons. However, a smaller number of students reported experiencing IBL strategies in physics (43%) and chemistry (42%) lessons. Additionally, it can be realized from the table that students' responses about biology were the classified into two major groups: about (38%) of the students reported that they have been subjected to FA and (31%) of them reported learning through IBL activities in every biology lesson. Whereas, third of the students (30%) reported that they never learned through FA or IBL activities in biology lessons.

After analyzing students' responses to particular items of FA, several results were revealed; the majority of students (70%) reported that they were asked questions every physics lesson, (63%) in chemistry and 51% in biology lessons. In addition, (66%) of the students agreed that they had the chance to ask questions every physics lesson, (61%) chemistry and surprisingly only one-third (31%) had chances to ask questions during biology lessons. In addition, the majority of students (70% in physics, 67% in chemistry and 53% in biology) reported that the teachers re-explain points that students did not understand. Looking at items about cooperative learning on average, about (50%) reported that they practice proper teamwork, as they discuss problems and activities with peers and learn from them in physics and chemistry, while in biology, about (44%) of the students reported proper cooperative learning. When it came to objective clarity, more than half of the students (61% in physics and

57% in chemistry) reported that they know what is required form each task and related it with the topic in every lesson. While less than half (43%) reported that in biology. As for the feedback and time given to think and reflect about answers, (56%) of students reported that they did, indeed have experienced that in every physics lesson and (47%) every reported that in every chemistry lesson.

Regarding IBL instructions, about half of the students (52%) reported that they receive detailed instructions to make investigations in chemistry lessons. (48%) reported that they can relate the conclusions with the main scientific concept in physics, and (46%) in chemistry. Considering students' responses regarding inquiry-based instructions revealed that one third of the students reported that they were able to connect scientific investigations to the main concept of every science lesson. On the contrary, one third of the students reported that they never experienced that in a biology lesson.

Teachers' Perceptions and Practices Regarding Formative Assessment and Inquiry-Based Learning.

The FAOIBA teacher questionnaire was designed to investigate teachers' perceptions and current practices regarding IBL and FA in science classrooms

The majority of teachers (74%) reported that they are using FA in their teaching practices every science classroom, while about one fourth (26%) reported using IBL activities once biweekly and (25%) reported using IBL activities every science lesson.

Responses show that most teachers (more than 70%) reported that FA items are frequently used during every science lesson. FA practices include asking questions, allowing students to ask questions, distributing questions equally, using students' ideas and suggestions during the lesson, allowing students to share ideas with peers and discuss how to solve a problem and explain how they reached a solution. Moreover, they reported controlling group work and encouraging students to take control of their learning and explain the target of each task in every science classroom. About half of the teachers reported that they explain the rubrics required to assess students every science lesson.

Results uncovered the actual practices of IBL activities. The top three components of IBL presented in table 8 include: first, less than half of the teachers (43%) reported that students are given systematic instructions before they conduct investigations in every science lesson. Second, about 36% of the teachers reported that each student has a role as investigations are conducted. Third, 36% of the teachers reported that students develop their own conclusions for investigations. Full data regarding teachers' perceptions and practices is demonstrated in appendix 7.

The one-way ANOVA statistical test results compared teachers' perceptions according to their specialization. Four items regarding FA were statistically significant: giving students opportunities to discuss ideas in the class, questions fairly distributed between students, the teachers addressing all misconceptions after a task is done and students being able to relate the task result with the lesson topic.

Correlation Test to Link the Best-Chosen Formative Assessment Items with The Best-Chosen Inquiry-Based Items for Both Teachers' and Students' Results.

Descriptive statistics were used to analyze data and find the Pearson correlation coefficient to measure the relationship between different IBL items and FA items in the teachers' questionnaire and students' questionnaire.

Pearson correlation figures revealed a positive relationship between the inquiry-based items and FA items as follows:

> Inquiry-based first item "Students formulate questions which can be answered by investigations" was positively related to FA item "students are given opportunity to discuss ideas in the class". The r value was = 0.29 which is considered weak relation.
>
> IBL 4th item "Students design their own procedures for investigations" was positively related to "Questions are fairly distributed between students" with relatively weak r value of 0.297
>
> IBL 6th item "The investigation is conducted by me in front of the class" was positively related to two items regarding FA: "I explain the target of each task and I provide my students with rubrics used to assess their task" with r value = 0.41 and 0.336 respectively.
>
> IBL 7th item "Each student has a role as investigations are conducted" was positively related to six items of FA. These items include: "Questions are fairly distributed between students. Students discuss with peers how to solve a problem. Students share their resources when doing an assignment. I encourage students to take control of their learning. I provide my students with rubrics used to assess their task, My feedback enhances my students to do actions to increase their performance, Students can relate the task result with the lesson topic" with r values that indicate moderate relation, ranging between 0.314- 0.81.
>
> IBL 8th item "Students determine which data to collect" was positively related with the FA items "Questions are fairly distributed between students, I provide my students with rubrics used to assess their task, My feedback requires students to practice the concepts they didn't master" with r values ranges from 0.310- 0.33, reflecting moderate relation.
>
> IBL 9th item "Students develop their own conclusions for investigations" was positively related to eleven items asking about formative assessment. These include: "Students are given opportunities to discuss ideas in the class.

> Questions are fairly distributed between students. Students discuss with peers how to solve a problem. I encourage students to take control of their learning. Students can relate the task result with the lesson topic, when designing the task rubrics, I consider my students' opinion, I provide my students with rubrics used to assess their task, my feedback enhances my students to do actions to increase their performance, my feedback helps students identify their strengths and weakness points". With relatively moderate relation ranged between 0.3- 0.44).

Qualitative Results

Notes were taken to document all the events that occurred in the classroom at that time, and a checklist was used to report the events related to formative assessment and inquiry-based learning strategies implemented in each lesson. The list below represents some situations in which FA of IBL strategies was observed.

Using Scientific Approach

Students worked independently in groups, formulated a hypothesis, set their procedure, and found the solution. The teacher was circulating to provide support when required and used questioning techniques that tackle students' higher order thinking skills and deepen their understanding of the scientific concept. Students were not provided with direct answers. Instead, they were always referred to investigate further. Independent group work enhanced students' SRL skills, as students were responsible for their own learning.

Effective Teachers' Observation

An example was observed of the FA of IBA, where topics were prepared and explained by students. These activities were assessed through effective observation by the teacher, as she was circulating to identify strength and weakness points of the student that is explaining, and other students in the learning group.

The same practice of FA was observed when the teacher was circulating among groups, asking them about their procedure, providing proper feedback to support students and guide them to identify the gaps in their understanding during a lab session about identifying the pH of household solutions.

Students' Group Work and Peer Collaboration

Peer collaboration was evident in one of the physics lessons, where groups of students were asked to prepare and explain a scientific concept through implementing an activity. One group had to explain the meaning of the magnetic field. Each student demonstrated the activity to three of their peers in the class and asked them to explain the regular arrangement of iron powder around the magnet. They were then able to define the magnetic field. Other groups explained different methods to magnetize objects and the concept of electromagnetism. In all the class activities, the responsible groups prepared activities and asked other students to explain the observations to conclude the required information.

An alternative example of peer collaboration was observed when students performed a lab lesson to identify the pH of unknown solutions. After testing the pH of several household solutions, groups followed guided instructions to identify the pH of known solutions, then a different unknown was prepared for each group. They were required to apply the experiment using three different methods and compare results to emphasize the conceptual understanding of pH and to determine the type and the pH of the unknown.

Assessment Worksheets

Assessment by using worksheets was observed in several science classes: in a grade 9 biology lesson about karyotyping and hereditable disorders, a grade 10 biology lesson to implement Tidal volume lab activity, in which students were asked to complete the worksheets and send it to their teachers to receive feedback.

The worksheets observed in the Grade 10 physics class required students to investigate elasticity in different types of springs. They were required to use Hooke's law and calculate the elasticity constant to their given spring, then compare their findings with the elasticity constant given from the producing company, which was unknown to the students. The teacher prepared five worksheets to guide and assess students' work throughout the lab session. Students worked in groups to complete the direct application of physics laws that facilitated conceptual understanding. By the end of the lesson, three of the five groups were able to correctly identify the elasticity constant. The other two groups required further mentoring and guidance from the teacher to find the correct figure. Students were asked to submit solved worksheets and were given a challenging question by the end of the lab to evoke their critical thinking skills (see appendix 10).

Students' Presentations to Communicate Their Learning Experience

In a lesson discussing karyotypes and genetic disorders, students were asked to arrange chromosomes and make a human karyotype, then interpret it to identify the genetic disorder that would result from it. Assessment took place through having the students prepare one PowerPoint slide about one of the disorders, which would include a question about the disorder, a general description of it, and its treatment options. Each group had to communicate their learning experience and explain their results to the rest of the class, and the teacher allowed peer discussion. Most importantly, the feedback presented at the end of the lesson identified the strength and weakness points for each group. In another session, students featured the same form of assessment, where students were asked to work in groups to design a model that represents breathing in lungs. They were then required to draw the main parts of the respiratory system and find out the function of each part. FA took place through asking each group of students to communicate their model and explain the main parts of the respiratory system and their functions. Each group explained different parts in the system, and by the end of the discussion, students were able to identify all the parts of the respiratory system and their functions.

Discussion to Support Exploration and Data Analysis

Comprehensive discussion was observed when students were studying methods of analyzing different types of motion graphs. The teacher used a motion detector to collect and graph data in the class, which enabled students to master graph analysis skills. One of the activities involved having groups of students discuss the procedure behind displaying a specific graph on the screen, and they were given the opportunity to move in front of the motion sensor and match the displayed graph according to their discussions. Each group was given time to think about their motion example and explain how to better match the graph in a second trial, which required students to discuss the solution in groups, and communicate their predictions with other groups. The teacher allowed open class discussion to find the correct conclusion and provided precise feedback to individual students within the groups.

Lab Reports as a Tool to Assess Inquiry Lab Work and Provide Proper Feedback

All the lessons observed in the laboratory required students to write a lab report that shows the main steps of their investigation, these reports were marked, and proper feedback was provided. Each lab report included the investigation question, hypothesis and variables used (dependent, independent, and control variables). Then, students were required to describe their procedure and arrange their data in tables. The last part of the lab report included the conclusion and some analysis questions to be answered. Teachers then check the lab reports and provide proper feedback with a score for each component of the lab report.

Using Online Platform to Assess Students' Conclusions

Online testing was used as a closure for a grade 9 biology lesson about karyotypes and genetic disorders, in which an online platform was used to investigate three case studies and build their karyotype, then conclude the type of genetic disorder they could have.

Explaining Concepts

Several examples observed reflected how students were given the opportunity to explain new concepts to their peers by using different strategies, such as: preparing hands-on activities to explain scientific concepts, developing a presentation to explain different symptoms and treatments of hereditable diseases, explaining their method and procedure through conducting lab activities. In all examples, students were required to speak up in the class, show an acceptable level of self-confidence, and respond to their classmates and teacher's questions.

Relations Between Formative Assessment Items with the Best-Chosen Inquiry-Based Items in Teachers' and Students' Results

After interpreting the results of the teachers' questionnaire, some items pertaining to IBL were positively related to FA items, meaning that teachers' practices were able to reflect that FA strategies could be used to assess IBL activities. This result can indicate that teachers' perceptions and educational practices reflect an improvement in their beliefs about teaching and learning in science education (Abell, and Lederman, 2010). However, students' results did not show any positive relations between IBL and FA practices. This indicates that the system still requires development, as students couldn't relate the FA and IBL in the teaching and learning practices implemented in their classroom environment.

Several FA and IBL items were found to be related; IBL 6th, 7th and 8th items were positively related with six items of FA, including, this is in line with the proposed model of integrating FA strategies with IBL cycles. Clark (2012) indicated that informing students of the factors that affect their assessment and explaining the rubrics that they will be evaluated according to will increase their ability to become accountable of their work and develop their self-learning strategies.

> Orientation: through teachers' explanation & informing students about the rubrics used to assess their performance Conceptualization: when students are given the opportunity to discuss with peers the methods of solving a problem Investigation: when every student has a role in the investigation. Teachers' feedback to adjust learning. Data collection relating the conclusion to the main topic explained. (Pedaste et al., 2015).

Another IBL item (9th): "Students develop their own conclusions for investigations" was positively related to eleven FA items 1, 4, 6, 7 and 8, in addition to practices such as considering students' opinions when designing rubrics, and the positive effect of teachers' feedback on student performance. This relation proves that using inquiry-based instructions is an ideal method to enhance students' thinking and provide them with effective feedback. Jalil, & Ziq, (2009) found that using experimental learning strategies would strengthen students' 'working memory' via exposing students to problems that require higher cognitive thinking skills. These strategies would also prepare students to pursue scientific discoveries.

The last inquiry item, "Students connect conclusions to scientific knowledge" was positively related to seven FA items, including: "Students are asked to explain how they reached a solution", "Using group work to enhance students to cooperate with peers", "Sharing resources when students are asked to complete an assignment" ,

"Proper team work implementation during group work", "Teachers' feedback enhances his students to do actions to increase their performance", "The teacher addresses all misconceptions after a task is done", "Students can relate the task result with the lesson topic". This result provides evidence of the importance of IBA in emphasizing teamwork and enhancing students' cooperative skills, in addition to evoking their metacognitive abilities when they are asked to reflect upon their work, as a part of the communication process. This is aligned with constructivist theories of education, developed after Piaget's ideas about learning through using sensory data collected from different methods of communication to build new concepts and knowledge (Pritchard, 2010).

Analysis performed on the students' questionnaire did not show any correlation between inquiry-based items and FA strategies implemented, which is normal; students did not experience IBL sufficiently to relate it to the FA practices they experience in every science classroom.

Qualitative Results Discussion

Results were collected from interpreting lesson observations that were focused on the situations in which FA of IBL activities was observed.

Practices observed in science classrooms reflected the use of the scientific approach, where students were required to form a hypothesis, apply an experiment, collect data and find conclusions related to the topic they are discussing. Teachers circulate between students and observe their practices to evaluate their progress and provide immediate feedback. This helps students develop their understanding of the required concepts. As per the model provided by Pedaste et al., (2015), the sequence of the inquiry cycle observed involves the main and sub inquiry phases including: orientation, hypothesis generation, experimentation, data interpretation and discussion, in addition, a communication sub phase is required throughout the lesson. Implementing this strategy in science classrooms would enable students to explain, connect and communicate scientific concepts, and get essential feedback that would provide them with a better understanding of the topic (Asay, & Orgill 2010), accordingly aiding the development of their academic performance.

From the perspective of teachers, preparation-wise, Towndrow, Tan, Yung, and Cohen, (2008) confirmed that it is important to reform practical science assessments and qualify teachers to better evaluate students' performance during laboratory work. On another note, Harrison, (2014) reported that teachers agreed that collecting evidence of students' understanding through effective observation, and recording their interactions and contributions to group discussions, is considered better than relying on limited written assignments, as it provides a clearer image about students' understanding and performance. In addition, Hattie, & Timperley, (2007) clarified that using feedback during the process stage would better improve students' understanding of different concepts.

Peer collaboration was observed during various inquiry cycles, as students were discussing and working together to apply different inquiry steps. Additionally, students were held responsible for their data, and were required to discuss their results amongst each other. The teacher's role in these phases was limited to being a complete facilitator, as they always directed students to analyze their results and guided them to adjust their procedure to get an acceptable result, without giving students direct answers. Using how and why questions would evoke students' critical thinking skills and enforce them to rethink their answers. This is a preferable feedback method provided by teachers, which would help students develop their learning accordingly. Paul, & Elder, (2004) identified that building students' critical thinking skills requires teachers to use different questioning strategies that would need students to gather information, interpret them to get reasonable conclusions, and eventually communicate their solutions to others. Most of the activities observed required students to explain concepts that make up the structure of the lesson, reflecting a student-centered approach. This is supported by Asay, & Orgill (2010), as they reported that one of the important features of a student-centered class is to require students to explain different concepts to their peers. Students were asked to prepare hands-on activities and were given the responsibility to explain scientific concepts to their peers in small groups. This practice would support students in building self-confidence and enforcing their understanding of all the concepts. They were required to communicate their findings and reflect upon their understanding, which was one of the best practices observed, as having students reflect upon their own work and think about the way they solved a problem would strengthen their metacognitive skills. As stated by Torrance, (2012) one important purpose of FA is to have students think about their method of solving problems and critique their own way of thinking. This can be achieved easily if students were asked to present their work and discuss it with their peers and the teacher. Relating this observation to Pedaste et al., (2015) who stated that students were responsible for the three sub-phases of inquiry cycles, including questioning, exploring, and experimenting to clarify scientific concepts. Hayes, and Devitt, (2008) proved that allowing students to form small groups to discuss a topic would develop their critical thinking skills, enabling them to develop their scientific literacy through IBL. Clark, (2012) confirmed that for students to gain SRL skills, they should be a constructive part of the learning process and interact with their peers and the teacher. Additionally, Clark, (2015) declared that in order to support a formative curriculum, students are encouraged to develop peer to peer communication, and should be trained to develop self–assessment skills that will enable them to measure their own understanding as learners, and locate their strength and weakness points, accordingly holding accountability of their learning.

When students were implementing a laboratory experiment, they were required to write a lab report that will be marked by the teacher according to specific rubrics. Each lab report should have included the question and hypothesis of the experiment, dependent and independent variables used, summary of the procedure, and data tables and data analysis. When students write the lab report, they would make sure that they understood the requirements of the lab experiment, and the rationale behind implementing it. Cartwright, and Stepanova, (2012) clarified the importance of writing a lab report, especially when students are required to answer questions that relate the lab work to the main topic explained in the theory class, thereby attaining the maximum benefit of the laboratory work. Similarly, assessment worksheets were used to evaluate students' understanding of scientific concepts. When worksheets were used after each inquiry step in the cycle, and were checked by the teacher directly, they served as a good tool for FA in class and enabled students to adjust their misconceptions. These can be considered as sub-phases that consist of the reflection and communication of students' understanding in the model presented by Pedaste et al., (2015). However, when worksheets are submitted by the end of the lesson, to be marked and returned to students after the lesson, they would not provide immediate feedback, thus requiring further follow-up from teachers to ensure that all students have mastered the information. Clark (2012) indicated that when evidence of understanding is collected at the end of the lesson, yet the feedback is not provided directly, it is called 'asynchronous', which could be used to prepare for the next lesson and address students' misconceptions, or as home work that would be checked the next day. Some teachers used online platforms to assess students' understanding of the scientific concepts during the class. This reflected a good example of measuring students' performance and providing immediate feedback for individual students, thereby guiding them to adjust their understanding. Clark, (2012) confirmed that immediate feedback is an important feature of effective FA. Using technology to achieve this was investigated by Keough, (2012) as he reported that using clickers in the classroom was preferable to students; it provided opportunities for class participation and required students to engage in class discussions. Yet, some weakness points caused by normal technical problems were reported, which were of the kind that may face technology integration in general. Similarly, Lee, Feldman, & Beatty, (2012) reported that several factors may affect teachers' utilization of 'Technology- Enhanced Formative Assessment' (TEFA) including software or hardware problems, time constrains, pacing, and teachers' and students' readiness to use the technology.

CONCLUSION

This research study explored students' and teachers' perceptions and practices of the FA of IBL activities in the teaching and learning environment in the UAE. Students' views regarding FA confirmed that most science classes exhibit a cooperative learning environment and create opportunities for various discussions among learners and between learners and their teachers. Likewise, students' views reflected the application of some guided inquiry learning activities, which is a step forward in implementing inquiry in teaching sciences. Students' results didn't show any correlation between FA and inquiry-based practices in the classroom.

Teachers' practices and perceptions reflected strong commands of using FA in physics and chemistry lessons. However, biology teachers seemed to exhibit less experience in utilizing different FA strategies in biology lessons. Furthermore, the teachers presented less confidence when responding to IBL items, as they were implementing inquiry cycles less frequently in their lessons. An important and interesting result was conveyed by teachers' responses, which presented a strong relationship between different items of IBA and FA. This was illustrated by the significant correlation between the various steps required in inquiry cycles, such as the requirement of explaining data, drawing conclusions and communicating results, along with cooperative learning strategies and effective feedback utilization to develop students' critical thinking and meta-cognitive skills.

The findings of this study also present various best practices of using FA strategies to assess IBL activities, such as using a scientific approach, effective teachers' observation, students' collaboration and peer work, the use of worksheets and lab reports, discussions, explanations and using online platforms to assess students' understanding.

To sum up, using FA strategies to evaluate inquiry cycles in science classrooms would enable teachers to collect authentic evidence regarding students' understanding of scientific concepts, and adjust teaching strategies to address misconceptions, considering various questioning techniques that would evoke students' critical thinking skills and require them to reflect upon their understanding. Accordingly, students will build SRL skills and become able to evaluate their own understanding, enabling them to gain the essential knowledge and skills needed to build the required scientific literacy, and contribute positively in building the future of their country.

Implications and Recommendations

This study has two types of implications: one in the educational field, at the school level, and the other in promoting further educational research.

Implications and Recommendations at School Level

The result of this study suggests different strategies to improve the use of FA when assessing IBL activities. First, the priority goes to supporting teachers to implement FA of IBL with confidence, and this would involve:

- Designing continuous professional development workshops that provide actual examples of implementing inquiry in science curricula, and clarify the real meaning of inquiry of science, and its relation to the nature of different sciences: biology, chemistry and physics.

- Providing teachers with training on different questioning skills, and how to evoke students' curiosity and critical thinking skills to solve problems
- Providing model lesson plans that include complete inquiry cycles with the suitable FA strategies, and help teachers develop similar lesson plans.
- Organizing lesson observations to evaluate teaching and learning practices in science classrooms and providing the required support to develop strategies for the FA of IBL.
- Designing curriculum documents and instructional guides that include learning outcomes targeting the development of inquiry skills.
- Designing various FA strategies to ensure that all students have mastered the needed skills to practice IBL.
- Considering time constrains and ensuring that the required topics can be taught through inquiry during the allocated time in the curriculum.

Implications on Educational Research

This study has investigated students' and teachers' perceptions of FA and IBL in science classrooms in the UAE. The significant relationship between inquiry cycles and FA that appeared in this study would require further research to identify different inquiry-based and FA items that are related and measure the effect of this relation on students' achievement, learning progress, self-satisfaction, self- efficacy and developing higher-order thinking to perform novel tasks and solve problems.

Further research is required to investigate the effect of using FA in IBL on the development of students' SRL skills and their ability to measure their conceptual understanding.

Study Limitations

The limitations of this study include the relatively small number of participants, as the students' questionnaire was sent to 2800 students, yet only 535 responded and only 51 from 80 teachers responded. The qualitative part of this study included 10 lesson observations distributed among physics, chemistry and biology lessons, which resulted in small samples for each subject, making it difficult to generalize the results and make definite conclusions based on this small number of teachers.

This study was performed within a series of schools following the same system; centralized curriculum documents are distributed to teachers to implement certain instructional strategies, which reduces teachers' freedom in changing the instructions and makes it difficult to find significant differences regarding implementing IBA.

REFERENCES

Abell, S. K. & Lederman, N. G (2010). *Handbook of Research on Science Education.* 2nd edn. Routledge.

Asay, D. & Orgill, M. (2010) Analysis of essential features of inquiry found in articles in the science Teacher 1998-2007, *The Science Teacher Education.* 21, 57-59.

Bell, J. (2010). *Doing your research project* 5th edn. McGrawHill education

Campbell, T., Abd-Hamid, N. H., & Chapman, H. (2010). Development of instruments to assess teacher and student perceptions of inquiry experiences in science classrooms, *Journal of Science Teacher Education,* 2,13–30.

Cartwright, E., and Stepanova, A., (2012). What do students learn from a classroom experiment: not much, unless they write a report on it, *The Journal of Economic Education,* 43(1), 48–57

Clark, I. (2012). Formative assessment: assessment is for self-regulated learning. *Educational Psychology* Review, 24, 205–249

Clark, I. (2015). Formative assessment: translating high-level curriculum principles into classroom practice, *The Curriculum Journal,* 26 (1), 91-114.

Cohen, L., Manion, L. & Morrison, K. (2000). *Research Methods in Education.* 5th edn. Routledge/Falmer

Cresswell, J. (2002). *Research design qualitative, quantitative, and mixed methods approach.* 2nd edn.

Cresswell, J. (2011). Educational research 4th edn. Pearson Education.

Dorman, P. J., Aldridge, M. J. and Fraser, B. J., (2006) Using students' assessment of classroom environment to develop a typology of secondary school classrooms *International Education Journal,* 7(7), 906-915.

DiBiase, W. and McDonald, J. R. (2015). Science teacher attitudes toward inquiry-based teaching and learning, The Clearing House: *A Journal of Educational Strategies,* Issues and Ideas, 88, 29–38.

Frankel, J., Wallen, N., Hyun, H. (2015) How to Design and Evaluate *Research in Education.* 9th edn. USA: Mcgraw-Hill

Forawi, S. A. & Liang, X. (2011). Developing in-service teachers' scientific ways of knowing, *International Journal of the Humanities,* 9, 265-270.

Harrison, C. (2014). Assessment of inquiry skills in the SAILS project. *Science Education International,* 25, 112-122

Hattie, J., & Timperley, H., (2007). The power of feedback. American *Educational Research Association,* 77 (1), 81-112.

Hanauer. D. I and Bauerle. C., (2012). Facilitating Innovation in Science Education through *Assessment Reform, Liberal Education series,* 34-41.

Irtiqaa Inspection Reports (ADEC) (2015). [online]. [Accessed 9 May 2015]. Available at: https://www.adec.ac.ae/en/education/keyinitiatives/pages/irtiqaa-reports.aspx

Keough, S. M., (2012). Clickers in the Classroom: A Review and a Replication, *Journal of Management Education*, 36(6), 822–847

The Knowledge and Human Development Authority (KHDAKnowledge and Human Development Authority (KHDA). (2015). Inspection reports [online]. [Accessed 17 April 2015]. Available at: http://www.khda.gov.ae/en/publications

Laban, P. A, (2012). A Functional Approach to Educational Research Methods and Statistics: Qualitative, Quantitative, and Mixed Methods Approaches. The Edwin Mellen Press

Lee, H., Feldman, A., & Beatty, I., (2012). Factors that Affect Science and Mathematics Teachers' Initial implementation of technology-enhanced formative assessment using a classroom response system, *Journal of Science Educational Technology*, 21, 523–539

Llewellyn, D. (2010). *Differentiated Science Inquiry*. Sage.

Ministry of Education in United Arab Emirates (2015). Education in the UAE [online]. [Accessed 17 April 2015]. Available at: https://www.moe.gov.ae/English/Pages/UAE/UaeEdu.aspx

Nagle, B. (2013). Preparing high-school students for the interdisciplinary nature of modern biology. CBE *Life Science Education*, 12, 144–147.

Nowak, K. H., Tiemann, N. R., & Belzen, A. U. (2013). Assessing students' abilities in processes of scientific inquiry in biology using a paper-and-pencil test, *Journal of Biological Education*, 47, 182-188.

National Authority for Qualification and Quality Assurance of Education and Training: Directorate of Government Schools Reviews Date of Review: 29 September 1 October 2014 http://www.qqa.edu.bh/en/Reports/Pages/default.aspx

Nehring, A., Tiemann, R. &Belzen, A.U. (2013). Assessing students' abilities in processes of scientific inquiry in biology using a paper-and-pencil test, *Journal of Biological Education*, vol. 47(3), 182-188

Paul, R. & Elder, L. (2004). *Critical Thinking: Concepts and Tools*, 4th edn. Foundation of Critical Thinking

Pedaste, M., Mäeots, M., Siiman, L. A., Jong, T., Riesen, S. A. N., Kamp, E. T., Manoli, C. C., Zacharia. Z. C., and Tsourlidaki, E. (2015). Phases of inquiry-based learning: Definitions and the inquiry cycle. *Educational Research Review*, 14, 47–61.

Pritchard, A. & Woollard, J. (2010). *Psychology for the Classroom: Constructivism and Social Learning*. Routledge.

Raleigh, M., (2012). *Outstanding school inspection: A study for Centre for British Teachers* (CfBT) Inspection Services.

Richards, K. (2003). *Qualitative Inquiry in TESOL.* 1st edn. Palgrave Macmillan.

Stone, E. M. (2014). Guiding students to develop an understanding of scientific inquiry: a science skills approach to instruction and assessment. CBE—*Life Sciences Education*, 13, 90–101.

Sieberg, J., (2008). *Measuring experimental design ability: a test to probe critical thinking*, A Thesis Submitted to the Graduate College of Bowling Green State University.

Tabari, R., (2014). Education reform in the UAE: an investigation of teachers' views of change and factors impeding reforms in Ras Al Khaimah schools, The Sheikh Saud bin Saqr Al Qasimi *Foundation for Policy Research* Working Paper Series.

Torrance, H. (2012). Formative assessment at the crossroads: conformative, deformative and transformative assessment. *Oxford Review of Education,* 8(3), 323–342.

CHAPTER 13

Inquiry-Based Learning in Context of TIMSS

Marwa Eltanahy

ABSTRACT

This chapter presents a comprehensive review of the literature from several angles related to inquiry-based learning (IBL) instruction. IBL is one of the most substantial key elements of many reform efforts in science education because of its impact on accelerating students' acquisition of scientific skills within the spectrum of 21st century skills as well as enhancing students' achievements through overcoming misconceptions, and acquiring the essential content knowledge which can be measured by standardized tests. Particular attention is paid to the historical and theoretical research backgrounds that shed light on IBL, followed by the nature of IBL and its effectiveness on education. Constructivism is then compared to IBL as an active learning instruction. The relationship between IBL and student acquisition of scientific skills is discussed as well. Factors affecting its implementation and IBL between theory and practice will be briefly mentioned. Finally, the relationship between TIMSS and strategies of inquiry model is interpreted.

INTRODUCTION

Inquiry-based learning (IBL) is one of the most substantial key elements of many reform efforts in science education because of its impact on accelerating students' acquisition of scientific skills within the spectrum of 21st century skills (Kazempour, 2012) as well as enhancing students' achievements through overcoming misconceptions, and acquiring the essential content knowledge (Kizilaslan et al., 2012) which can be measured by standardized tests (TIMSS, 2011). This chapter delivers a review of the literature from several angles related to IBL instruction. Particular attention is paid to the historical and theoretical research backgrounds that shed light on IBL, followed by the nature of IBL and its effectiveness on education. Constructivism will then be compared to IBL as an active learning instruction. The relationship between IBL and students' acquisition of scientific skills is also discussed. Factors affecting its implementation and IBL between theory and practice are briefly mentioned. Finally, the relationship between TIMSS and strategies of inquiry model is interpreted.

Scientific Inquiry

Inquiry is commonly defined as the process of understanding and doing science in the way scientists follow to discover the nature of life, while Science is a "unique mix of inquiry and argument" (Yore et al. 2004, p. 347). Three main concepts intensively used in science education were differentiated as Inquiry, Science Inquiry and Scientific Inquiry. The first term, Inquiry, refers to the process of seeking knowledge through posing a question. The second term which is Science Inquiry is described as the implementation process where students can follow its procedures through learning activities and exploration by conducting experiments to solve specific questions (Llewellyn, 2011). Finally, Scientific Inquiry is the learning process where science students can investigate the nature of the real world through diverse ways that rely on evidence-based explanation (Al-Naqbi, 2010). Additionally, it requires a combination of the content matter with general science process skills, critical thinking, creativity, and problem solving to successfully develop scientific knowledge (Lederman, 2009).

Recently, a positive correlation has been found between understanding nature of science (NOS) and inquiry teaching skills (Forawi, 2010). Although Lederman et al. (2014) mentioned that scientific inquiry is often used as a synonymous term of NOS, Next Generation Science Standards (NGSS) supports a contrast between these two terms that makes them independent. Basically, NOS embodies various aspects of science, including the general features that make it different from other disciplines, while scientific inquiry characteristics derive from acceptable scientific knowledge, social-cooperative group work and its significant impact (Irzik & Nola, 2011). Ultimately, science education reform explicitly emphasized the importance of the development of understanding NOS that requires two essential kinds of experiences consisting of learning the history of science and conducting scientific investigation consistently (Nott & Wellington, 2000) which refers to implementing scientific inquiry. Previous research has revealed that inquiry-oriented strategy is an indirect teaching approach develops better conceptions of the nature of science (Forawi, 2003). Furthermore, inquiry learning is an effective instructional strategy that enhances students' scientific skills (Eltanahy & Forawi, 2019). Consequently, NOS recommended IBL as an effective classroom instruction because it is considered as a fundamental component of scientific literacy (Wan & Wong, 2013).

Inquiry-based Learning in Science Education

There has been a sound definition of inquiry learning as an active process which is strongly related to scientific inquiry approach, while there are no boundaries or specific definitions of inquiry teaching. The popularity of the concept of inquiry has increased in supporting pedagogical science as well as its practices. Yager and Ackay (2010) argues that though being familiar with this term, the majority of science teachers are unaware of the core meaning of inquiry and its actual implementations in their classes. Figure 1 illustrates the continuum of teaching instruction (Llewellyn, 2011).

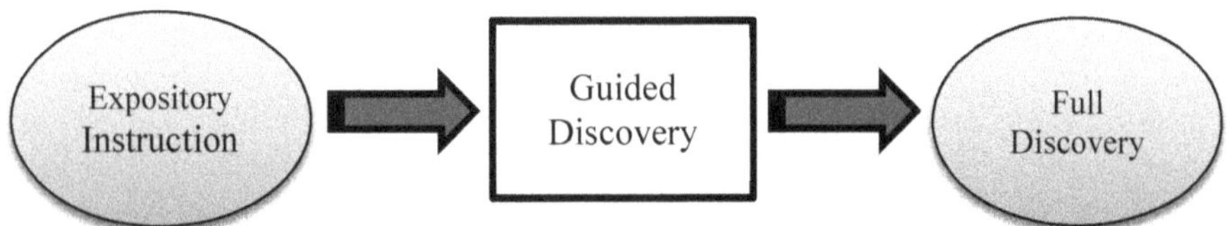

Figure 2: Teaching Instruction Diagram

Different levels of inquiry were distinguished according to the amount of instruction that was given to the learners (Buck et al., 2008). In addition, researchers classified inquiry as four major types according to whether each is the responsibility of the student or the teacher (Llewellyn, 2011) as shown in table 1.

Table 1: Types of Inquiry-based Learning Strategy

Area of inquiry	*Demonstrated inquiry*	*Structured inquiry*	*Teacher initiated inquiry*		*Self-directed inquiry (Open)*
			Guided inquiry	**Coupled inquiry**	
1. ***Posing a question***	Teacher	Teacher	Teache r	*Students select from a bank of questions*	Student
2. ***Planning procedures***	Teacher	Teacher	Student	Student	Student
3. ***Analyzing results***	Teacher	Student	Student	Student	Student
4. ***Drawing conclusion***	Teacher	Student	Student	Student	Student

Although the essential features of inquiry practices vary according to both cultural and social conditions (Abd-El- Khalick et al., 2004), there is only one single way to conceptualize IBL, which is student-centered approach based on hands-on activities that effectively enhance the acquisition of critical-thinking skills as well as science content knowledge.In addition, not all hands-on activities given to students refer to inquiry learning, e.g. building an atom model, because it does not depend on a research question (Bell et al., 2005). The main focus in this section is the features of IBL in the science classroom and its relationship to the main categories of knowledge that are measured through standardized exams such as TIMSS.

Student-centered activities such as guided inquiry practices based on constructivist methodology require concrete learning experiences rather than abstract presentations (Orlich et al., 2013). In addition, Ronis (2008) refers to integrating inquiry practices in science classes as an equivalent approach to problem-based learning (PBL) which simultaneously operates three significant types of learning: cognitive learning, collaborative learning and content learning, which support the acquisition of high-order thinking skills, social constructivism skills and better use of knowledge gained respectively (Capraro et al., 2013).

International-Standarized Tests and Science Education Reform

Thirty-two international comparative studies have been initiated in education by the International Association for the Evaluation of Educational Achievement (IEA) which are classroom-based studies to address a wide range of academic subjects like mathematics, science and reading literacy, to assess the educational outcomes such as students' achievements based on the context of teaching and learning process (Drent et al., 2013). There has been a growing concern among policy makers participating in these international reports and studies started by only 12 countries involved in the First International Math Study (FIMS 1963- 1967) and reached 69 countries from the different regions in the world (Rindermann, 2007).

The importance of these studies is not exclusive to ranking participating countries according to their average achievements' scores, but they also provide an in-depth vision into the diverse aspects of educational systems all over the world. Moreover, countries participating in these studies for more than one cycle of Trends in International Mathematics and Science Study TIMSS (every 4 years) or Program for International Student Assessment PISA (every 5 years) can analyze the provided information across assessment for educational development purposes such as elaboration of textbooks in order to enhance the successful acquisition of scientific habits of minds, learning skills and scientific knowledge (Dagher & BouJaoude, 2011). Therefore, international exams like (TIMSS) and (PISA) are considered as the benchmark and the significant driving force for the current educational reform in many countries.

Background of the Discussion

In the past decades, many educators recognized the importance of students' active role in learning such as Lane (1857-1925), Dewey (1870-1952) and Montessori (1870-1952) who refer to the nature of inquiry-based education. Their views were drawn from the earlier ideas of Rousseau (1712-1778), Pestalozzi (1746- 1827), and Froebel (1782-1852) that represent the roots of an approach that seeks to stimulate students' imagination and curiosity (Harlen, 2013). In the 21st century, inquiry has been gradually adopted in science education, and its importance has been progressively emphasized in many educational aspects such as curriculum development, teaching and learning science (Al-Nabqi, 2010; Bryant, 2006).

Inquiry practice is the main focus of science education reform, which allows students to construct their knowledge and accommodate these new experiences in a suitable way to their natural brain development (NRC, 1996). Furthermore, it supports the acquisition of modern learning skills such as critical thinking, and problem solving to be implemented in real-life situations which are considered as major benefits in applying the inquiry process in schools. In addition, many researchers found that guided inquiry as a student-centered active method is an effective approach in science generally as it enhances students' conceptual understanding (Hofstein et al., 2005).

National Science educational standards interpreted three main angles of scientific inquiry. They are process skills, content, and strategies of teaching and learning science which refer to consistent practice of inquiry investigation, nature of this inquiry method based on standards, and instructions used to enhance students' conceptual understanding respectively (Wang, 2011). As such, implementing effective scientific inquiry activities in schools is one of the significant goals in learning science where teachers are encouraged to practice and apply this approach in their classes in order to produce a new generation of students who are scientifically literate. Accordingly, implementing inquiry-based science education can efficiently support the noble goals of eduaction through designing effective science curriculum that supports the development of students' critical and creative thinking skills.

Statement of the Problem

Inquiry-based learning practices address many of the challenges that generally face the learning process (McKinley, 2012). In addition, it is seen as a tool to support science education reform through integrative methods that address all the students' educational levels and provide them with sufficient opportunities to solve a variety of authentic problems through implementing guided-inquiry instruction (Lehman et al., 2006). That is why, the consistent demand for advanced and innovative workplace skills has increased the importance of IBL at the workplace in terms of developing scientific literacy.

The aim of IBL is to shift the traditional learning paradigm into more constructivist instruction that relies on a social, and collaborative environment to support independent learning (Magee & Wingate, 2014). Previous studies have emphasized that IBL is an active learning approach that successfully engages learners in real investigation to enhance their cognitive skills such as critical thinking and problem solving (Etherington, 2011; Harlen, 2013). Moreover, experts contend that the consistent experience of inquiry-based activities can significantly motivate students in learning effectively to produce better achievement in standardized tests that measure scientific skills as well as content knowledge (Akınoğlu & Tandoğan, 2007). However, teachers' inquiry practices are affected by the way they have been taught. Accordingly, it is essential to clarify that successful inquiry features require them to act as facilitators who scaffold the students' activities and support independent learning (Asghar et al., 2013) and these features are related to knowledge measured by international exams. That is why, the relationship between instructional principles of IBL strategy and knowledge categories measured by standardized exams like TIMSS should be discussed to clarify how this rapport can impact students' results in TIMSS.

Conceptual Framework

In the light of standardaized tests, Figure 2 demonstrates the conceptual framework that is designed to illustrate the main theories that guide the relationship between IBL and TIMSS exams

Inquiry instruction as a constructivist learning aproach focuses on three main domains to enhance studnets' scientific skills that are measured by international exams.. Firstly, Content domain which emphasies the integration of technology in learning science. Secondly, the Inquiry domain, which is based on teaching strategy to support cognitive learning, questions used to support content learning, and students' interaction to support collaborative learning. Finaly, Cognitive domain, which is based on students' scientific skills gained by inquiry implementation. Content and Cognitive doamins are the same domains used in the TIMSS structure to measure students' scientific skills in three main categories named, knowing, applying and reasoning.

Historical and Theoretical Background of IBL Instruction

John Dewey (1910) proclaimed that students could understand and experience the process of science more efficiently by giving more emphasis on attitudes of mind and thinking skills rather than being passive receivers when the main focus is only on scientific facts. Therefore, developing thinking and formulating habits of minds by engaging students in learning are articulated as the main objectives of inquiry instruction. Thus, being an active learner is the primary factor that motivates students to freely construct their knowledge in a supportive environment (Dewey, 1938).

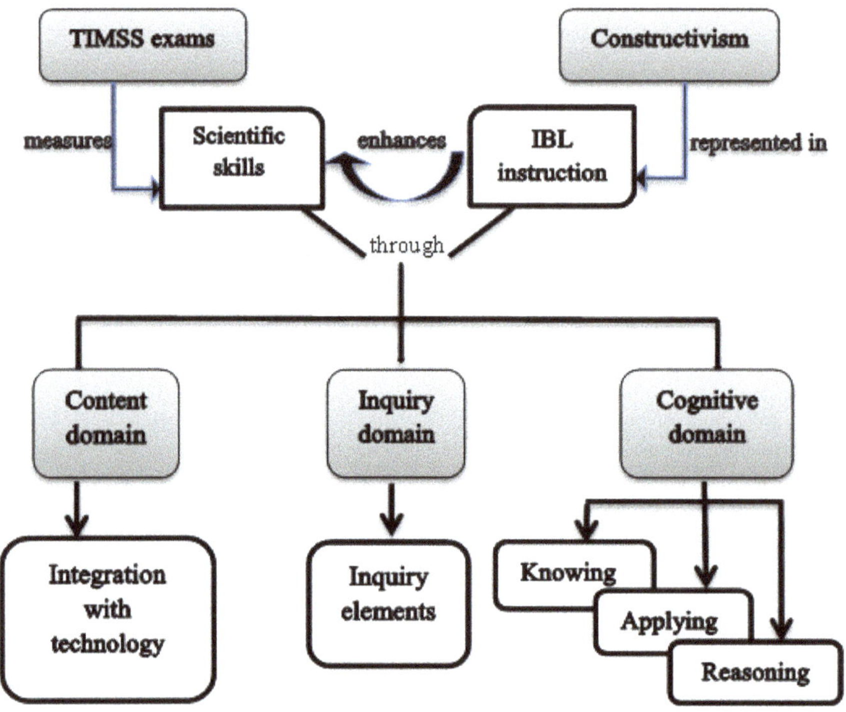

Figure 2: A conceptual framework shows the incorporation of IBL and International Assessment Theories

Accordingly, calling for reform in education was essential to meet the demands of the new century to produce students who are scientifically literate through developing methods of teaching science as enquiry (Schwab, 1960) in order to enhance scientific skills such as observing, inferring, classifying and controlling variables (Martin, 2010). In addition, science teachers were encouraged to use the school laboratory to help their students to be fully engaged in the process of scientific investigation which is considered as an invitation to enquiry learning, within which inquiry-based instruction was recommended because of its potential to empower learners when they work as experts to clarify their own data (Tabak & Baumgartner, 2004). Although, science teachers supported the inclusion of this new instruction in teaching science and opposed the didactic techniques, their actual teaching of science does not represent inquiry as they are more comfortable implementing structured inquiry to control the group work (Eltanahy & Forawi 2019).

National Science Educational Standards (NSES) advocated that IBL activities that accomplish scientific literacy benchmarks are pertinent for enhancing student achievements (NRC, 2000). The American Association for the Advancement of Science claimed that science-literate students would be generated by 2061 when IBL is implemented consistently in science classes (AAAS, 1993). Basically, successful IBL involves the development of students' cognitive abilities when they act as scientists to answer specific question during investigation which help them to gain better scientific knowledge and enhance their inquiry abilities (Bybee, 2000).

NRC identified the main features of inquiry that should be the responsibility of students in science classes, which are posing problem-oriented questions, collecting evidence, developing explanation, evaluating justifications and communication. Subsequently, a list of statements were designed by NRC (1996) to illustrate the emphasis shift that guide all science teachers in general to the essential elements of inquiry learning as shown in table 2.

The literature on IBL discusses several kinds of instruction based on significant qualifiers that emphasize its nature, scaffolding level, learning emphasis depending on either existing knowledge or constructing knowledge, and finally, the scale of inquiry which could be within the class or course, or represented in the whole course or degree (Spronken-Smith & Walker, 2010). Hence, scaffolding process refers to the guidance and support provided to students at the beginning, and then gradually decreased until they can take the charge of their learning as decreasing scaffolding level leads to increasing students' independent learning abilities and enhancing their skills.

Nature of IBL Instruction & its Benefits (Theoretical View)

Inquiry instruction is a powerful method within the learning process that depends on student-center approach as it supports the inquisitiveness of learners (Blanchard et al. 2008) who are expected to pose critical questions in a collaborative setting. Investigation is required to answer these questions through seeking evidence and discussing critical reasoning. NSES seek to promote IBL instruction that enables educators to build on students' natural curiosity through understanding science discipline as human endeavor. In addition, it increases their motivation and stimulates their interest (Gibson & Chase, 2002; NRC, 2005) to pursue a scientific career when they show better acquisition of both essential scientific concepts (Minner et al., 2010) and experimental skills (Drayton & Falk, 2001). Figure 3 illustrates the nature of inquiry learning as demonstrated by Etherington (2011).

Table 2: The Main Aspects that Support Implementing IBL Instruction (NRC, 1996, P.113)

Changing Emphases to Promote Inquiry	
Less emphasis on	**More emphasis on**
Treating all students alike and responding to the group as a whole	Understanding and responding to individual students' interests, strengths, experiences, and needs
Rigidly following curriculum	Selecting and adapting curriculum
Focusing on student acquisition of information	Focusing on student understanding and use of Scientific knowledge, ideas, and inquiry processes
Presenting scientific knowledge through lecture, text, and demonstration	Guiding students in active and extended scientific inquiry
Asking for recitation of acquired knowledge	Providing opportunities for scientific discussion and debate among students
Testing students for factual information at the end of the unit or chapter	Continuously assessing student understanding
Maintaining responsibility and authority	Sharing responsibility for learning with students
Supporting competition	Supporting a classroom community with cooperation, shared responsibility, and respect
Working alone	Working with other teachers to enhance the science program

Figure 3: Nature of Scientific Method of Inquiry (Etherington, 2011, p.38)

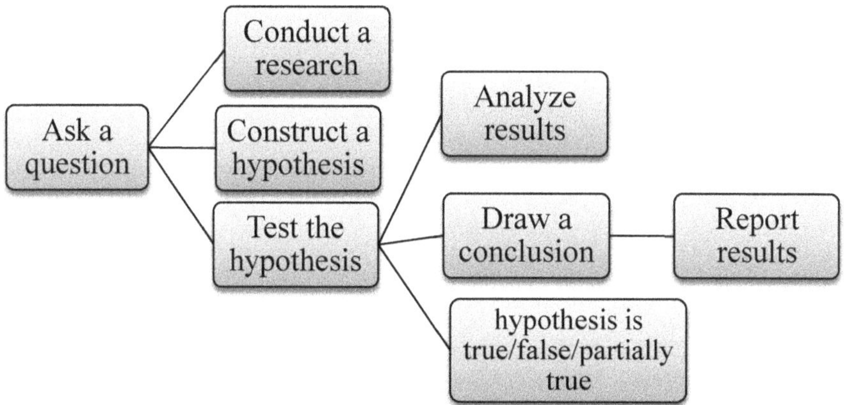

IBL model focuses on engaging students in conducting a research to answer a specific question through constructing their hypothesis and testing it, then analyzing the results to draw a logical conclusion (Banerjee, 2010). IBL instruction has been regarded in the international educational field as an effective learning strategy that strongly enhances science education (Mäeots & Pedaste, 2014). A vast body of literature affirms that a wide range of educational goals are clearly affected by applying IBL practices such as students' motivation and disposition towards learning science, understanding scientific concepts, and acquiring problem solving skills (Wilke & Straits 2002).

Hence, inquiry learning is considered as a student-driven process. Understanding and investigating students' perceptions and interests towards learning science by IBL instruction is a critical issue that affects the direction of the inquiry process (Magee & Wingate, 2014). Students who reliably apply inquiry practices in their class usually acquire the habits of working as scientists faster. These habits enhance their inquiry learning skills rather than just reading from the given text (Latta et al., 2007).

IBL Instruction as Constructivist and Active Learning Strategy (Practical View)

The most appropriate teaching and learning approaches should facilitate students' acquisition of 21-century skills and competencies through independent learning groups (DES 2003; NASRV 2008). Supporting this viewpoint, Airasan and Walsh (1997, cited in Orlich et al. 2013, p. 82) explain that "[C] onstructivism is a theoretical model about how learners come to know". Moreover, it is a pedagogical concept that refers to discovery learning where students acquire new knowledge that builds on their prior information when they conduct their own experiences (Balım 2009; Mayer 2004). Additionally, Forawi (2014, p. 41) contends that "[C] onstructivism is the dominant paradigm of learning in science, and a large amount of science education research has been carried out from a constructivist perspective".

Ronis (2008) argues that there is a close match between inquiry model and the theory of constructivism. Thus, IBL is an example of constructivist approach where learners should investigate the surrounding phenomena by evaluating the collected data to make a logical conclusion (Blanchard et al., 2010) rather than the teacher passing the information passively to them. Therefore, constructivist philosophy encourages self-directed learning strategies that allow students to have the active role of monitoring the learning process through planning procedures and evaluating evidences (Kang et al. 2009) which enhance their problem-solving abilities (Hmelo-Silver, 2004). There are several points of agreement that justify to what extent inquiry-oriented approach and constructivist approach are significantly connected (Richardson 2003). Table 3 shows the commonalities between them.

Previous studies have referred to any constructivist approach as an outstanding contribution that positively supports the quality of science education (Nezvalova, 2008). IBL instruction refers to diverse ways that provide students with the opportunities to connect their content knowledge to real-life applications through organized thinking (Slavin, 2012). Consequently, teachers are encouraged to apply IBL techniques consistently so as to fulfill the basic goals of literacy, and effectively maximize students' understanding of a wide range of scientific concepts in order reduce their misconceptions. Thus, shifting the paradigm from the didactic-traditional atmosphere into inquiry-constructivist learning becomes essential in science education. It is still impossible to do this shift without transforming the epistemology of both teachers and students to fully understand how knowledge should be constructed to provide them with a meaningful learning experience (Lamanauskas, 2010). This is why professional development programs are required to improve teachers' pedagogical knowledge about IBL and its implementations. Inquiry teaching needs sufficient time where teachers behave in the same energetic manner while playing a supportive role (Powell & Kalina, 2009).

Table 3: Commonalities between IBL & Constructivist Approach (Richardson, 2003)

Points of Agreement between IBL & constructivism	
	Student is the main focus
	flexible pace of instruction
	Students should search for implications, produce conclusions and evaluate their ways to solve real problems
	Humans classify objects provided by nature
	It is not a sole learning model

IBL Pedagogical View

There has been a reasonable tendency to consider that teachers' beliefs which represent espoused theory is the potential factor that direct teachers' practices underlying theory of action (Kane et al., 2002). Basically, it is crucial to explore both perceptions and practices of teachers to determine the notion of two educational theories named "espoused theory and theory-in-use"(Jones, 2009, p. 177) that underpin the ongoing learning experience. Roehrig (2004) claims that in spite of the constant emphasis on the implementation of IBL instruction in science education reform, translating this reform vision from theory to actual practice is still difficult because teachers are affected by the way they were taught, which formed their beliefs (Roth et al., 2006) that may render them resistant to modification or change. Moreover, many studies asserted that science teachers frequently use appropriate new concepts or educational terms as a reform jargon to refer to existing teaching instructions or learning activities such as 'hands-on', 'students build on their own knowledge', and 'teachers as facilitator' even though those terms do not match their actual implementations which has led to an urgent call for teachers' professional development programs to help them feel more confident in IBL applications (Bryan, 2003).

Although there is an obvious disconnection between the didactic ways of teaching observed in science classes and teachers' perceptions and descriptions of their inquiry teaching instruction (King et al., 2001), a recent study concluded that "the challenge of bridging the theory-into-practice gap is not actually an eternity away" (Leaman & Flanagan, 2013, p. 59), but it can be directed to the reform path by avoiding or modifying any vulnerable tasks that negatively affect the learning process.

In Standard B of National Science Educational Standards (NSES), it is stated that "teachers of science guide and facilitate learning" (NRC, 1996, p. 32) which means that teachers should not provide their students with the scientific knowledge as a straightforward answer present in their textbook. Instead students should be encouraged to debate collaboratively in order to develop more scientifically oriented questions. Furthermore, the question under investigation is called the efficacy of IBL because its technique can be designed in a way that does not underpin the inquiry goals which provide students with meaningful learning experience (Pine et al., 2006). Therefore, the stress on underlying scientific content should be decreased, while behavioral and skillful outcomes through IBL activities should be highlighted. Researchers' observations of IBL classes found that teachers' priorities were in getting students to enjoy doing experiments and using a variety of materials more than cognitively, thus engaging them to understand in-depth the required scientific concepts (Furtak & Alonzo, 2009). Accordingly, prioritizing activity over developing cognitive abilities and understanding content-oriented objectives is considered as a misinterpretation of IBL goals and leads to an undesirable mismatch between science reform-oriented expectations and teachers' priorities in their actual practices which certainly threaten the acceleration of the reform achievement and its success.

The essential capabilities that are expected from high school students have been identified by five science Threshold Learning Outcomes (TLOs) as a foundation to articulate the required scientific skills of a science graduate, which are demonstrating good understanding of science as inquiry, representing breadth and depth of the content knowledge, and communicating knowledge to real life. These skills were categorized by the Science Students Skills Inventory (SSSI) in six significant areas, which are team work, oral communication, scientific writing, scientific knowledge, quantitative skills and ethical thinking (Matthews & Hodgson, 2012). Five of these skills are developed effectively by enhancing interactive learning activities in feasible classes that implement IBL instruction (Aurora, 2010) especially in laboratory investigations. More recently, Hodgson et al. (2014) argued that traditional teaching instructions provide students with the required content knowledge without developing any of their learning skills, while practical laboratory classes based on IBL activities utilize all the six nominated scientific skills as perceived by 93% of the students in his study.

Basically, learning science is a physical attempt to positively understand the natural phenomena in the surrounding environment through observation, questioning and investigation that requires acquiring some essential scientific skills. Learning the nature of science is a major part in learning science through implicit application of

IBL instruction coupled with explicit teaching of NOS (Forawi, 2014). Correspondingly, Bybee and Powell (2014) classified the science process skills that are the substantial outcomes of consistent IBL practices into five types. The first type is the acquisitive skills which refer to abilities of students to collect the required data. The second type is the organizational skills which indicate how students are able to put their data in systematic order to be available for analysis stage. The third type of skills is the creative skills which identify the abilities of students to develop new ways of approach to think more critically. Communicative skills are the fourth type that distinguishes the abilities of students to transfer and explain their information successfully to others. Finally, manipulative skills signify the abilities of students to handle the scientific tools, machines or any laboratory instruments that should be used during their investigation.

The consistent emphasis on theoretical scientific knowledge through traditional teaching techniques and neglecting hands-on activities integrated with technology applications are the main reasons behind the problems affecting the quality of science education in many countries (Dagher & BouJaoude, 2011). That is why IBL as an experiential learning instruction has also been endorsed as a student-centered pedagogy (Hmelo-Silver 2004) which has the maximum benefits for students' learning by enhancing both the development of their high-order thinking skills and their outcomes (Spronken-Smith & Walker, 2010).

Harlen and Qualter (2009) suggested three main dimensions of students' progression in scientific skills based on IBL strategies from simple to more elaborated skills, followed by the effective use of these skills in familiar and unfamiliar situations, and end up with consciously making reasonable predictions which imply development of skills into the more advanced stage of metacognition skills. Akınoğlu and Tandoğan (2007) advocated that employing the highest level of IBL in science education which is the guided and open inquiry with problem-based active learning strategies will positively enhance students' achievement in standardized tests that measure both learning content and skills. In addition, increasing the level of problems complexity leads to increasing the acquisition of skills that enrich students' cognitive development (Capraro et al., 2013). Ronis (2008) referred to IBL as a brain-compatible methodology that provides students with real opportunities in order to connect educational theories to effective practical experiences (Schwartz et al., 2005) which enable them to positively interact with types of reasoning and achieve high levels of Bloom's taxonomy within the cognitive process (Liu & Lin, 2009). Thus, IBL activities promote students' critical thinking and problem solving skills because this instruction utilizes questioning techniques that launch students in the direction of analyzing and evaluating their data to make proper decisions and find a conclusion to solve an authentic problem (Snyder & Snyder, 2008), as well as increasing focus on practical opportunities enhances students' intrinsic motivation and improves their metacognitive skills (Zimmerman, 2007). Conclusively, IBL practices as a scientific methodology enhance students' cognitive acceleration and develop their scientific reasoning skills (Hugerat et al., 2014). Thus, investigating ill-structured problems stimulates students' innovation and creativity through brainstorming to find alternatives that efficiently solve the required tasks (Gurses et al., 2007).

The technique of IBL is interpreted based on complex and interrelated contextual influences (Pea, 2012) such as teacher competence, learner autonomy, assessment and technology integration. Mostly, teachers' competence is the most significant factor that underpins positive implementation of IBL instruction (Onwu & Stoffels, 2005) because teachers are considered as central-decision makers who prepare and adapt IBL activities in their classroom (Colburn, 2000). Accordingly, science teachers not only require a deep understanding of their subject matter to introduce correct scientific concepts to their students, but they should also be aware of the scientific process to effectively facilitate the learning development, and guide the students to formulate successful investigation (NRC, 2005). Lack of the combination between content knowledge, theoretical knowledge and pedagogical knowledge would lead to unsatisfactory experience of learning science and low quality of education (Kim & Tan, 2011). Consequently, providing teachers with sufficient and authentic professional development opportunities has a positive relationship with the level of IBL practices that they apply in their classes (Rogan & Aldous, 2005) because generally most of them have never experienced IBL instruction in their own education stage. That is why investigating science teachers' perceptions of applying IBL in their teaching method is advantageous to support their practice. Moreover, using the term inquiry investigations or practices is beneficial in the educational field to "stress that engaging in scientific inquiry requires coordination both of knowledge and skills simultaneously" (NRC, 2012, p. 41).

Learner autonomy has been flagged as one of the dynamic principles that reinforce expressive activities in IBL environment where students' practical work and self directed learning drive the vehicle of the educational situation in the classroom with less control from teachers and more learning responsibility from students. Furthermore, high stakes of standardized assessments, including summative examinations play a practical role in improving inquiry-based practices in science classes (Blanchard et al., 2010). Moreover, there has been an argument that there is an intrinsic link between acquiring scientific process skills and enhancing students' content knowledge (Chiappetta & Adams, 2004).

NRC (2005) encourages teachers to integrate technology as a valuable tool in the scientific inquiry process which strongly contributes to the enhancement of constructivist, student- centered practices (Seimears et al., 2012), and enriches teachers' pedagogical knowledge (Hechter & Vermette, 2013). Recently, the integration of technological skills into science education has become an extensively popular trend (Tsai & Chai, 2012). According to the literature of science-based technologies, this integration facilitates the learning process and helps students to understand more key scientific concepts in short time as well as engage them in real scenarios for further investigations (Capraro et al., 2013).

Previous studies emphasized that technology integration into scientific academic content is another key element (Almekhlafi, 2006) required to solve problems by operating IBL to develop students' metacognitive skills (Sherry et al., 2001) as a result of its progressive impact in both teaching and learning practices (Almekhlafi & Almekdadi, 2010). In this regard, Science education reform identifies the inclusion of technical applications in learning science as a primary goal that directly enhance students' motivation to gain more knowledge (Seimears et al., 2012).

The Relationship between TIMSS & Strategies of Inquiry Model

Previous research categorized scientific content knowledge into six main kinds which are "canonical, procedural and experimental, nature of science, real-life issues, classroom safety, and meta- cognitive" (Furtak & Alonzo, 2009, p. 427). Given the prominence of IBL in science education reform to all grade levels, researchers have been inspired to consider the relation between knowledge categories and features of the inquiry model. Figure 4 shows the strategic cycle of inquiry model and the relevant skills that are developed through the implementation and reflection on the learning process.

Figure 4: The strategic cycle of Inquiry Model

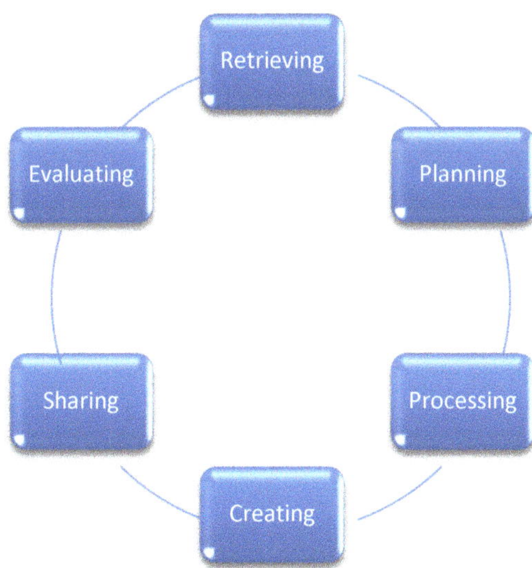

1. Retrieving

    - Observe phenomena
    - Identify available information and group them
    - Locate information and identify a gap to pose a question
    - Develop an information retrieval plan

2. Planning

    - Identify an interesting area of inquiry
    - Identify accessible sources of information
    - Identify target audience and participants
    - Establish criteria and rubric for evaluation
    - Outline an inquiry plan

3. Processing

    - Identify the purpose and focus of inquiry
    - Select, collect pertinent information and record them
    - Create possible dots of connections and inferences
    - Analyze the data and identify gaps
    - Review, revise, modify and update the inquiry plan

4. Creating

    - Organize information to make new meaning
    - Understand and consider audience' perceptions and needs
    - Create a product to represent the new meaning
    - Review and update the inquiry plan

5. Sharing

    - Communicate with the target audience
    - Introduce and present the new ideas and conclusion
    - Discuss and consider suggestions to improve
    - Demonstrate and support the conclusion with evidences
    - Review and update the inquiry plan

6. Evaluating

- Evaluate the conclusion, results and the product
- Evaluate the inquiry process and plan
- Review and modify the inquiy model applied
- Apply the new learning in a different situation

Accordingly, IBL is a nonlinear process. It has a recursive, highly individual and flexible nature. Hence, inquiry is defined by NRC (2000) as "consisting of the activities and thinking processes of scientists, as well as knowledge of what scientists do and what science is, embedded within a foundation of deep conceptual understanding" (Furtak & Alonzo, 2009, p. 428). Based on this definition and considering all the six categories of knowledge that are measured by TIMSS, they have parallel representations with fundamental inquiry goals which aim to produce independent learners who acquire not only scientific skills, but also the disposition to employ those abilities along with deep understanding of self-assessment strategies and their own responsibilities towards learning in order to thrive in a digital environment full of complex information (Alberta, 2004). In addition, the progressive inquiry principles were designed to explicate the dynamics of scientific inquiry process, and adapted to represent inquiry strategies (Kong & Song, 2014).

Based on the previous literature, table 4 is designed to highlight the link between knowledge categories and inquiry principles in relation to the essential scientific skills that can be achieved.

Table 4: The Relationship between Inquiry Strategies and Categories of Knowledge meansued by TIMSS

(TIMSS) Knowledge Categories	Meaning	Strategies of Inquiry Model Related	Instructiona l Principles of IBL
Canonical knowledge	propositional information found in a textbook such as facts, laws, theories and explanations	**Retrieving:** gain background & connect to self prior information to pose a question e.g: (Forawi 2014)	Building on what students' already know
Procedural &	steps of scientific practices and	**Planning:** identify the topic area and sources of collecting	Using authoritative sources

(TIMSS) Knowledge Categories	Meaning	Strategies of Inquiry Model Related	Instructiona l Principles of IBL
experimental knowledge	choose materials to execute scientific procedures	data for inquiry investigation	constructively in a sequence of stages
Nature of science knowledge	what scientists do conduct investigation to discover natural phenomena which represent the value of science	**Processing:** choose pertinent information to make connection and inferences	Encouraging diverse ideas at critical point relys on evidence-based assessment
Real-life issues	Connection between content learning and life situations such as social or personal	**Evaluating:** transfer learning to new situations with critical analysis & logical decision e.g: (Mäeots & Pedaste 2014)	Working in authentic problems
Metacognitive knowledge	Strategies for learning to skillfully develop new thinking	**Creating:** organize & edit the information to establish new product	Engaging actively and reflecting on that experience.
Classroom safety knowledge	Handling classroom science materials safely in group work	**Communicating:** Share and present to audience	Providing collaborative opportunities in social interaction

The assessement of students' performance in science is scaled to 500 as an international average, and standard deviation of 100. These achievements are based on four criteria which are called international benchmarks that explain each level of students' understanding and knowledge. Table 5 explains how each science international benchmark in TIMSS refers to inquiry types.

Table 5: The Relationship between Science Internatinal Benchmark and Inquiry Types

International Benchmark	***Description***	***Type of IBL required***
Advanced 625	- understand and apply basic elements of investigation to draw conclusions and solve real life problems. - communicate & explain knowledge in a written report.	Open inquiry
High 550	- Some scientific skills are demonstrated - Interpreting data in graphs, diagrams …etc. - Writing conclusion in short intensive response.	Guided inquiry
Intermediate 475	- Interpreting information from graphs to draw conclusion. - Brief explanation.	Structured inquiry
Low 400	- Interpreting simple graphs. - Applying basic knowledge to similar situation.	Demonstrated inquiry

The Results of Previous Studies

Many studies have highlighted the effectiveness of IBL applications (Anderson, 2002; Ketpichainarong et al., 2010) as increasing students' behavioral, intellectual engagement and providing them with lifelong learning skills (Richardson & Liang, 2008). These latter qualities are seen as vital in developing students' scientific conceptual understanding and supporting logical ways of creating this scientific knowledge and its nature. As a result, a new generation of successful decision makers to their future investigations will be produced (Harlen, 2013) that achieve the major goals of scientific literacy through enhancing their cognitive skills such as problem-solving and critical thinking (Wu & Hsieh, 2006).

Ramnarain (2014) argued that science teachers believe that IBL practices provide students with more enjoyable science classes and develop their skills in the experimental process during consistent inquiry investigations to construct new knowledge based on their prior information. Researchers emphasized that teachers play a critical role to encourage their students to understand the nature of scientific process that is constantly changing and dynamic (Khishfe & Abd-El-Khalick, 2002) through IBL techniques, which means that teacher training and development programs should frequently include more teachers' practices of inquiry-based laboratory which increase their attitude towards learning science, and develop their scientific process skills that are essential to professionally lead their students to apply IBL activities to support their creative skills (Yakar & Baykara, 2013).

Al-Naqbi (2010) found that although, students are working in cooperative groups to implement IBL activities in the science class, it is difficult to effectively develop their inquiry abilities because insufficient opportunities are given to them to work independently, and all their investigations are under intensive support from their science teachers. Thus, other obstacles such as inadequate class time, limitations of students' skills and prior knowledge, teachers' frequent intervention, and shortage of scientific tools act as the main barriers towards implementing successful IBL experiences (Cheung, 2007).

Conclusion

Successful implementation of IBL requires preparation of teachers through professional development programs that highlight the key features of IBL. Great focus should be given to the main instructional principles of IBL and the most relevant startegies to its model in order to enhance teachers' practices. Furthermore, the relationship between different types of knowledge categories that are measured by TIMSS and IBL principles that are implemented in the classroom should be identified and emphasized as a base for future assessment. Class time should be manipulated for further investigations in order to provide students with more authentic learning opportunities to improve their scientific skills.

REFERENCES

Abd-El-Khalick, F., BouJaoude, S., Duschl, R., Lederman, N. G., Mamlok-Naaman, R., Hofstein, A. & Tuan, H. (2004). Inquiry in science education: International perspectives. *Science Education*, 88(3), 397–419.

Anderson, R. D. (2002). Reforming science teaching: What research says ahout inquiry. *Journal of Science Teacher Education*, 13 (1), 1-12.

Akınoğlu, O. & Tandoğan, R. O. (2007). The Effects of Problem-Based Active Learning in Science Education on Students' Academic Achievement, Attitude and Concept Learning. *Eurasia Journal of Mathematics, Science & Technology Education*, 3(1), 71-81.

Almekhlafi, A.G. (2006). The effect of computer assisted language learning (CALL) on United Arab Emirates English as a foreign language (EFL) school students achievement and attitude. *Journal of Interactive Learning Research*, 7(2), 121-142.

Almekhlafi, A. G. & Almeqdadi, F. A. (2010). Teachers' Perceptions of Technology Integration in the United Arab Emirates School Classrooms. *Educational Technology & Soci*ety, 13 (1), 165–175.

Al-Naqbi, A. K. (2010). The degree to which UAE primary science workbooks promote scientific inquiry. *Research in Science & Technological Education*, 28(3), 227-247.

American Association for the Advancement of Science (AAAS). (1990). *Science for all*. New York: Oxford University Press.

American Association for the Advancement of Science (AAAS). (1993). *Benchmarks for Science Literacy*. New York: Oxford University Press.

Asghar, A., Ellington, R., Rice, E., Johnson, F. & Prime, G. (2013). Supporting STEM Education in Secondary Science Contexts. *Interdisciplinary Journal of Problem-based Learning*, 6(2).

Aurora, T. S. (2010). Enhancing Learning by Writing Laboratory Reports in Class. *The Journal of Faculty Development*, 2(1), 35–36.

Balım, A., G. (2009). The Effects of Discovery Learning on Students' Success and Inquiry Learning Skills. *Eurasian Journal of Educational Research*, 35, 1-20.

Banerjee, A. (2010). Teaching science using guided inquiry as the central theme: A professional development model for high school science teachers. *Science Educator*, 19(2), 1-9.

Banchi, H. & Bell, R. (2008). The many levels of inquiry. *Science and Children*, 46(2), 26–29.

Bell, R. L., Smetana, L. & Binns, I. (2005). Simplifying inquiry instruction. *The Science Teacher*, 72(7), 30–33.

Blanchard, M. R., Southerland, S. A. & Granger, E. M. (2008). No Silver Bullet for Inquiry: Making Sense of Teacher Change Following an Inquiry-Based Research Experience for Teachers. *Science Eduation,* 93(2), 322 – 360.

Blanchard, M., Southerland, S., Osborne, J., Sampson, V., Annetta, L. & Granger, E. (2010). Is Inquiry Possible in Light of Accountability?: A Quantitative Comparison of the Relative Effectiveness of Guided Inquiry and Verification Laboratory Instruction. *Science Education*, 94 (4), 577-616.

Bryan, L. A. (2003). Nestedness of beliefs: Examining a prospective elementary teacher's belief system about science teaching and learning. *Journal of Research in Science Teaching*, 40, 835–868.

Bryant, R. (2006). Assessment results following inquiry and traditional physics laboratory activities. *Journal of College Science Teaching*, 35(7), 56–61.

Bybee, R.W. (2000). Teaching science as inquiry. In J. Minstrell & E.H. van Zee (Eds.), *Inquiring into Inquiry Learning and Teaching in Science*. New York: Aas Project 20161, 20-46.

Capraro R. M., Capraro M. M. & Morgan J. R. (2013). STEM Project-based Learning. *An Integrated Science, Technology, Engineering,and Mathematics (STEM) Approach.* 2nd Edn. Rotterdam: Sense Publishers.

Cheung, D. (2007). Facilitating chemistry teachers to implement inquiry-based laboratory work. *International Journal of Science and Mathematics Education*, 6(1), 107–30.

Chiappetta, E. L. & Adams, A. D. (2004). Inquiry-based instruction: Understanding how contentand process go hand-in-hand with school science. *The Science Teacher*, 71(2), 46–50.

Dagher, Z. R. & BouJaoude, S. (2011). Science education in Arab states: bright future or status quo? *Studies in Science Education*, 47(1), 73–101.

Department of Education and Skills [DES]. (2003). *The future of higher education*, London: The Stationary Office Limited.

Dewey, J. (1910). Science as subject matter and as method. *Science*, 31(787), 121-127.

Dewey, J. (1938). *Experience and Education.* New York: Collier Books.

Drayton, B. & Falk, J. (2001). Tell-tale signs of the inquiry-oriented classroom. *NASSP Bulletin*, 85(623), 24-34.

Drent, M., Meelissen, M. R. & Van der kleij, F. M. (2013). The contribution of TIMSS to the link between school and classroom factors and student achievement. *Curriculum studies*, 45(2), 198-224.

Eltanahy, M. & Forawi, S. (2019). Science Teachers' and Students' Perceptions of the Implementation of Inquiry-Based Learning Instruction in a Middle School in Dubai. Journal of Education, 199(1), 13-23.

Etherington, M. B. (2011). Investigative Primary Science: A Problem-based Learning Approach. *Australian Journal of Teacher Education*, 36(9), 36-54.

Forawi, S. A. (2003). The effects of contributing factors to understanding the nature of science. *The Urban Researcher*, 15-31.

Forawi, S. A. (2010). Investigating the relationship between the nature of science and guided inquiry instruction. *Journal of Applied Research in Education*, 14(1), 8-16.

Forawi, S. (2011). Inquiry instruction and the nature of science: how are they interconnected? *Journal of Science Education*, 12, 11-14.

Forawi, S. (2014). Impact of Explicit Teaching of the Nature of Science on Young Children. *The International Journal of Science, Mathematics and Technology Learning*, 20, 41-49.

Furtak, E. & Alonzo, A. (2009). The Role of Content in Inquiry-Based Elementary Science Lessons: An Analysis of Teacher Beliefs and Enactment. *Research in Science Education*, 40(3), 425-449.

Gibson, H. L. & Chase, C. (2002). Longitudinal impact of an inquiry-based science program on middle school students' attitudes toward science. *Science & Education*, 86(5), 693-705.

Gurses, A., Acikyildiz, M., Dogar, C. & Sozbilir, M. (2007). An investigation into the effectiveness of problem-based learning in a physical chemistry laboratory course. *Research in Science & Technological Education*, 25(1), 99-113.

Harlen, W. & Qualter, A. (2009). *The teaching of science in primary schools.* 5th ed. London: David Fulton.

Harlen, W. (2013). Inquiry-based learning in science and mathematics. Review of science, *Mathematics and ICT education*, 7(2), 9-33.

Hechter, R. P. & Vermette, L. A. (2013). Technology integration in K-12 science classrooms: An analysis of barriers and implications. *Themes in Science & Technology Education*, 6(2), 73-90.

Hodgson, Y., Varsavsky, C. & Matthews, K. E. (2014) Assessment and teaching of science skills: whole of programme perceptions of graduating students. *Assessment & Evaluation in Higher Education*, 39(5), 515-530.

Hofstein, A., Navon, O, Kipnis, M. & Mamlok-Naaman, R. (2005). Developing students' ability to ask more and better questions resulting from an inquiry-type chemistry laboratories. *Journal of Research in Science Teaching*, 42, 791-806.

Hmelo-Silver, C. (2004). Problem-Based Learning: What and How Do Students Learn? *Educational Psychology Review*, 16 (3), 235-266.

Hugerat, M., Najami, N., Abbasi, M. & Dkeidek, L. (2014). The cognitive acceleration curriculum as a tool for overcoming difficulties. *Journal of Baltic Science Education*, 13(4), 523 – 534.

Irzik, G. & Nola, R. (2011). A family resemblance approach to the nature of science for science education. *Science & Education*, 20(7), 591-607.

Jones, A. (2009). Generic attributes as espoused theory: the importance of context. *Higher Education*, 58, 175-191.

Kane, R., Sandretto, S. & Heath, C. (2002). Telling half the story: A critical review of research on the teaching beliefs and practices of university academics. *Review of Educational Research*, 72(2), 177–228.

Kang, W. C., Jordan, E. & Porath, M. (2009). Problem-Oriented Approaches in the Context of Health Care Education: Perspectives and Lessons. *Interdisciplinary Journal of Problem-based Learning*, 3(2), 43-62.

Kazempour, M., Amirshokoohi, A. & Harwood, W. (2012). Exploring students' perceptions of science and inquiry in a reform-based undergraduate Biology. *Journal of College Science Teaching*, 42(2), 38-43.

Ketpichainarong, W., Panijpan, B. & Ruenwongsa, P. (2010). Enhanced Learning of Biotechnology Students by an Inquiry-Based Cellulose Laboratory. *International Journal of Environmental and Science Education*, 5 (2), 169-187.

Khishfe, R. & Abd-El-Khalick, F. (2002). Influence of explicit and reflective versus implicit inquiry-oriented instruction on sixth grades' views of nature of science. *Journal of Research in Science Teaching*, 39(7), 551-578.

Kim, M. & Tan, A. L. (2011). Rethinking difficulties of teaching inquiry-based practical work: stories from elementary pre-service teachers. *International Journal of Science Education*, 33(4), 465 - 486.

King, K., Shumow, L. & Lietz, S. (2001). Science education in an urban elementary school: Case studies of teacher beliefs and classroom practices. *Science Education*, 85(2), 89–110.

Kong, S. C. & Song, Y. (2014). The Impact of a Principle-based Pedagogical Design on Inquiry-based Learning in a Seamless Learning Environment in Hong Kong. *Educational Technology & Society*, 17 (2), 127–141.

Lamanauskas, V. (2010). Integrated science education in the context of the constructivism theory: Some important issues. *Problems of Education in the 21st Century*, 25, 5-9.

Latta, M. M., Buck, G., Leslie-Pelecky, D. & Carpenter, L. (2007) Terms of inquiry. *Teachers and Teaching: theory and practice*, 13(1), 21-41.

Lederman, J. S. (2009). Teaching scientific inquiry: Exploration, directed, guided, and opened-ended levels. *National geographic science: Best practices and research base*, 8–20.

Lederman, J. S., Lederman, N. G., Bartos, S. A., Bartels, S. L., Meyer, A. A. & Schwartz, R. S. (2014). Meaningful Assessment of learners' understanding about scientific inquiry-The views about scientific inquiry (VASI) Questionnaire. *Journal of research in science teaching*, 51(1), 65-83.

Lehman, J., George, M., Buchanan, P. & Rush, M. (2006). Preparing Teachers to Use Problem- centered, Inquiry-based Science: Lessons from a Four-Year Professional Development Project. *The Interdisciplinary Journal of Problem-based Learning*, 1(1), 76-99.

Leaman, L. H. & Flanagan, T. M. (2013). Authentic Role-playing as Situated Learning: Reframing teacher education methodology for higher-order thinking. Studying Teacher Education: *A journal of self-study of teacher education practices*, 9(1), 45-61.

Liu, E. Z. F. & Lin, C. H. (2009). Developing evaluative indicators for educational computer games. *British Journal of Educational Technology*, 40(1), 174-178.

Llewellyn, D. (2011). *Differentiated Science Inquiry*. California: Corwin.

Mäeots, M. & Pedaste, M. (2014). The role of general inquiry knowledge in enhancing students' transformative inquiry processes in a Web-based learning environment. *Journal of Baltic Science Education*, 13(3), 19-31.

Magee, P. A. & Wingate, E. (2014). Using inquiry to learn soil: A fourth grade experience. *Science activities*, 51, 89-100.

Martin, L. A. (2010). Relationship between teacher preparedness and inquiry-based instructional practices to students' science achievement: evidence from TIMSS 2007. Doctoral Thesis. Indiana University of Pennsylvania.

Matthews, K. & Hodgson, Y. (2012). The Science Students Skills Inventory: Capturing Graduate Perceptions of their Learning Outcomes. *International Journal of Innovations in Science and Mathematics Education*, 20 (1), 24–43.

Mayer, R. E. (2004). Should There Be a Three-Strikes Rule Against Pure Discovery Learning? *American Psychologist*, 59 (1), 14-19.

Minner, D., Levy, A. J., & Century, J. (2010). Inquiry-based science instructiondwhat is it and does it matter? Results from a research synthesis years 1984 to 2002. *Journal of Research in Science Teaching*, 47(4), 474-496.

National Assembly of the Socialist Republic of Vietnam [NASRV] (2008). Education Law, No. 38/2008/QH11-2008. Hanoi: Education Press.

National Research Council (NRC) (1996). *National science education standards*. Washington, DC: National Academy Press.

National Research Council (NRC) (2000). *Inquiry and National Science Education Standards*. Washington, DC: National Academy Press.

National Research Council (NRC) (2005). *America's lab report: Investigations in high school science*. Washington DC: The National Academy Press.

National Research Council (NRC) (2012). *A Framework for K-12 Science Education*. Washington DC: National Academies Press.

Nott, M. & Wellington, J. (2000). A programme for developing understanding of the nature of science in teacher education. In W. F. McComas (Ed.), *The nature of science in science education*. Netherlands: Springer, 293-312.

Onwu, G. & Stoffels, N. (2005). Instructional functions in large, under-resourced science classes: perspectives of South African teachers. *Perspectives in Education*, 23(3), 79-91.

Orlich, D.C., Harder, R.J., Callahan, R.C., Trevisan, M.S., Brown, A.H. & Miller, D.E. (2013). *Teaching Strategies: A guide to effective instruction*. 10th ed. USA: Wadsworth Cengage learning.

Pea, C.H. (2012). Inquiry-based instruction: Does school environmental context matter? *Science Educator*, 21(1), 37-43.

Pine, J., Aschbacher, P., Roth, E., Jones, M., McPhee, C., Martin, C., Phelps, S., Kyle, S. & Foley, B. (2006). Fifth graders' science inquiry abilities: A comparative study of students in hands-on and textbook curricula. *Journal of Research in Science Teaching*, 43(5), 467–484.

Powell, K. C. & Kalina, C. J. (2009). Cognitive and social constructivism: Developing tools for an effective classroom. *Education*, 130(2), 241-250.

Ramnarain, U. D. (2014). Teachers' perceptions of inquiry-based learning in urban, suburban, township and rural high schools: The context-specificity of science curriculum implementation in South Africa. *Teaching and Teacher Education*, 38, 65-75.

Richardson, G. M. & Liang, L. (2008).The use of inquiry in the development of preservice teacher efficacy in mathematics and science. *Journal of Elementary Science Education*, 20(1), 1- 16.

Richardson, V. (2003). Constructivist pedagogy. *Teachers College Record*, 105(9), 1623-1640.

Rindermann, H. (2007) The g-factor of international cognitive ability comparisons: the homogeneity of results in PISA, TIMSS, PIRLS and IQ-tests across nations. *European Journal of Personality*, 21(5), 667–706.

Roehrig, G. H. (2004). Constraints experienced by beginning secondary science teachers in implementing scientific inquiry lessons. *International Journal of Science Education*, 26(1), 3–24.

Rogan, J. & Aldous, C. (2005). Relationships between the constructs of a theory of curriculum implementation. *Journal of Research in Science Teaching*, 42(3), 313-336.

Ronis D. (2008). *Problem-Based Learning for Math & Science.* Integrated Inquiry and the Internet. 2nd Edn. Thousand Oaks, CA: Corwin.

Roth, K. J., Druker, S. L., Garnier, H. E., Lemmens, M., Chen, C., Kawanaka, T., et al. (2006). *Teaching science in five countries: Results from the TIMSS 1999 Video Study* (NCES 2006-011). U.S. Department of Education, National Center for Education Statistics. Washington, DC: Government Printing Office.

Schwab, J. J. (1960). Enquiry, the science teacher, and the educator. *The Science Teacher*, 27, 6-11.

Schwartz, D. L. Barnsford, J. D.& Sear, D. A. (2005). Efficiency and innovation in transfer. *Transfer of learning from a model multidisciplinary perspective*, Stanford University, 1-52.

Seimears, C. M., Graves, E., Schroyer, M. G. & Staver, J. (2012). How constructivist-based teaching influences students learning science. *Educational Forum*, 76(2), 265-271.

Slavin, R. (2012). *Educational Psycology: Theory and Practice.* 10th ed. New Jersey: Pearson Education Inc.

Snyder, L. G. & Snyder, M. J. (2008). Teaching critical thinking and problem solving skills. *The Delta Pi Epsilon Journal,* 2, 90-99.

Spronken-Smith, R. & Walker, R. (2010). Can inquiry-based learning strengthen the links between teaching and disciplinary research?. *Studies in Higher Education*, 35(6), 723-740.

Tabak, I. & Baumgartner, E. (2004). The teacher as partner: Exploring participant structures, symmetry, and identity work in scaffolding. *Cognition and Instruction*, 22(4), 393-429.

TIMSS (2013). *TIMSS 2011 School Report. Trends in international mathematics and science study.* Dubai: Knowledge and human development authority.

Tsai, C. C. & Chai, C. S. (2012). The "third"-order barrier for technology integration instruction: Implications for teacher education. Building the ICT capacity of the next generation of teachers in Asia. *Australasian Journal of Educational Technology*, 28(6), 1057-1060.

Wan, Z. H. & Wong, S. L. (2013). As an infused or separated theme? Chinese science teacher educators' conceptions of incorporating Nature of Science instruction in the courses of training pre-service teachers. *Science Education International*, 24(1), 33-62.

Wilke, R.R. & Straits, W. J. (2005). Advice for teaching inquiry-based science process skills in the biological sciences. *The American Biology Teacher*, 67(9), 534-540

Wu, H. & Hsieh, C., (2006). Developing Sixth Graders' Inquiry Skills to Construct Explanations in Inquiry-based Learning Environments. *International Journal of Science Education*, 28(11), 1289-1313.

Yager, R. E., & Ackay, H. (2010). The Advantages of an Inquiry Approach for Science Instruction in Middle Grades. *School Science and Mathematics*, 110(1), 5-12.

Yakar, Z. & Baykara, H. (2013). Inquiry-Based Laboratory Practices in a Science Teacher Training Program. *Eurasia Journal of Mathematics, Science & Technology Education*, 10(2), 173-183.

Yore, L, Hand, B., Goldman, S., Hildebrand, G., Osborne, J. & Ti-eagust, D. (2004). New directions in language and science education research. *Reading Research Quarterly*, 39, 347-352.

Zimmerman, C. (2007). The development of scientific thinking skills in elementary and middle school. *Developmental Review*, 27(2), 172-223.

CHAPTER 14

Science Project-Based Learning

Mona Mohamed

ABSTRACT

This chapter presents an in-depth descriptive review of selective empirical research studies that are related to implementing the project-based learning (PBL) approach in science. It also provides main findings and discussion of the significant issues related to implementing project- based learning in science instruction. The review covers the following topics: the definition of project-based learning, the effective implementation of the science PBL in science, the common features of PBL compared to the inquiry-based learning, the models of PBL and their instructors' and students' roles and perceptions.

INTRODUCTION

The Definition of Project Based Learning as an Educational Approach

Project-based learning is considered one of the effective student-centered methods of learning which is established on three constructivist pillars: learning should be done through a specific context, learners accomplish the learning goals through the effective participation in the learning process, and learners understand by interacting effectively and sharing knowledge (Cocco, 2006). Project-based learning is a model that brings life to the classroom. It is definitely based on challenging questions or problems that involve students in design, problem-solving, decision making, or investigative activities so it gives students the opportunity to learn effectively (Jones et al., 1997; Marx et al., 1994). In the project-based learning approach, students build up and direct their own learning through developing their creativity and solving problems cooperatively. The project-based learning is an approach based on students' working individually or in small groups in order to produce products (Ergül & Kargın 2014). This approach supports students in acquiring manual skills in order to learn more by performing original activities (Chen, 2004). Providing the project-based learning environments lead to many advantages for students. It has gradually become widespread especially in lessons where daily life is related more such as science and technology (Ergül, &Kargın, 2014)

The Effectiveness of Applying Project-Based Learning in Science According to Empirical Studies in the Context of Different Countries:

Many studies used a variety of instruments to measure the impacts of project- based learning on the students' skills, attitudes and attainment. A number of studies have explored the effectiveness of project-based learning concerning science education in different countries. These are summarized in the section that follows.

In Oman, Al-Balushi and Al-Aamri (2014) carried out a quasi-experimental study with the students of the twelfth grade. That study aimed to explore the effects of environmental science projects on students' environmental knowledge and attitudes towards science. The experiment was conducted on two classes randomly. The students were divided into an experimental group and a control group. The findings showed that there was a positive relationship between the students' results and engaging in the project- based learning. This means that, the experimental group significantly demonstrated more positive results in the post tests compared to students who received traditional instructions that were mainly based on information recall. The study found that the students' enthusiasm in the experimental group in using new technology to design their products led to more positive results in the post-tests. It could be inferred that the students that engaged in the project-based learning curriculum demonstrated positive benefits in developing content knowledge and thinking skills. That was a quasi-experimental study using a pretest-posttest design. In America, a quasi-experimental study was conducted by Hsu, Van Dyke, Chen and Smith (2015) aiming to investigate the effects of project- based learning environment on the seventh graders. The study aimed to explore the development of argumentation skills and the construction of science knowledge in a graph-oriented computer based on using project-based learning environment. The study indicated the significant impact of implementing project-based learning environment on building students' science knowledge. In another American empirical study, Geier et al. (2008) indicated that 7th and 8th graders that participated in project-based inquiry science units demonstrated a development in science content understanding and achieved better process skills. In addition, more students at project-based schools significantly managed in passing the General Certificate of Secondary Education (GCSE) at the end of the three-year study than those students receiving the traditional instruction. In the context of Taiwan, studies showed that applying project-based learning in terms to STEM (science, technology, engineering and mathematics) curriculum design led to increasing the students' enjoyment, engagement in the project and the ability to combine theory and practice effectively (Lou, Liu, Shih & Tseng, 2011). That study was an in-depth study that aimed to investigate 84 students' cognition, behavioral intentions and attitudes in the project-based STEM environment and involved text analysis and questionnaire survey as the main data collection tools. In Taiwan,

ChanLin's (2008) noted that project-based learning developed the students' skills related to synthesizing, elaborating knowledge and engaging in scientific exploratory work with the use of technology. In Greece, it was also found that project-based learning could be considered as an effective method of instruction with low-achieving students because it enabled the students with low abilities to achieve positive learning outcomes (Koutrouba & Karageorgou, 2013). Similarly, Karaçalli and Korur (2014) conducted a quasi-experimental study in Turkey with science students. The study asserted the positive significant impact of the project- based learning approach on the students' academic achievement and retention of knowledge.

Facilitating Factors and Project Based Learning

Facilitating factors include using technology, teaching resources, providing professional development training and providing equipped buildings. Many empirical studies investigated the relationship between facilitating factors and the implementation of project-based learning instructions. On the basis of empirical findings, modern digital technology is considered as a major factor that enabled students to comfortably participate in the process of designing and developing their projects as they can document the whole process and easily share their creations in a digital format (Patton, 2012). Also integrating technology effectively in the project-based learning helped both of the low and high achievers in constructing knowledge in the project-based learning environment (Erstad, 2002). However, Bell (2010) pointed out that children need to be guided and supported in using technology safely and effectively to acquire creativity skills that technological involvement can offer. The students' group work played a significant role in enhancing collaboration during the project-based learning (Cheng, Lam & Chan, 2008). Furthermore, the team's members should be a high-quality group that shows positive interdependence, individual accountability, equal participation and social skills. However, it was noticed that project-based learning was affected by challenges associated with social class differences, gender and attainment hierarchies to affect power relations among some students in the project team. These challenges could lead to unequal learning. Moreover, Al-Balushi and Al-Aamri (2014) indicated that project-based instruction is not more demanding than traditional instruction in terms of resources and time. Hence, this approach can be implemented with few resources, inside the school building and within the time allocated for the study of particular topics.

Regarding the professional development, Crossouard (2012) claimed that teachers need to be better supported in order to develop more sensitivity towards the social and gendered hierarchies that can often be implicit in pupils' discourse, particularly in relation to peer assessment interactions. Moreover, issues of social equity can become part of the pedagogic focus and the language used in the classroom in order to explore social relations.

The Common Effects of Inquiry Based Learning and Project Based Learning on the Science Learning Outcomes:

According to Barron (1998), research indicated that inquiry-based learning and Project-Based Learning strategies developed communication, problem-solving, and critical thinking skills and improved student achievement (Barron et al., 1998). In the context of learning science, project-based learning gave the students a chance to solve interdisciplinary problems by themselves. Also, it enabled them to respond to activities outside the school environment (Holubova, 2008). It helped students in developing students' perceptions of achievement and reaching instructional goals. Also, it enhanced understanding of learning, studying habits and interactions with others.

Furthermore, Panasan, and Nuangchalerm (2010) concluded that applying project-based and inquiry-based learning activities were appropriately efficient and effective. The students in two groups demonstrated similar learning achievement, science process skills and analytical thinking. Therefore, science teachers could implement both of these teaching methods in organizing activities according to the learners' needs. Project-based learning (PjBL) and problem-based learning are closely related learning methods that share some overlapping features. Both of them are designed to evaluate the student learning, focus is on the application, and integrate the previous acquired knowledge (Uziak, 2016). Both of these educational methods depend on using technology because students use computers to design and produce projects. Additionally, problem-based learning is also considered very successful in lower levels of education, both primary and secondary (Uziak, 2016).

Problem-based learning is based on the problem that is encountered by the students, so it focuses on research and inquiry. In other words, it starts with a problem that demands a solution. On the other hand, project-based learning aims to make "the end product". This means that, the process requires an assignment to be accomplished. That requires certain tasks leading to the production of the final product.

The empirical study indicated that science teachers should understand that constructivist theory can provide meaning to teaching and learning by beginning lessons with what students know and understand (Panasan, & Nuangchalerm, 2010). This means that, teachers are encouraged to implement project-based or problem-based learning activities in order to enhance the students' achievement. Both of these methods helped students in constructing knowledge through solving real world problems based on information obtained during experimentation (Panasan, & Nuangchalerm, 2010).

Many other empirical studies indicated that using project-based learning affected the students' science achievement positively. According to Ergül and Kargın (2014), it was noted that the use of the project-based learning method increased students' science success and supported the students' performance. However, sufficient substructure should be prepared in the selection of the project-based learning method for both the teacher and the student. In other words, every student should be assigned a role according to his or her own ability, when projects are presented.

According to Jalinus and Nabawi (2017), project-based learning PjBL method provides students the opportunity to solve the problems of the real world by creating innovative works through their project work. Using the project- based learning PjBL provides opportunities and experiences for students to solve the problem. Therefore, the students would be ready with experiences and abilities to make decisions about what effort needs to be done to solve the problem in the next project. That would increase the power of thinking ability in solving the problem implications and the students' competency skills. In other words, any problems found can be resolved properly so that the process and the work of students would be improved. Based on these obtained results, implementing project-based learning models can enhance the problem-solving ability and competency skills of students. It was found that project-based learning enabled the students to enhance the soft skills and hard skills, while facing the harsh competition of work. The method developed students' knowledge and skills through making a product which is based on the real world. The students' work was adopted from the real work of the industrial world in accordance with the competency skills being acquired by the students. To conclude, project -based approach must be applied in classrooms in order to produce graduates who are ready to face the world of work. Similarly, Uziak (2016) asserted that integrating project based learning in the applied science in the field of engineering enabled students to be prepared for their professional careers because it helps students to acquire problem-solving and lifelong learning abilities, rather than memorizing prescribed content and simply spoon feeding them. Project-based learning provides students with required skills to face real-life problems. It gives them an effective chance to synthesize the knowledge through participating in the project work. Therefore, it supports the development of life-long skills and students' autonomy. It can be concluded that project-based learning is considered as an effective educational way to fulfill the work place needs.

The Model of Project Based Learning and the Role of Effective Instructors

According to Jalinus and Ramli (2016), there is a developed syntax project -based learning (PjBL) model for vocational education with seven-steps learning process. These steps are: the formulation of expected learning outcomes, understanding the concept of teaching material, skills training, designing project theme, making the project proposal, executing the tasks of project, and presentation of the project report. Additionally, the roles of teachers according to this model are suggested as: the formulation of expected learning outcomes, explaining and discussing with students about the learning outcomes of the course, explaining and discussing with the students about the relevance of competence in relation to the real world, and discussing with students about problem-solving or emerging challenges that are relevant to the scientific community (Jalinus, Nabawi, & Mardin, 2017).

In terms of project- based learning, teachers are not only facilitators, but they also provide students with the required skills. For example, teachers should demonstrate the operation of the required machinery, guide students, give the rational questions, evaluate the work done by the student, and help in designing the project report. At the stage of designing the project report, teachers instruct students to present appropriately the project report and the product (Jalinus, Nabawi, & Mardin, 2017).

Furthermore, instructors demonstrate a major role in forming a study group of students, distributing the study materials for the task discussions and student presentations, instructing students to present the assigned material and guiding students to carry out discussion (Jalinus, Nabawi, & Mardin, 2017).

Teachers should supervise and direct students while applying the project- based approach. For instance, effective instructors discuss problems or challenges that developed in the community, discuss and establish the theme of the project task as an attempt to solve real world problems, divide the tasks on students' group and guiding students to make the project proposal. Effective instructors assist students in making the project proposal. For instance, they instruct students to propose responsibilities of their project, instruct the student to make a proposal project task, and help students in making the proposal framework (Jalinus, Nabawi, & Mardin, 2017).

The proposal framework is usually composed from background information, objectives, working drawings and estimates of production.

Foundations and Criteria of Project Based Learning:

According to Thomas (2000) there are five criteria for identifying characteristics of project-based learning. Firstly, there is a centrality between projects and curriculum. Secondly, the projects should focus on questions or problems that "drive" students to encounter and struggle with the central concepts and principles of a discipline. Thirdly, the central activities of the project must involve the construction of knowledge by students. Fourthly, projects are student-driven to some significant degree. Fifthly, projects are realistic or authentic, not school-like. In the same context, Dori and Tal (2000) suggested that projects and task assignments should include elements of inquiry, argumentations, and authentic, everyday life context. The study results showed that projects that fostered the combination of Web-based exploration of new concepts and theories, multidisciplinary learning, and practicing the traversal of the four levels of chemistry understanding, result in high achievements in learning chemistry.

According to Barak and Dori (2000) the empirical findings indicated that students who participated in the project- based learning performed significantly higher than their classmates not only on the posttest, but also on the course final examination. Incorporating chemistry project- based learning into higher education enhanced and improved student understands of chemical concepts, theories, and molecular representations. The construction of computerized models and Web-based inquiry activities promoted students' ability.

Project based learning (PjBL) is considered as an effective educational approach because it aims to enhance the student engagement and helping them develop deeper understanding of important ideas (Blumenfeld & Krajcik, 2005). In other words, the main purpose of using PBL is to engage students, not by lecturing, but by assigning them a goal and allowing them to work by applying academic work to real situations. It can be noticed that PjBL is against the traditional classroom learning. Krajcik and Blumenfeld (2005) established five central doctrines that represented the foundations for any PjBL lesson plan. These foundations are: the driving question situated Inquiry related to real life, collaboration, using technology and creation of artifacts. Firstly, the driving question can be considered as the main motivation for students to learn. The main aim of the driving question is to attract the students' attention to an interesting and engaging topic. Either the teacher or the students can choose a specific question that motivates discovery about the topic. However, it should also be interesting enough to the students a free opportunity to come up with creative solutions and to answer the related questions in different disciplines (Thomas, 2000). Secondly, in terms of science courses, using a situated inquiry motivates students to engage students in activities that more closely reflect and represent their interactions with the world to truly learn and retain information (Thomas, 2000). This means that, PjBL should be contextualized by linking it to the natural world in order to reflect the natural integration and world complexity. This educational approach is significantly different from conventional learning that provides short term learning activities in the classroom, but it does not often provide the students with the suitable opportunity to make connection between these classroom activities and the real world. Thirdly, project-based learning enhances collaboration between the students through encouraging discussion and group work in order to facilitate learning. This means that it helps in creating a "community of learners" (Blumen & Krajcik, 2005). Applying this educational approach would motivate students to interact and work in groups in order to answer the driving question. The teacher's role should be focusing on coaching, guiding and providing resources. Fourthly, technology must be integrated into classroom while applying the project-based approach. Teachers should distance themselves from the normal classroom format and instead implement a project- based lesson. Computers can also offer students real-time access to learning tools on the internet, which allows students to research and learn at their own pace, as well as software to "present information in dynamic and interactive formats" (Blumen & Krajcik, 2005). Finally, Students learn better when they create 'artifacts', or physical representations of what they learned about the driving question (Blumen & Krajcik, 2005). Students can connect together the knowledge that they have gained through creating a concrete model, game, or other physical item. Creating an artifact could represent the practical answer to the driving question.

The Project Based Learning is Determined by Students' Needs and Teachers' Demands

The traditional classroom teaching assumes that all learners have the same learning style and pace. However, modern learning theories suggest that learners have multiple learning styles and different speeds of acquiring the knowledge (Uziak, 2016). Project based learning considers the different learning paces and team learning. On the other hand, teachers had to implement project- based learning to transfer skills in order to fulfill the demands of the work place. Additionally, teachers adapted the project-based learning to suit the student- centered environment in which tasks are being solved through conducting group discussions (Uziak 2016). Moreover, this approach effectively has enabled teachers to establish an environment of self-directed learners who also receive motivation from the relevance of the project (Uziak, 2016). While using the project -based learning, students are the central part of the learning process. This environment forces the student to perform activities involving research, decision-making and writing (Uziak, 2016). This educational approach managed in attracting the student's attention because the more the task reflects reality, the more the students feel motivated (Uziak, 2016). The approach intensified learning due to engaging students in real-life projects and involving them in active inquiry.

Perspectives About the Advantages and Disadvantages of Project Based Learning in Relation to Teamwork

In project-based learning, students work on academic tasks in small groups. The task can be in the form of an investigation or research on a specific topic. The topic being studied usually integrates concepts from multiple disciplines or fields of study (Blumenfeld et al., 1991). Group members collaborate with one another to produce a collective outcome over a designated period of time. In project-based learning, knowledge is constructed through social interactions. This instructional strategy is part of a revolutionary paradigm shift from traditional to constructivist approaches of teaching and learning (Katz & Chard, 1989). Nowadays, schools are required to teach not only academic subject area knowledge, but also generic skills to students. These skills include collaboration, communication and problem-solving skills. This means that teachers need to shift from implementing teacher-centered approaches to student-centered approaches in order to quip students with the required skills (Cheng & Chan, 2008). It is described as a teaching strategy that aims to enable students to connect knowledge, skills, values and attitudes and to construct knowledge through a variety of learning experiences (Cheng & Chan, 2008).

Despite the popularity of project-based learning in this time of education reform, some researchers have reservations about its effectiveness. For example, Cowie and Berdondini (2001) did not fully advocate the positive claims made by proponents of teaching strategies that involve cooperative group work. They argued that it is difficult for group members to resolve interpersonal problems during cooperative activities. Project-based learning is a teaching strategy that requires collaboration among students in small groups and its positive effects on student learning are determined by the nature of group composition and the quality of group processes.

Concerning grouping in project-based learning, there has been on- going debate on how students should be grouped in learning. Cooperative learning proponents (Slavin, 1995) believed that under heterogeneous grouping, low achievers can get assistance, encouragement and stimulation from high achievers; while high achievers can improve their cognitive abilities and presentation skills through explaining and elaborating concepts to low achievers. Another study by Webb (1982) supported the effectiveness of heterogeneous grouping for both high and low achievers. The researcher argued that when high achievers were grouped homogeneously, they would interact less effectively as they assumed that everyone in the group should have understood the materials. Also, it was argued that when low achievers were grouped homogeneously, they would have insufficient ability to help each other to learn. Consistent with these arguments, there have been studies showing that high, average and low achievers gained equal benefits in heterogeneous grouping (Slavin, 1991; Stevens & Slavin, 1995).

On the other hand, not all studies advocated the assumed superiority of heterogeneous grouping. For example, Robinson (1990) found that while low achievers benefited from heterogeneous grouping, high achievers did not. Fuchs, Fuchs, Hamlett, and Karns (1998) also found that high achievers collaborated less effectively and produced work of lower quality when they worked in heterogeneous rather than in homogeneous grouping. Similarly, Hooper and Hannafin (1988) indicated that the achievement of high achievers in homogeneous groupings increased by approximately 12% when compared with high achievers in heterogeneous groupings. In contrast, the achievement of low achievers in heterogeneous groupings increased by approximately 50% when compared with low achievers in homogeneous groupings. Another study by Hooper and Hannafin (1991) also found that high achievers completed the learning task more efficiently in homogeneous than in heterogeneous groups, while low achievers had more interaction and completed the learning task more efficiently in heterogeneous rather than in homogeneous groups. Furthermore, the superiority of homogeneous grouping was inconsistent across students with different levels of ability. Low achievers performed better in heterogeneous than in homogeneous groups; medium achievers performed better in homogeneous than in heterogeneous groups; and high achievers performed equally well in either homogeneous or heterogeneous groups.

In general, findings about the effects of homogeneous and heterogeneous grouping were varied and inconsistent across studies. As suggested by Webb and Lewis (1988), the nature of intragroup cooperation might be important in determining the effects of grouping. Lou, Abrami, and Spence (2000) have attempted to identify a number of factors that account for the variability in the findings on the effects of grouping. In the current research, we expected that the quality of group processes or group functioning, that is, the quality of interaction among group members, might be a strong predictor of students' learning outcomes in project-based learning.

Cheng and Chan (2008) found that the quality of group processes was a positive predictor, but student achievement was a negative predictor of the discrepancy between collective- and self-efficacy. In general, students would have higher collective efficacy than self-efficacy if they experienced a high quality of group processes. Low achievers would in general have higher collective efficacy than self-efficacy, but high achievers would in general have lower collective efficacy than self-efficacy. Nevertheless, this unfavorable condition for high achievers disappeared once the quality of group processes was taken into consideration. They found an interaction effect between group processes and student achievement on the discrepancy between collective- and self-efficacy. When the quality of group processes was high, both high and low achievers reported higher collective efficacy than self-efficacy. At the group level, the study indicated that none of the group-level variables was related to the discrepancy between collective- and self-efficacy.

Conclusion and Discussion

Based on the reviewed literature, the empirical studies that were conducted in America, Taiwan, Greece, Turkey, and Oman asserted that implementing the project-based learning approach demonstrated positive benefits in building content knowledge and developing thinking skills. Additionally, this approach helped in improving the students' performance especially concerning passing the General Certificate of Secondary Education (GCSE). In terms of STEM (science, technology, engineering and mathematics) curriculum, the project -based teach enabled students to combine theory and practice effectively. It was also found that this approach improved the outcomes of the low achievers. The project- based learning is effective in developing the students' scientific skills and high thinking skills. In other words, this educational approach enabled students to participate significantly in the process of designing and developing their projects. Furthermore, it can be concluded that project-based learning is affected by using technology, teaching resources, professional development training and providing equipped buildings.

Additionally, Project-based learning (PjBL) and inquiry-based learning share many overlapping features. Both of them aim to evaluate the student learning, focus is on the application, and integrate the previous acquired knowledge (Uziak, 2016). Project based learning is an effective educational approach because it develops students' knowledge and skills through making a product which is based on the real world. In short, this educational approach must be implemented in science classrooms in order to produce graduates who are ready to face the work places. In other words, this practical approach contributes in developing life-long learning skills that equip students for the work market.

Moreover, teachers can be described as effective partners in the project -based learning process. Teachers and students cooperate together in discussing problems, establishing the theme of the project task as an attempt to solve real world problems, and dividing the tasks on students' group. It can be concluded that project -based learning is a teaching strategy that enhances and fosters teamwork and effective collaboration between instructors and students. However, some researchers argued that working in groups while implementing the project- based learning could have negative effects on the low and high achievers. This means that, high achievers interacted less effectively as they assumed that everyone in the group should have understood the materials. Also, when low achievers were grouped homogeneously, they had insufficient ability to help each other to learn.

The project- based learning approach is more effective than traditional instruction in increasing academic achievement with lower achieving students, enhancing teamwork, increasing long-term retention, developing 21st-century skills, and increasing satisfaction of students. Project based learning (PjBL) is intentionally adopted and integrated into classrooms as an effective approach of learning through the multiple learning stages. The (PjBL) approach is increasingly applied while learning science because of its effectiveness. For example, it increases the students' engagement level and self -directed learning. In other words, this approach deals with students as independent learners who lead their learning process. These independent learners usually participate collaboratively in real life opportunities in order to achieve specified learning goals. Additionally, the project- based learning approach encourages students to explore real life experiences and challenges. It increases students' motivation and engagement because it helps learners to deal with the curriculum through an authentic problem -solving process. Hence, it enhances the problem -solving skills and the critical thinking methods. The main aim of this educational approach is to build the knowledge content and the ability to transfer learning across different contexts.

Also, it supports "learning by inquiry" because it allows the students to ask questions, to make decisions and to make reflection on their projects. In other words, it enables students to evaluate their own work as independent learners. Furthermore, the project-t based learning is an educational approach that enhances" learning by doing". It engages learners in finding solutions for real life problems and making a project product. Hence, learning gains a profound meaning that extends beyond classrooms as students are equipped with the critical skills to function more effectively in the practical world. Furthermore, the project- based learning approach aims to support the 21st century skills. In other words, this approach enables students to improve high thinking skills, like: solving complex problems, making critical analysis, creativity and evaluating information. Moreover, project -based learning is an educational approach that fosters the group work skills. It forces students to work in groups collaboratively in order to come up with creative solutions, ideas and final projects. On the other hand, instructors who implement project-based learning activities play the roles of facilitators and coaches rather than simply transferring knowledge. They teach their students how to think critically and to make hypotheses in order to obtain information. Hence, this educational approach transformed teachers into co-learners as their students take on a different learning project. This educational approach represented a real move from the result to the process. In other words, the lecturer had been transformed from the classroom main actors and dictators to advisors. To conclude, project- based learning can be considered as a shift of a lecturer from an actor on the stage to a guide on the side.

Recommendations and Suggestions:

Based on the reviewed literature, there are some suggestions about the Implementation of the Project-Based Learning Method (PjBL). The Project-Based Learning Method should be integrated effectively in the science curriculum and lesson plans frequently in order to increase students' attainment and achievement. It aims to assist students in acquiring higher-level thinking skills and making positive contributions in attaining acquisitions. Furthermore, the project- based learning fosters lifelong learning, active learning, learning by doing, acquiring research skills and cooperative learning. This approach encourages students to work in groups effectively aiming to solve problems while implementing projects. According to this educational method, teachers do not decide and predetermined every learning step, but students lead the learning process through engaging in learning independently.

Additionally, this educational approach must be applied widely in order to build learning experiences, while applying the web projects throughout life rather than learning only during the school time. In the same context, the project- based learning approach builds effective research skills through using the real tools that are used in the workplace. It also builds critical thinking skills needed in today's workplace. In other words, students are forced to collect data, explore, create, physically experiment things, and organize information and obtaining information through the Internet. Hence, applying these skills could help students in acquiring knowledge in real deep contexts, rather than the oversimplified textbook techniques. Moreover, this approach supports active learning through applying "learning by doing". In other words, students work in a hands-on mode with the physical science world while implementing well -organized web project. They access to information and get to communicate with new characters from the real world. Therefore, they make a closer relationship to the real-world context of problems and projects. Furthermore, this technique strengthens cooperative learning and active students' engagement. Online projects enable students to acquire new acquaintances and skills through cooperative learning. This helps students to communicate with multiple types of people around the world. While implementing web projects, students learn working and collaborating not only with peers, but with instructors and experts in different fields.

Furthermore, effective instructors should have strong time management while implementing the project- based learning. For instance, they should cooperate with the student groups to assign clear timelines and project plans. Moreover, setting interesting research questions would facilitate the students' missions and encourage the thoughtful work. Also, the instructors and the students must agree on the project rubrics and the work plans before stating the project. Furthermore, instructors must create a positive learning environment that asserts the student's responsibility for his learning process. This positive environment motivates students to learn how to learn. However, teachers are responsible for creating the suitable grouping pattern that encourages effective participation. Also, teachers should play effective roles in monitoring the students 'work and recording evidences of achievement. The assessment process could be through group grades and collecting formative evaluation information from students about the project. Additionally, teachers should provide technological resources for projects. For instance, they should use the Internet effectively in order to guide students to the suitable relevant web sites. Effective instructors should motivate students to cooperate with experts and partners outside the classrooms.

The project-based learning PjBL enables students to become responsible for their own learning. They work as teachers of their peers. They access to new information and apply an effective mode of learning which is called as "learning by discovery" and "Side by Side learning". This approach enhances the interaction and the relationships between students and teachers. They both interact effectively during the process of project designing and implementation. To conclude, project-based learning (PjBL) should be adapted in the modern changing world because it provides an effective alternative to traditional education by shifting the focus of education from what instructors teach to what students learn. Future studies should investigate and explore the challenges that have been facing teachers and students while implementing the project- based learning in the context of science education.

REFERENCES

Al-Balushi, S. M., & Al-Aamri, S. S. (2014). The effect of environmental science projects on students' environmental knowledge and science attitudes. *International Research in Geographical & Environmental Education, 23*(3), 213-227.

Baran, M., & Maskan, A. (2010). The effect of project-based learning on pre-service physics teachers' electrostatic achievements. *Cypriot Journal of Educational Sciences*, 5, 243-257

Blumenfeld, P. C., Soloway, E., Marx, R. W., Krajcik, J. S., Guzdial, M., & Palincsar, A. (1991). Motivating project-based learning: Sustaining the doing, supporting the learning. *Educational Psychologist, 26, 369–398.*

Boubouka, M., & Papanikolaou, K. A. (2013). Alternative assessment methods in technology enhanced project-based learning. *International Journal of Learning Technology, 8(3), 263-296.*

Bell, S. (2010). Project-based learning for the 21st century: skills for the future. *The Clearing House: A Journal of Educational Strategies, Issues and Ideas, 83(2), 39-43.*

ChanLin, L.J. (2008). Technology integration applied to project-based learning in science. *Innovations in Education and Teaching International, 45*(1), 55-65.

Chen, L. (2004). Cooperative Project-Based Learning and Student's Learning Styles on Web Page Development. *Educational Technology Systems, 32(4), 363-375.*

Cheng, R. W., Lam, S., & Chan, C. (2008). When high achievers and low achievers work in the same group: The role of group heterogeneity and processes in project-based learning. *British Journal of Educational Psychology, 78*(2), 205-221.

Cocco, S. (2006). *Student leadership development: the contribution of project-based learning.* Unpublished Master's thesis. Royal Roads University, Victoria, BC.

Cowie, H., & Berdondini, L. (2001). Children's reactions to cooperative group work: A strategy for enhancing peer relationships among bullies, victims and bystanders. *Learning and Instruction, 11, 517–530.*

Crossouard, B. (2012). Absent presences: the recognition of social class and gender dimensions within peer assessment interactions. *British Educational Research Journal, 38(5), 731-748.*

Dori, Y.J. & Tal, R. T. (2000). Formal and informal collaborative projects: Engaging in industry with environmental awareness. *Science Education, 84, 95–113.*

Ergül, N.R., & Kargın, E.K. (2014). The effect of project- based learning on students' science success. *Procedia-Social and Behavioral Sciences, 136, 537-541*

Erstad, O. (2002). Norwegian students using digital artifacts in project-based learning. *Journal of Computer Assisted Learning, 18*(4), 427-437.

Fuchs, L. S., Fuchs, D., Hamlett, C. L., & Karns, K. (1998). High-achieving students' interactions and performance on complex mathematical tasks as a function of homogeneous and heterogeneous pairings. *American Educational Research Journal, 35, 227–267*

Geier, R., Blumenfeld, P.C., Marx, R.W., Krajcik, J.S., Fishman, B. Soloway, E., & Clay-Chambers, J. (2008). Standardized test outcomes for students engaged in inquiry-based science curricula in the context of urban reform. *Journal of Research in Science Teaching*, *45*(8), 922-939.

Holubova, R. (2008). Effective teaching methods project-based learning in physics. *US-China Education Review*, *12*(5), 27-35.

Hooper, S., & Hannafin, M. J. (1991). The effects of group composition on achievement, interaction, and learning efficiency during computer-based cooperative instruction. *Educational Technology, Research and Development, 39, 27–40*

Jalinus, N, & Nabawi, R.A. (2017). Implementation of the PjBL model to enhance problem solving skill and skill competency of community college student. *Journal Pendidikan Vokasi*, *7*(3), 304-311.

Jalinus, N., Nabawi, R.A., & Mardin, A. (2017). *The Seven Steps of Project Based Learning Model to Enhance Productive Competences of Vocational Students.* In International Conference on Technology and Vocational Teachers (ICTVT 2017). Atlantis Press.

Karaçalli, S., & Korur, F. (2014). The effects of project-based learning on students' academic achievement, attitude, and retention of knowledge: the subject of "electricity in our lives". *School Science and Mathematics*, *114*(5), 224-235.

Kokotsaki, D., Menzies, V, & Wiggins, A. (2016). Project-based learning: a review of the literature. *Improving Schools, 19 (3), 267-277.*

Koutrouba, K., & Karageorgou, E. (2013). Cognitive and socio-affective outcomes of project-based learning: Perceptions of Greek second chance school students. *Improving Schools, 16(3), 244-260.*

Krajcik, J. S., & Blumenfeld, P. C. (2005). *In the Cambridge Handbook of the Learning Sciences; Sawyer, R. K., Ed.*; Cambridge Handbooks in Psychology; Cambridge University Press: Cambridge, 317–334.

Lam, S.-F., Cheng, R. W.-y., & Choy, H. C. (2010). School support and teacher motivation to implement project-based learning. *Learning and Instruction, 20(6), 487-497.*

Lou, Y., Abrami, P. C., & Spence, J. C. (2000). Effects of within-class grouping on student achievement: An exploratory model. *Journal of Educational Research, 94, 101–112.*

Lou, S.J., Liu, Y.H., Shih, R.C., & Tseng, K.H. (2011). Effectiveness of on-line STEM project-based learning for female senior high school students. *South African Journal of Education 27, 399-410.*

Lou, Y., Abrami, P. C., & Spence, J. C. (2000). Effects of within-class grouping on student achievement: An exploratory model. *Journal of Educational Research, 94, 101–112.*

MacNeal, G.K., & Shukan, J.L. (2018). *Applying Project Based Learning to Middle School Science Education.* (Doctoral dissertation, Worcester Polytechnic Institute).

Patton, M. (2012). *Work that matters: the teacher's guide to project-based learning.* London: Paul Hamlyn Foundation.

Panasan, M., & Nuangchalerm, P. (2010). Learning outcomes of project-based and ınquiry-based learning activities. *Journal of Social Sciences, 6 (2), 252-255*

Ritz, J.M., & Fan, S.C. (2015). STEM and technology education: International state-of-the-art. *International Journal of Technology and Design Education*, *25*(4), pp.429-451.

Robinson, A. (1990). Cooperation and exploitation? The argument against cooperative learning for talented students. *Journal for the Education of the Gifted, 14, 9–27*

Sanders, M. (2009). STEM, STEM education, STEM mania. *The Technology Teacher, 68(4), 20–26.*

Slavin, R. E. (1987). Ability grouping and student achievement in elementary schools: A best evidence synthesis. *Review of Educational Research, 57, 293–336.*

Slavin, R. E. (1991). Are cooperative learning and untracking harmful to the gifted*? Educational Leadership, 48, 68–71.*

Stevens, R.J., & Slavin, R.E. (1995). The cooperative elementary school: Effects on students' Robinson, A. (1990). Cooperation and exploitation? The argument against cooperative learning for talented students. *Journal for the Education of the Gifted, 14, 9–27*

Thomas, J. W. (2000). A review of research on project-based learning executive summary. San Rafael, CA: The Autodesk Foundation. Retrieved from: http://www.k12reform.org/foundation/pbl/research

Uziak, J. (2016). A project-based learning approach in an engineering curriculum. *Global Journal of Engineering Education*, *18*(2), 119-123.

Webb, N. M. (1982). Peer interaction and learning in small cooperative groups. *Journal of Educational Psychology, 74, 642–655.*

PART 5

CHAPTER 15

Student Conceptions of the Nature of Science

Sufian Forawi

ABSTRACT

The nature of science has evolved in the last century due to the complex and dynamic nature of the scientific knowledge. The United Arab Emirates (UAE) is an active member of the Gulf Cooperation Council (GCC) and may become a strong partner for the European Union (EU) in a number of domains. In its recent country report debriefing, the European Union (EU, 2012) introduced the report by stating that the UAE has the most open and diversified economy in the region. This chapter provides extensive literature review of the nature of science (NOS), its connection to science education, student understanding, pedagogy and inquiry instruction, and professional development. It also presents major results of a study on UAE students' conceptions of NOS. Finally, it concludes by critical discussion and implication of multicultural contexts.

INTRODUCTION

The free, publicly provided education has been a central tenet of the social contract in every country in the Middle East and North Africa since independence (Forawi, 2015). Post-independence governments have significantly expanded their education systems, driven by rapidly expanding youth populations, the need to build nationhood and to establish political legitimacy and popular support for new regimes through making education a fundamental right of citizenship. It is well known that population growth in Arab countries is among the highest in the world, which makes providing basic education a major challenge (Bahout & Cammack, 2020). However, education systems in the region, with a few exceptions, now provide basic education to most children.

One of the most pressing needs in the Gulf Cooperation Council (GCC) countries is to develop citizen competencies to lead the long-awaited development of their nations. While oil development is seen as a key factor for financial and social development, a lasting effect of such development can only be achieved if citizens are adequately prepared in all different walks of life and especially in education and the sciences. The United Arab Emirates (UAE) is an active member of the GCC and is a strong partner for the EU in a number of domains. In her first visit as EU chief officer, Mogherini stated that the UAE, EU work hand in hand on bilateral issues in an excellent manner (Gulf News, 2016).

Recent reforms stress an increased emphasis on the nature of science (NOS) and scientific inquiry. Student understanding of the NOS and scientific inquiry has, in fact, been an agreed upon by most scientists and science educators for the past 90 years (Forawi, 2014). While arguments abound as to what constitutes the NOS, that there is still an acceptable level of generality regarding the NOS that is accessible to K-12 students and relevant to their daily lives. Many researchers asserted that science education had a crucial role in the development and success of nations (Wegerif, Li, & Kaufman, 2015). The way science is taught in the classrooms had a significant impact on students understanding of what is science and how it works. Therefore, science teachers and their students need to have adequate understanding of the NOS. Furthermore, such understanding should couple with proper instructional practices. According to Eltanahy and Forawi (2019), well-informed science teachers who combine implicit and explicit teaching approaches of the NOS proved to be essential for students to develop proper understanding of science and how it works. The National Science Education Standards (NSES) along with the recently developed Next Generation Science Standards (NGSS) emphasized the NOS in their guidelines and performance indicators. While the UAE is incorporating the standard-based education and the international benchmarking of science education, no studies have been conducted in this regard in the country and region.

Nature of science is characterized by subjectivity, creativity, and tentativeness. nature of science as the epistemology of science and scientific knowledge. NOS is subject to human imagination and creativity which is presented in a set of representations and values of scientific knowledge. Accordingly, nature of science represents the values and methods of scientific development (Forawi, 2014). This chapter provides extensive literature review of NOS, its connection to science education, student understanding, pedagogy and inquiry instruction, and professional development. It also presents major results of a study on UAE students' conceptions of NOS. Finally, it concludes by critical discussion and connections of findings and review to UAE context.

Nature of science is embedded in social and cultural ethos. Thus, science is considered a way of thinking that leads to the development of scientific knowledge. Besides, nature of science is the developmental process of science. Lederman (2004, p.303) defines Nature of Science as: Nature of science is the epistemology of science, science as a way of knowing, or the values and beliefs inherent to scientific knowledge or the development of scientific knowledge. There is a connection between NOS and epistemological beliefs (EB), as explained by Michel and Neumann (2014). The epistemological beliefs have a positive impact on developing students' reasoning skills, metacognitive abilities and their learning strategies. Therefore, the EB have a strong effect on student knowledge, especially when they overlap with aspects of NOS. NOS affects positively the student use of problem-solving strategies and scientific reasoning. Consequently, for that reason, many called for explicitly teaching of NOS aspects using generic NOS activities (Brase, 2014; Forawi, 2014). Students can establish an authentic access to NOS once NOS and science concepts are meaningfully interwoven. Various approaches have been discussed to introduce NOS aspects such as modeling (Oh & Oh, 2011), contextualized NOS instruction (Michel & Neumann, 2014), and NOS-oriented metacognitive prompts. For instance, contextualized NOS instruction can be realized through open or guided inquiry, class discussions, and hands-on activities (Akerson et al., 2010).

Brain Psychology and NOS

The history of the brain psychology went back to ancient medical practitioners, such Aristotle, Galen and Rufus who had provided a general physical description of the brain- that its basic structures such as the pia mater and dura mater (the soft and hard layers encasing the brain) and less about function that mental activity seems to reside in it rather than the heart as wrongly claimed by Aristotle. In the Middle Age, the Islamic medical philosopher Avicenna wrote to describe 'the brain' as the house of the "faculty of fantasy," later referring to the five senses. Not till the Renaissance, during the Sixteenth and early seventeenth-century, that anatomists, such Willis and Steno who contributed a great deal to the physical description of the brain terms such as cerebrum, cerebellum and medulla were commonly used but made few significant advances in their understanding of its function and learning.

Now, the development of new imaging technology, (e.g., PET, MRI, CAT, and EEG) now allows us to observe children's brains in action. An even newer area of study, cognitive science, combining psychology, computing science, and neurophysiology, offers interesting and possibly important insight into patterns of learning. We now know that all the neurons of the adult brain are present at birth. During the first six months of life, the synaptic connections between neurons appear at an amazing rate and by the age of six months the child's brain contains more synapses than an adult brain. Neural connections or networks are activated and stabilized as the brain learns. As long as there are undedicated neural connections available, learning is easy. Stimulation from the environment causes 'learning' either by stabilizing existing networks in the brain or by forging new ones. These appear to be the fundamental physical mechanisms by which cognitive capacities develop. This new understanding of brain growth provides for the first time an appreciation of how biology and the environmental content and context are inextricably linked in the very tissue of the developing brain.

The Rutgers Center for Cognitive Science relate how they, along with developmental psychologists and preschool staffs, collaborated on the development of the Preschool Pathways to Science (prePS) approach that takes into account results of recent developmental research. They believe that recent research shows that "preschool children have some potent cognitive competencies and related learning potentials. These include early arithmetic abilities and skills, implicit understanding of cause and effect sequences, pre-literacy 'writing,' and some science knowledge." (Gelman & Brenneman, 2004, p. 150)

A domain-specific, constructivist perspective on learning scientific materials, suggests that since the concepts in the mind and in the world are connected then learning experiences should be conceptually linked as well. Language and concepts related to a domain are connected and since children learn new words at an astonishing rate during these years (preschool) and because proper vocabulary is part and parcel of conceptual growth, appropriate terms related to the concepts, processes, and the language of science should be employed from the onset of a child's science education background. Observe, predict, test, measure, classify, communicate, collaborate, record, theorize, verify, explain, identify, relate, create, and develop are but a few of the vocabulary words of science. Each term plays a part in engagement in the scientific process and, perhaps more importantly, some of the words represent concepts characteristics of the NOS.

Additionally, topics requiring brain functioning, such as adolescent risk-taking behavior have shown connection to scientific knowledge. Researchers (in medical decision-making and public health have come to realize that simply remembering and repeating a number verbatim does not capture how most people represent risk in their minds and brains. The riddle of adolescent risk taking may not be solved any time soon, but we are seeing a great leap in scientific knowledge about this topic, with broad implications for enhancing public health and well-being. This is true nowadays with the Coronavirus pandemic that is seeping most counties in the world, alerting people to such health care, wellbeing, and science awareness.

The Empirical Evidence Based Scientific Knowledge

A scientific theory is a principle derived from facts that explains the collective evidence from observations. Scientific theory is the best explanation of available evidence from empirical investigations at a given time. A law describes the relationship between variables or observed phenomena. So, the laws as descriptive statements for observable phenomena while theories are indirect explanations for those occurrences. On the other hand, a theory provides explanations based on data collected from repeatedly proven observations. Brace (2014) believes that a scientific theory is good in nature when it provides testable predictions. Scientists make predictions for future observations of a particular scientific problem based on scientific theories.

The observations of the natural world provide rich data that enhance science. Park et al. (2014) insist on the urgency of empirical evidence to underpin scientific theories. Such experimental confirmation makes the scientific theories reproducible by other scientists. Science is based on empirical data, yet it is objective. Kuhn's work (1996) formed a solid base for science reformation and philosophy of science. According to Kuhn, the existence of science is tightly connected to observation and empirical experiences. In the first place, scientific inquiry starts with data collection. This data form a body of pure facts which constitute predictions- matching natural observations and the paradigm that lead to articulation of scientific theory or its paradigm.

Different researchers confront the same phenomena in many different ways. Thus, a variety of descriptions and interpretations are available for one specific natural model. Further investigations of results in rigid articulations which strengthen the conditions related to the issue. Kuhn (1996) identifies three factual scientific investigations. First type focuses on collecting precise, reliable and accurate data for the predetermination of facts. Second type compares the collective evidences to the theoretical framework in an attempt to form a cluster. Third type is related to the residual ambiguities and permitting the solution of problems which had previously drawn the attention of other scientists.

All scientific theories are testable. Lederman et al. (2002) state that scientific knowledge can be tested empirically, and it requires the support of empirical evidence. AAAS (1993) declares that the most important criteria in evaluating scientific theory is the empirical evidence, observations, and experiment testability. Irzik and Nola (2011) insist that people underdetermine the significance of empirical evidence of theories. Empirical evidence is required to confirm or disconfirm the scientific theory and further tests its probabilistic or falsifiability status (Kuhn 1996). When theories conflict each other, then empirical evidence is the only used tool to determine their validity and acceptability. Hickey (2005) states that empirical evidence validates theories while the fame of a scientist or society's value does not. Kuhn (1996) insists that the results of observations and experiments should be in complete agreement with a theory.

Creativity plays a major role in providing new explanations for observed phenomena. Although natural world circumstances are observed by many people. Yet, the level of interpretation to explain the observed facts may allow the formation of a new paradigm. Though empirical data is limited, many interpretations and justifications may exist for one single observational outcome. Many theories may explain the same evidence equally well. Scientists may interpret collected artifacts from empirical experience differently which results in diversified outcomes. Thus, science demands imagination and creativity and that scientific discoveries happen as a result of creativity and innovation. Thinking beyond what is usual and common results in new scientific discoveries. Additionally, Forawi (2011) insists that scientific knowledge is a developmental process of human imagination. Creative designs of scientific experiments are required as well as logical explanations.

Following a systematic scientific method assists in testing the scientific theories. Thus, a scientist has to follow a specific investigative approach to reach a substantive conclusion acceptable by the science community. Many scholars (Kuhn, 1996; Macaulay, 2015; Louzek, 2016; Fang, 2014) insist that scientists construct their own scientific beliefs based on their own attitudes of mind and use of scientific methods. Thus, the scientific knowledge becomes embedded in the cultural and social beliefs of the scientist (Gondwe & Longnecker, 2015; Toomey, 2016) and under the direct influence of the imagination and creativity of the scientist (Kuhn, 1996).

Scientific Knowledge

The word tentative means subject to change. Science and scientific knowledge are considered tentative since they are subject to change due to a new discovered input. Thus, the nature of science is tentative and changeable. Park et al. (2014) state that "tentativeness of science means that scientific knowledge is durable but not permanent" (p.1171). Scientific knowledge is reliable though it can be proven as an absolute and final sense at a specific time. It changes due to many factors over time. Scientific knowledge witnesses a change if empirical data emerge. In fact, scientists reinterpret data differently due to advanced analysis tools or social and cultural principles changes. Louzek (2016) believes that scientific knowledge is not owned by anyone, however, it is a result of an infinite number of examinations done by scientists.

Some researchers believe that scientific laws and ideas are absolutely true, and they are never tentative to change. For instance, Popper (1963) insists on rejecting the confirmation of any scientific theory even if it is supported by empirical evidence. He states that scientific theory is not confirmed absolutely; yet if an experiment does not falsify a theory then the theory becomes validated. According to Popper, scientific theories are vulnerable and strongly affected by the falsification. Once one single item of counter evidence is detected, then the science communities become doubtful about the truthfulness of the theory.

Kuhn (1996) investigated the history of science and focused on the paradigm shift theory to present his view about the tentative nature of scientific knowledge. The history of science witnessed continuous changes in scientific theories and various paradigm shifts. According to Kuhn (1996), the same theoretical assumptions that form the base of the work of a number of scientists may lead to paradigm formation. In reality, science has been done differently from how it should have been done.

Abnormal observations and experimental results that threaten a paradigm of normal science attract scientists to focus their investigation on testing the dominant paradigm. To be more precise, the newly conducted experiments lead to scientific revolution as well as the establishment of a new paradigm which replaces the existing paradigm. As a matter of fact, new paradigm cannot coexist with existing paradigm. When a new paradigm is established, it takes over the current one. Scientific knowledge continuously develop, it is neither permanent nor unalterable (Kuhn, 1996). Many other researchers (Forawi, 2014; Lakatos, 1970) confirm that scientific knowledge is neither fixed nor true.

Realists and anti-realists have different points of views about the tentative nature of science. They both believe that scientific knowledge is tentative and not fixed. However, Realists believe that a theory changes when new data is available (Park et al., 2014). Correspondingly, Anti-realists consider the tentative nature of scientific knowledge a normal process. Park et al. (2014) state that anti-realists consider a new theory as an alternative point of view about natural events. Empirical investigations provide empirical data to measure accurately the observed phenomena. Thus, empirical data is definitive but subject to human imagination and accuracy. Empirical evidence sets the base for a new discovery (Louzek, 2016; Park et al., 2014).

Once observational data is obtained, scientists start conducting thorough investigations to consolidate their findings. Correspondingly, investigations allow for possible theory formation to explain the observed phenomena (Louzek, 2016; Park et al., 2014; Irzik, Nola, 2011). Observations and experimentations yielded to the foundation of the new discovery. Toomey (2016) considers the experiment evidence an essential requirement to confirm that the new theory is true. Hence, an absolute fact is a result of investigation, observation or experiment, experimental testability, and possible formation of conclusion, law or theory.

Implicit NOS

Science is a methodical approach to study the natural world. Science continuously seeks answering questions using experimental tests, interpretation through logic and observations. As defined by Macaulay (2015, p.77) that science is a discipline, which is systematic and comprehensive. It is difficult to give a universally accepted definition of science. Additionally, various methods are considered as scientific. It is therefore well-known that one scientific uniform method is not used in the sciences. Macaulay (2015) considers the scientific method investigations dependent on the nature of the studied object. Thus, experiments and examination are done using a scientific method to obtain a supporting evidence to defend the theory. Hence, as defined by Macaulay (2015, p.79):

> "the term "scientific method" refers to not really the method which the scientists use in their discoveries or experimentation. Rather, it refers to an attitude of the mind. It refers to the sense of rigor, rationality, objectivity, thoroughness, incisiveness (or detail), logic, consistency and coherence with which their inquiry is carried out."

The scientific method is a way of using comparative critical thinking to learn about nature. Only testable or falsifiable things are considered science. The scientific method is neither about the steps nor about the sequence of procedures. It does not matter which type of investigation comes first. The essence of a scientific method is the elements used to carry on with the investigations. Macaulay (2015) states that the scientific method is a definite route of carrying out scientific reasoning. Furthermore, the variety of utilized forms add a lot of value to the scientific method. Some of those processes are observations, predictions, evaluation, analysis, and experimentations.

Engaging students in inquiry activities foster the development of adequate understanding of NOS (Torres et al., 2015). Students learn best when they are actively involved in their learning. inquiry fosters student learning about the process of science and nature of science. Inquiry approach can serve as a medium of instruction to consolidate the NOS understanding. For instance, models are powerful tools used by scientists to develop scientific knowledge. The involvement of students in constructing models as a representation of a reality develop scientific knowledge. Modelling activities in science classes contribute to the understanding of scientific inquiry and many other aspects of NOS (Torres et al., 2015; Oh, and Oh, 2011; Giere, 2010). Modelling activities gives a meaningful dimension for learning. Oh, and Oh (2011) consider modeling a useful method to demonstrate how things work and to explain complicated scientific concepts.

The process of scientific inquiry does not follow fixed key steps sequence. Thus, there isn't fixed set of steps that all scientists follow. Yet, the scientific investigations have to involve relevant evidence, logical explanations and the use of logical reasoning. Various scientific inquiry methods can be used, and all may lead to successful results. Lawson (2009) proposed the hypothetical-deductive thinking as a scientific method. His proposal is based on the need for a hypothesis to give direction to data collections and thus leading deductively to specific conclusions. Although one universal definition of scientific method does not exist, yet science is recognized as one distinct discipline uniquely separated from all other disciplines.

Irzik and Nola (2011) call for a family resemblance approach to NOS that is based on investigating similarities and differences among scientific disciplines. The family resemblance approach presents "nature of science as fixed and timeless". This approach is known as the consensus view of nature of science. Scientific literacy is a critical component of learning that prepare students to resolve scientific, societal and personal issues (Smith et al., 2012). Science education focuses on the learning to do science, the learning about science and the learning of science.

It is widely believed that unless science teachers recognize scientific enterprise it will be difficult for them to assist their students to gain a sound understanding of it. In one of our studies (Forawi, 2014), two major criteria advocated by the major science education reform agendas, the conceptions of the nature of science and the use of inquiry instruction were investigated. Both of these deserve more attention, especially for pre-service elementary teachers as well as in-service teachers and students. One result of that study indicated that participating pre-service elementary teachers' conceptions of the NOS have generally shown a slight instrumentalist view. Pre-service elementary teachers developed a better understanding about views of the nature of science while using "hands-on" inquiry teaching approach. This result provided a good rationale for the present study that explicit teaching of NOS coupled with implicit inquiry instruction can improve learners' NOS conceptions.

The implicit NOS view presented above echoed Hodson's (1998) model which presents three pillars of science education an in the figure below.

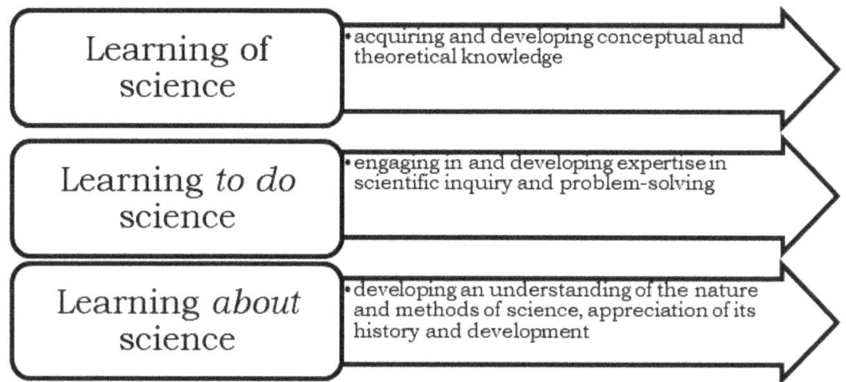

Figure 1: Hodson's (1998) Dimensions of Science Education

Science education is seen as a result of the proper development of scientific literacy and NOS understanding. Many scholars conclude that students possess as inadequate view of NOS (Forawi, 2014; Torres et al., 2015). NOS is not emphasized in science textbooks (Forawi, 2010) and teachers do not give too much attention to the construction of their students' NOS principles in designing the learning activities (Buaraphan, 2012). NOS understanding has a direct impact on the learning process. Understanding NOS permits the development of critical thinking, intellectual excitement and argumentative skills. Thus, students become engaged in deep discussions. Students grasp the key features of scientific phenomena through their engagement in deep discussions related to NOS (Lederman et al., 2002). Finally, carrying on inquiry process allow students to design explorations but they might be unable to gain insightful ideas about NOS. NOS views do not necessary develop by an inquiry-based science class (Abd-El-Khalick, 2013). Therefore, Smith, et al., (2012) identify three approaches to teach NOS: Implicit approach, explicit and reflective approach, and historical approach.

As a result, NOS instruction can better be discussed under the combination of the three approaches as earlier suggested by our research, interaction among teachers NOS, inquiry instruction (implicit NOS) and textbooks (explicit NOS) (Forawi, 2003). Akerson et al. (2010) stated that contextualized NOS instruction can take many forms such as reviewing historical case studies, class discussion, open or guided inquiry and hands-on activities. A teacher plays the role of a moderator for discussions in such an instructional setting. Clough (2006) states that in a contextualized NOS, the teachers' role is to moderate discussions about scientific actions, generic NOS activities and historical or modern approaches. The only evidence for the success of the contextualized NOS approach is the way students use NOS understanding to learn science content. The discipline of NOS seeks to find answers related to defining science and scientific knowledge, the origin of the discovery of scientific laws and theories as well as the inventions done by scientists. McComas et al., (1998) in Torres et al. (2015, p.4) defines NOS as:

> NOS describes how science works, what science is, how scientists operate and how science relates with society, merging aspects of history, sociology, philosophy and psychology of science.

UAE Students' Conceptions of NOS: The Study

The following sections presents development of the questionnaire and its results of the study that is based on an internal funded grant project. The complete study followed a quantitative design approach by use of newly, developed questionnaire on the nature of science (NOSQ). The postpositivistic philosophy foundation for such design relates to postpositivism. This questionnaire makes a needed contribution to instrumentation of the nature of science in the UAE and the Middle East context. Population of the study is based on K-12 students in the diverse UAE country. Convenience sampling technique is followed in this study to include selected elementary (302), middle (408) and high school (1870) students from the seven Emirates, Abu Dubai, Dubai, Sharjah, Ajman, Ras Al-Khaimah, Umm Al-Quwain, and Al Fujairah. The present study incorporated only the high school sample of 1870 students, 860 boys and 1010 girls.

Development of Nature of Science Questionnaire (NOSQ)

The Nature of Science questionnaire (NOSQ) consisted of three parts. Firstly, part A focused on the demographic information. This part included six items about: gender, grade level, date of birth, mother tongue language, nationality and language spoken at home all the time. Secondly, part b constituted of items with 5-point Likert scales with six sections: scientific theory, scientific knowledge, scientific inquiry, scientific learning, science connections, and science values and culture. Items on the scales were anchored at (SA)=strongly agree, (A)=agree, (N)=don't have an opinion, (D)=disagree, (SD)= = strongly disagree. There were three versions of the NOSQ: Early Years version, middle school version and high school version. The items in part b were customized to be age-appropriate. Thus, there were 24, 30 and 48 items in the early-years version, middle school version, and high school version respectively. Thirdly, part c included two open ended questions to allow students to reflect their belief about science and scientists.

The six aspects of NOS were used into designing the scales in the NOSQ. 1- Scientific Theory: Students study about scientific theories to understand natural phenomena with characteristics of changeability, imagination, and experimental evidence. 2- Scientific Knowledge: Students explore scientific knowledge through understanding its tentativeness, subjectivity to experimental tests, ethical judgment, moral judgment, and reliability. 3- Scientific Inquiry: Students are actively engaged in scientific investigations to draw generalizations. In the same context, students formulate their conceptions about scientists' investigations to obtain accurate results, reveal reality and draw generalizable conclusions. 4- Science Learning: The value of science learning is to engage students in an inquiry activity that stimulate their own thinking to find the relevance of science with their daily life experiences. 5- Science Connections: The learning environment in the science class sets the scene for students about science learning and the learning process. Universal laws, real life experiences and human curiosity mark science learning. 6- Science Values and Culture: Perceiving the connection between scientific researches, social and cultural values reflect how students understand the dynamics of the scientific research.

After establishing the six scales, a pilot study was conducted to validate the items of the NOSQ. Few revisions were made based on validity construct checks by expert for clarity, appropriateness and importance. Reliability with use of Cronbach was check and instrument found to be reliable (0.723). Arabic translation was done for the government school sample using the backward method between the Arabic and English languages.

Results and Discussion

The main result of the study indicated statistical significant difference with the sample size selected and use of t test, T = -19.390, P-value = 0.000. This means that participating K-12 students had an overall agreed conception of science based on the Nature of Science questionnaire (NOSQ). The agreed conceptions of the newly, contextualized, and developed NOS questionnaire, closer to the average 4, shows that UAE students demonstrating contemporary understanding of the nature of science.

There is a statistically significant difference in overall participating students' conceptions of science indicating that to the importance of aspects of NOS based on the questionnaire. Additionally, female students were found to indicate a higher mean than male students (t = 6.614, p-value = 0.000), as in table below.

Table 1: Students' Conceptions of NOS and their Clusters

Statement	**Mean**	**Standard deviation**	**Test**
Overall sample	3.7	0.67	One sample t test T = -19.390 P-value = 0.000
Overall sample by gender	M = 3.59 F = 3.79	M = 0.698 F = 0.627	Independent sample t test T = - 6.614 P-value = 0.000
Overall sample by nationality	UAE = 3.66 GCC = 3.90 Other Arab Asia = 3.81 Other Arab Africa = 3.97	UAE = 0.69 GCC = 0.52 Other Arab Asia = 0.59 Other Arab Africa = 0.49 Europe & America = 0.42 Asia = 0.71	One-way ANOVA F = 5.973 P-value = 0.000

Statement	Mean	Standard deviation	Test
	Europe & America = 3.94 Asia = 3.50 Africa = 3.84	Africa = 0.56	
Overall sample by mother tongue	Arabic = 3.69 Africa = 3.50 Europe & America = 3.78 Hindi = 3.89 Asian = 4.00	Arabic = 0.67 Africa = 0.71 Europe & America = 0.51 Hindi = 0.56 Asian = 0.58	One-way ANOVA F = 1.436 P-value = 0.220
Overall sample by language home	Arabic = 3.69 English = 3.88 Hindi = 3.94 Asian = 4.00	Arabic = 0.67 English = 0.54 Hindi = 0.61 Asian = 0.00	One-way ANOVA F = 3.252 P-value = 0.021
Overall sample by grade	12 = 3.64 High = 3.66 Univ = 3.70	12 = 0.74 High = 0.70 Univ = 0.70	One-way ANOVA F = 0.387 P-value = 0.679

understanding that varies from one person to another. Considering the history of the development of understanding, science topics from childhood to adolescence illustrate clearly the tentative nature of scientific knowledge. Although scientific knowledge develops the belief that scientific knowledge increases and progresses based on new scientific discoveries or deep analysis of existing information. Thus, tentatively is a crucial component that allows individuals to recognize the nature of science. Yet, gender difference warrants further investigation.

There is also a third statistical significant difference in the overall responses of the NOSQ by nationality, using ANOVA test, F = 5.973, P-value = 0.000. The order shows the following descending nationalities by the use of Tukey HSD: Other Arab Africa – Europe & America – GCC – Asia – Other Arab Asia – UAE – Africa. Kuhn (1996) claims that personal considerations of a scientist affects his/her determination to solve and confront issues. It is the culture which sets the level of complexity of scientific processes. Social restrictions might reject new attempts and restrict the major margins. Toomey (2016) believes that society and science are interrelated. Science is directly or indirectly affected by the values and beliefs of the people in a particular society. For one thing, society is triggered by science and scientific discoveries. Philosophers have different points of view about the relationships between science, cultural and social influence (Gondwe, Longnecker, 2015; Popper, 1963; Kuhn, 1996; Toomey, 2016). Universalists believe that scientific knowledge remains the same across history and culture. Whereas realists consider scientific knowledge a product of a particular social and cultural context. As in the present study, students' nationality seem to influence their understanding of what science is, how scientific knowledge develops, and how science works.

Interestingly, there is a significant difference found for overall participating students' NOSQ scores by the language spoken at home (F = 1.436, P-value = 0.220). Order of descending languages by using the Tukey HSD is as follows: Asian, Hindi,

English, and Arabic, despite the large difference in sample size for the Arab participants. Students become skillful when they are able to analyze, synthesis and apply the abstract knowledge they learned to solve word problems. Having the scientific knowledge alone is not enough. Lederman (2007) states that although many factors influence student's learning and achievements but curriculum structure and instructions has the greatest impact. Additionally, the present study contributed another factor that is the students' language spoken at home. Further investigation is warranted to explore the impact of language spoken at home and culture of families to their children understands of the nature of science.

There are no statistical significant differences found of the NOSQ, regarding participants' grade levels and mother tongue. These indicate that UAE students' conceptions of science don't differ by different high school grade levels, 10 to 12. Also, despite the great diversity in terms of about 200 nationalities that live in the country, participating students' spoken language at home did not influence the overall statistical significance in the study.

Conclusions and Implications

For a century, the nature of science has been acknowledged as a common objective for science education and seemed as central to developing scientific literacy in recent reforms. There is not one specific definition of the nature of science (NOS). However, it generally refers to the representations of epistemological and values of scientific development. Additionally, several of the NOS characteristics are stated by many researchers as indicated in this chapter. Also, the history and nature of science has been a major chapter of K-12 science texts.

The instrumentalist represents scientific knowledge as a product of human imagination and creativity. The instrumentalist view allows us to make inferences and construct arbitrary models to explain the behavior of physical phenomena." The realist view considers theories to be taken at face value and despite their transcendence of empirical data. However, the instrumentalist view corrects this epistemological difficulty that realist runs into it, by acknowledging the fact that the scientific theories have an irreducible role, but not a straightforwardly descriptive one. Theories, according to the instrumentalist view, have an important role and are scaffolding for the only descriptive parts of science, namely, those statements which are directly checkable by observation. Theories, then, are either empirically adequate or empirically inadequate; they are not, however, either true or false descriptions of the world. The instrumentalist view is based on the empirical, observable data. It has been noted earlier by Albert Einstein (1934, p.14) that "pure logical thinking cannot yield us any knowledge of the empirical world; all knowledge of reality starts from experience and ends in it." Science is 'self-critical,' or able to be criticized, but many non-scientists don't realize that most empirical propositions, if in demonstrable error, can be corrected, because scientists have a 'reference point' that is open to scrutiny and therefore highly validated.

To connect the extensive review of NOS studies and the study results on UAE students, the author provide a missing framework in this section. Let us return to the idea from the previous discussion of Gelman's (2004) brain psychology that 'concepts in the mind and in the world are connected,' and 'learning experiences should be conceptually linked as well.' Science education practices have historically been guided by what the child did not know or could not conceptualize. Small bits of information about many subjects were introduced in seemingly no particular order and without clear explanations of the concepts behind the facts yet children were expected to learn the information, internalize it, and over the years organize it into a comprehensive, seamless, and understandable picture of the world of science (Forawi, 2014).

Lederman (1999) stated, in his exemplary article, that the subject matter without context, discusses science education at the pre-college and undergraduate levels. He explains the Nature of science and scientific inquiry and says:

> 'understanding the NOS and scientific inquiry provide a guiding framework and context for scientific knowledge. Without an understanding of how scientific knowledge is derived and the implications the process of derivation has for the status and limitations of the knowledge, all students can ever hope to achieve is knowledge without context. Context is necessary for student to understand what the knowledge means.' (Lederman, 1999, para. 3)

Combining the thoughts of Gelman and Lederman, seems logical not only attest to the fact that preschool and early elementary science students might be capable of understanding the basic principles of the NOS, to K-12 students' better understanding of NOS and the need to benefit from its inclusion in science curriculum. Similarly, the UAE can benefit from such research and findings to improve their science education and its curricula. A basic understanding of the explicit NOS might actually be the perfect foundation for a science education a perfect frame for the puzzle. This what I suggested in a paper presented at the Oxford Round Table in 2006 and continued to be appropriate in the present context. The main concept behind the model is to incorporate the NOS discourse, as appropriate, starting from early childhood grades, PreK to grade 3 and continue to provide complex explicit aspects of NOS as progress to high school. This should provide a strong foundation on which science knowledge, skills, and dispositions are developed in meaningful ways. Additionally, the NOS discourse may implicitly be provided based on the nature of grade/level curriculum and student readiness. This satisfies the current understanding of constructivist as the dominant psychological theory of learning and further pave the way to more developments, i.e., brain and neuropsychology learning.

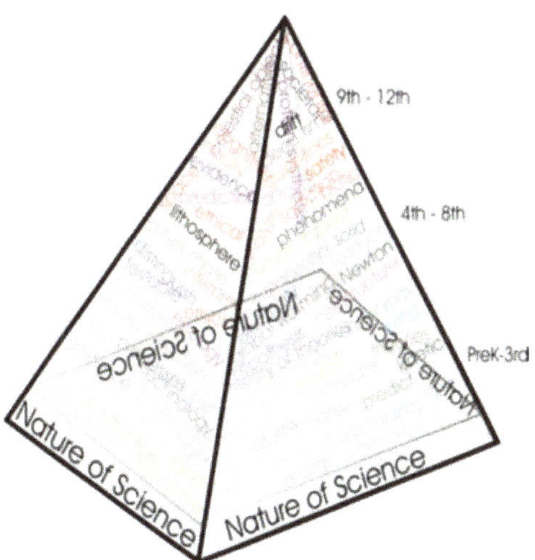

Figure 2: The Upright NOS Model. (Forawi, 2006)

The challenge of teaching the NOS to students has continued to be existed in the literature review. Recommendations to teachers presented to have an 'informed position' to include both implicit and explicit discourse of NOS (Forawi, 201) and that NOS should be a significant item in science classroom and textbooks. We must continue to develop curriculum models that focus on the NOS and insist that textbook authors weave NOS lessons through the content chapters instead of relating it to the introduction. Science curriculum is a momentous factor in impacting students' understanding of the nature of science. As one implication, it becomes imperative that the development of science curricula and textbooks should incorporate research findings, such as the ones found by this study, if the desire is to promote students' learning and understanding to include the NOS.

From the result of this study, students' understanding of the nature of science can no longer be linked separately to only the teachers' conceptions of the nature of science, or the instruction choice, or the type of textbooks used, but to the effective combination of such factors and the student's won cognitive ability. As suggested by Gelman, et al., (2004) that since the concepts in the mind and in the world are connected then learning experiences should be conceptually linked as well, the NOS then need to be made explicit in science curriculum, especially at the early childhood level. Unless teachers and science developers incorporate views of the nature of science into the content of science texts and instruction, many of the benefits anticipated may be lost.

Culture and society affect the development of human enterprise. Science exists in every culture. Students in UAE held diverse views of the nature of science that are similar to reported conceptions, yet show difference regarding ethic, cultural and language background. Shizha (2006) insists that there exists a unique science for each culture. Science is under the influence of social and cultural ethos. Kuhn (1996) states that scientists and science development are results of culture. Every culture has its own unique structure. Gondwe and Longnecker (2015) state that every culture possesses its own knowledge and beliefs concerning the natural world. Confirming Gondwe and Longnecker claim, Toomey (2016) insists that the boundaries that determine where science ends and where culture starts does not exist. Humans existing in different cultures have many ways of constructing reality. Toomey (2016) states that the process of scientific knowledge production is directly affected by the perceptions and beliefs of the members of the society. Cultural values and beliefs are embedded in the human thoughts of its population. In particular, some scientists tend to prioritize their cultural believes over their scientific thinking. Main results of the study indicated that high school students in UAE tend to have contemporary views of science as measured by the newly developed NOSQ and that female students responded higher than their counterparts. These results are supported by El-Sayary, Forawi, and Mansour (2015) who reported on science education reform in STEM in high schools in the UAE. One research recommendation would be the need to investigate closely factors affecting UAE students' conceptions. Also, replicating the study with other different student samples, e.g., college. A practical recommendation would be to train teachers to explicitly provide NOS instruction to improve their students' conception and learning.

REFERENCES

Abd-El-Khalick, F. (2013). Teaching with and about nature of science, and science teacher knowledge domains. *Science & Education, 22*(9), 2087-2107.

Akerson, V. L., Weiland, I., Pongsanon, K., & Nargund, V. (2010). Evidence-based strategies for teaching nature of science to young children. *Journal of Kirsehir Education Faculty, 11*(4), 61–78.

American Association for the Advancement of Science (AAAS) (1993). *Benchmarks for science literacy.* Oxford University Press.

Bahout, J. & Cammack, P. (2020). Arab Political Economy: Pathways for Equitable Growth. Carnegie Endowment for Peace. https://carnegieendowment.org/2018/10/09/arab-political-economy-pathways-for-equitable-growth-pub-77416

Brace, G. (2014). The nature of thinking, shallow and deep. *Frontiers in Psychology, 5*, 1-7

Buaraphan, K. (2012). Embedding Nature of Science in Teaching About Astronomy and Space. *Journal of Science Education and Technology, 21*, 353-369.

El-Sayary, A, Forawi, S. &, Mansour, N. (2015). STEM education and problem-based learning. In R. Wegerif, L. Li, & J. Kaufman (Eds.). *The Routledge international handbook of research on teaching thinking* (pp.357-363). Routledge Publisher.

Eltanahy, M & Forawi, S. (2019). Science Teachers' and Students' Perceptions Regarding the Implementation of Inquiry Instruction in a Middle School in Dubai. *Journal of Education*,1-11.

Forawi, S. A. (2015). Science Teacher Professional Development Needs in the United Arab Emirates. In N., Mansour & S. Al-Shamrani (Eds.). *Science education in the Arab Gulf States: Visions, sociocultural contexts and challenges* (pp. 25-32). Sense Publishers. Rotterdam, the Netherlands.

Forawi, S. (2014). Impact of Explicit Teaching of the Nature of Science on Young Children. *The international Journal of Science, Mathematics and Technology Learning, 20*, 41-49

Forawi, S. (2011). Inquiry instruction and the nature of science: How are they interconnected. *Journal of Science Education, 1*(12), 14-17.

Forawi, S. A. (2010). Impact of Teachers' Conceptions of Science and use of Textbooks on Students. *The International Journal of Learning, 17*(5). 281-294.

Forawi, S. A. (2006). The upright pyramid: Is there room for the nature of science at the early childhood level? *The Oxford Round Table Proceedings, 11*, 65-72.

Forawi, S. A. (Winter 2002/2003). Effects of contributing factors to students' understanding of the nature of science. The *Education Researcher, 18*, 2, 15-31.

Gelman, R., & Brenneman, K. (2004). Science learning pathways for young children. *Early Childhood Research Quarterly, 19*, 150-158.

Giere, R. N. (2010). An agent-based conception of models and scientific representation. *Synthese, 172*, 269-281.

Gondwe, M., & Longnecker, N. (2015). Scientific and Cultural Knowledge in Intercultural Science Education: Student Perceptions of Common Ground. *Research in Science Education, 45*(1), 117-147.

Hodson, D. (1998). *Teaching and Learning Science: Towards a personalized approach.* Buckingham: Open University Press.

Irzik, G., & Nola, R. (2011). A family resemblance approach to the nature of science for science education. *Science & Education, 20*, 591-607.

Kuhn, T.S. (1996) *The Structure of Scientific Revolutions.* University of Chicago Press.

Lakatos, I. (1978). *The Methodology of Scientific Research Programmes: Philosophical Papers Volume 1.* Cambridge University Press

Lawson, A. E. (2009). On the hypothetico-deductive nature of science—Darwin's finches. *Science & Education, 18*, 119-124.

Lederman, N. G. (2007). Nature of science: Past, present, and future. In S. K. Abell & N. G. Lederman (Eds.). *Handbook of research on science education.* Mahwah, N.J: Lawrence Erlbaum Associates.

Lederman, N.G. (1999). Teachers' understanding of the nature of science and classroom practice: Factors that facilitate or impede the relationship. *Journal of Research in Science Teaching, 36,* 916-929.

Lederman, N.G., Abd-El-Khalick, F., Bell, R.L, & Schwartz, R. (2002). Views of nature of science questionnaire (VNOS): Toward valid and meaningful assessment of learners' conceptions of nature of science. *Journal of Research in Science Teaching, 39*(6), 497-521.

Loužek, M. (2016) 'The economic approach to science', *Prague Economic Papers*, 1–12.

Macaulay, K. (2015). The limitations of science: A philosophical critique of scientific method. *IOSR Journal of Humanities and Social Science, 20*(7), 77-87.

McComas, W.F., Clough, M.P., & Almazroa, H. (1998). The nature of science in science education: An introduction. *Science and Education, 7*, 511-532.

Michel, H., & Neumann, I. (2014). Nature of Science and Science Content Learning: Can NOS Instruction Help Students Develop a Better Understanding of the Energy Concept? *Paper presented at the international conference of the National Association of Research in Science Teaching (NARST), Pittsburgh, Pennsylvania. March 31st, 2014*

Gulf News (2016). European Union mission inaugurated in Abu Dhabi. https://gulfnews.com/uae/government/european-union-mission-inaugurated-in-abu-dhabi-1.1932577

National Research Council (1996). *National science education standards*. National Academy Press.

Oh, P. S., & Oh, S. J. (2011). What teachers of science need to know about models: An overview. *International Journal of Science Education, 33,* 1109-1130.

Park, H., Nielsen, W., & Woodruff, E. (2014). Students' conceptions of the nature of science: perspectives from Canadian and Korean middle school students. *Science & Education*, *23*(5), 1169-1196.

Popper, K. (1963). Conjectures and refutations. In. M. Curd & J.A. Cover (Eds.). *Philosophy of Science: The central issues* (pp.3-10). W.W. Norton & Company.

Sadler, T. D., Chambers, F. W., & Zeidler, D. L. (2004). Student conceptualizations of the nature of science in response to a socioscientific issue. *International Journal of Science Education, 26*(4), 387–409

Shizha, E. (2006). Legitimizing indigenous knowledge in Zimbabwe: A theoretical analysis of postcolonial school knowledge and its colonial legacy. *Journal of Contemporary Issues in Education, 1*(1), 20–35.

Smith, K. V., Loughran, J., Berry, A., & Dimitrakopoulos, C. (2012). Developing Scientific Literacy in a Primary School, *International Journal of Science Education, 34,* 127-152.

Steno, N. (1669). The brain, the masterpiece of creation, is almost unknown to us. https://web.stanford.edu/class/history13/earlysciencelab/body/brainpages/brain.html

Toomey, A.H. (2016) 'What happens at the gap between knowledge and practice? Spaces of encounter and misencounter between environmental scientists and local people', *Ecology and Society*, *21*(2).

Torres, J., Moutinho, S., & Vasconelos, C. (2015). Nature of Science, Scientific and Geoscience Models: Examining Students and Teachers' Views. *Journal of Turkish Science Education*, *12*(4), 3-21.

CHAPTER 16

Science Cooperative Learning Strategies

Mina Radhwan

ABSTRACT

This chapter presents a comprehensive review of the literature related to cooperative learning strategies and the context of science. Cooperative learning is one of the most significant teaching strategies to have a substantial positive effect on developing student learning skills. The chapter has a closer look at the use of science cooperative learning strategies that play a vital role in promoting students' academic knowledge and social skills. Particular attention is paid to the historical and theoretical research backgrounds that shed light on the meaning of cooperative learning, followed by the effects of cooperative learning strategies in science education. The pedagogy of cooperative learning and its influence on effective analysis of learning habits is discussed as well. Finally, the benefits of using cooperative learning strategies in science classes are interpreted to help teachers in their learning process.

INTRODUCTION

Autonomous, cooperative learning (CL) represents the essential factors that improve the latest teaching methods (Teng Z et al., 2015). Cooperative learning is one of the most accomplished methods to play an important role in the area of instructional modernization (Slavin, 1999). Since 1980, it has been an active pedagogical method and its importance has increased especially in the 21st Century as one of the appreciated instruments in educational institutions during modern times (Johnson et al., 2007) CL affords welfare for both schoolchildren and teachers (Shimazoe and Aldrich, 2010). Tsay and Brady (2010, p.78) confirmed that cooperative learning has a significant effect in students' learning through their interaction with each other. In other studies during the last fifteen years, carried out by (Cohen, 1994; Gillies, 2006; Johnson & Johnson, 2009; Slavin, 1996; Tsay & Brady, 2010) they reinforced the role of cooperative leaning in developing positive self-image, community skills and academic success along with the schoolroom atmosphere. Moreover, CL has multi-dimensional advantages, for instance improving students' interactive personal skills in addition to developing low-achievers' self-confidence by sharing different responsibilities (Joyce et al., 1992). Additionally, schoolchildren's mental capabilities will increase towards making better progress along with motivation (Kim et al., 2012) as well as their societal progress (Gillies, 2004; Jordan & Le Metaias, 1997).

Cooperative Learning (CL)

CL is an education that is grounded on a two people or a small-group approach towards schooling that grasps learners responsible for individual as well as team achievement. CL arranges for distinctive learning capabilities designed for students in addition to offering an alternative to competing models of teaching (Johnson and Johnson, 2014). While (Doymus, 2008; Hennessy and Evans, 2006; Johnson et al., 2007; O'Leary and Griggs, 2010) agree that cooperative learning is one of the learning techniques where the students work as a group with the purpose of achieving their targets. Working cooperatively enhances students' thoughts towards knowledge through their engagement with peers or in small groups (Lafont et al., 2007; O'Leary and Griggs, 2010). Slavin (1995) considered cooperative learning to be one of the most important strategies to represent an ideal of collaborating groups, building knowledge and clarifying any misunderstanding in students' thoughtfulness. According to Cheng (2010), CL is a method where a small group of individuals argue and communicate about their own personal ideas in order to accomplish the goal of the lesson. In this way, each student has to have a role in the activity in addition to respecting others' opinions. Therefore, under these conditions students will be able to structure knowledge in different ways by improving their work in a cooperative manner. Akinbobola, (2006) confirmed the previous definitions and he assumed that the schoolchildren with dissimilar levels of skill could participate in achieving the objectives of the topic.

There are many benefits of cooperative learning rather than solely the educational outcomes. One of them is promoting students' social skills positively in addition to the cognitive benefits (Willis, 2007). According to Vygotsky's opinion (1987), human participation will help in understanding the progression of a learner through the use of their mental and socio-cultural system. These processes, especially the social ones, have a great effect on the students' development through their interactive cooperative skills. In this situation, there will be a significant value of social constructivism in supporting the implementation of working collaboratively, which will lead to developing the students' success attainment and creativity along with their communications skills in science classes (Johnson & Johnson, 1994; Hacklinget al., 2007).

Working cooperatively represents a model of social life through cooperation where the individuals of the community are continuously interacting with each other in order to survive, for that reason these social skills will be valuable in solving some of the problems that students might face in the future. As a result of these social skills, students will be able to define themselves, take advantages from others' opinions and develop their personality along with their national consciousness positively so they can serve their country as effective citizens (Özer Aytekin and Saban, 2013). Additionally, CL strategies have a significant role in stimulating the aims of second language acquisition. This can be achieved via the use of the five necessary elements of cooperative learning activities (Johnson and Johnson, 1999).

Another benefit of using CL strategies which was demonstrated by Meng (2010) who explained the importance of the Jigsaw cooperative learning strategy in understanding many different science experiments as well as the students' interest towards English as a subject. Meng (2010) confirmed that this strategy improved the students' ability to read English correctly and to stimulate their motivation to do so. This cooperative learning strategy is one of the best active techniques that represents a student-centered, teacher-assisted, affirmative interdependent communication even for the students who study foreign language in Universities. The same strategy had been used by other science teachers where they confirmed the benefits of using CL in their chemistry classes. These teachers concluded that there is a recognizable improvement in the students' academic level through using this strategy compared to the traditional method of teaching (Yasemin et al., 2010). Moreover, in a previous study which was done by Gradel and Edson (2010) in which they highlighted the importance of cooperative learning strategies in addressing both education and knowledge challenges that take place in schools. CL is not only important for students, but for teachers as well, because one of the important studies which has been carried out by (Angela and Rylee, 2013) whereby they support the essential use for a deep implanting of CL pattern language in many of the tutor preparation and professional development programs as well as focusing on the continuous challenge of translating a learning theory into actual training in a large amount of schools.

Cooperative learning has a great effect, not only for schoolchildren and adolescents, but even for college students. Thus a number of authors have tried to enhance CL in some classes of engineering because it has the ability to improve learning implications (Smith et al., 2005; Terenzini, 2001; Pimmel, 2001; Haller et al., 2000; Hsiung, 2010; Kaufman et al., 2000; Ohland et al., 2005; Prince, 2004). According to Slavin (1990) there are many methods and strategies that represent CL such as Group Investigation (GI), Student Teams-Achievement Divisions (STAD), Academic Controversy (AC), Problem-based learning (PBL), Think–pair–share, Teams-Games Tournament (TGT), Team Assisted Individualization (TAI) in addition to Jigsaw, Cooperative Integrated Reading and Composition (CIRC) and more. All these approaches help the students to work together as a group towards a better understanding.

The Effects of Cooperative Learning Strategies in Science Education

Cooperative learning strategies play a vital role in promoting student's knowledge as well as their societal skills which are related to conventional/traditional whole tutorial methods of education (Adeyemi, 2002; Kolawole, 2008; Adesoji and Ibraheem, 2009).

CL administers greater knowledge outcomes in science classes through improving learners' thinking and constructing their thoughts about science content via building up and spreading the topic's knowledge (Lin, 2006). In addition to stimulating learner engagement as well as improving their scientific thinking progressions, particularly for young schoolchildren who are under eighteen years old (Souvignier & Kronenberger, 2007). Many studies illustrate the advancement that students achieved in science classes through their cooperation and collaboration together. These accomplishments include schoolchildren's creative thinking and community skills (Johnson & Johnson, 1994; Hackling et al., 2007).

In other research, carried out by Saka (2010) indicates that cooperation skills between the students which take place before the teacher's intervention have the capability to improve student's aptitudes in learning science by themselves through their arguments and cooperation. Learning through arguments and conversations lead practitioners on the way to success in science knowledge procedure (Chin & Osborne, 2008). In addition to providing deep learning via developing the students' abilities as well as their thoughts, then using these skills in different steps in the scientific methods and other approaches. These results, which are related to working cooperatively and to guided discussion in science classes are confirmed in many different studies such as those carried out by several authers (Hogan et al., 2000; Zohar & Dori, 2003; Liang & Gabel, 2005; Schaal & Bogner, 2005; Dymond & Bentz, 2006; Hourigan, 2006; Cammarata & Tedick, 2007; Fitzgerald et al., 2008).

Andrew and Alexandria (2015) confirmed the schoolchildren's positive outcomes by using Think-Pair-Share cooperative learning strategy. Using this method, students are able to discuss and solve different types of questions with sufficient time, which will help the students to improve their learning skills, (especially the low achievers) in addition to reducing their anxiety in answering challenging questions. Furthermore, CL has the aptitude to be responsible for providing more opportunities to improve better collaborative abilities, interactive skills plus enhancing the pupils' achievement in the direction of a higher level (Bobbette, 2012). CL strategies are considered as one of the most effective ways in understanding physics topics. Schoolchildren's skills are developed by using many of these strategies along with the help from stakeholders in coaching the students especially at secondary school. This method showed a noticeable improvement for the students in physics classes (Adetunji, 2010).

Demirci (2010) proved as a result of his research study that teaching by using cooperative learning methods progressed the students' attitude as well as their accomplishment towards science. Demirci's results have been confirmed with other studies by Candaş Karababa's (2009) and Doymuş (2009). Additionally, these instructional strategies will also increase the students' motivation successfully towards a better result in the future (Huang et al., 2010). The human societal survival, which builds positively through working in groups will be beneficial for every individual in a group and this will lead to a better performance for the students in different topics (McLeish, 2009). The students' interaction helps them in sharing the knowledge of the lesson, consequently the reciprocal dialogues between the students will support them in clarifying their thoughts in different ways of thinking (Gillies, 2004).

The Pedagogy of Cooperative Learning

Pedagogy is one of the main factors that are correlated to the quality of teaching and learning along with the faculty recruitment and the quality of education (Abaalkhail and Irani, 2012). Quality pedagogy has a great effect on the requirements of quality education because it measures the necessities that are of paramount importance. Henard and Roseveare (2012) clarified quality teaching as a practice of pedagogical skills in order to generate the appropriate educational results for the learners. These methods subsequently produce active and positive teaching outcomes. In pedagogy, there is significant attention given towards a particular schooling methodology, which many educators are searching for in order to get the best out of learning. This can happen through the construction of students' understanding when they are vigorously and enthusiastically involved in a topic (Gradel and Edson, 2010) and (Mestre and Cocking, 2002). CL symbolises one of the teaching and learning strategies. It started three hundred years ago and has since developed through the years to become one of the most important techniques that many tutors are searching for. According to Sharan (1980), CL is presented as a number of different instructional and educational plans, in which a small number of students are sitting and cooperating in a group with the purpose of understanding a topic. In this process, students will give attention to the knowledge of each member of the group, in addition to collaborating together in the direction of working for a specific shared target (Sharan, 1980; Johnson & Johnson, 1999). Slavin (1980, p.315) defined cooperative learning as, "the term that refers to classroom techniques in which students work on learning activities in small groups and receive rewards or recognition based on their group's performance." From another point of view, Cooper and Mueck (1990) specified CL as a guarantee system that targets the essential education environment through a particular instructional policy where small groups of different levels of students are working together to grasp a common objective.

Roon et al., (1983) clarified that cooperative learning is one of the most important education structures where students' aims for success are positively associated. This strategy has the ability to promote creative thinking where the students produce new thoughts and schemes as well as solutions (Johnson and Johnson, 1989). In addition to the positive effect that had been evidenced by Roon through using CL strategies in the courses that he implemented in his laboratory for his students that he taught in their first year (Roon et al., 1983). It is an investigator's opinion that CL has the capacity to increase the students' intellectual capability. Felder (1996) realized that many of the learners come to be so trained in the direction of working in teams and the gained data that was implemented adapted into other courses. In the same study, Felder concluded the high average of the gained knowledge through the use of cooperative learning strategies. He confirmed via his research study that this procedure had the goal of altering the ethics of the students' work through relying on pedagogies of engagement between the students, which is considered as one of the important factors for CL strategy where the teacher's role was represented in developing the learners' experience (Felder and Brent, 2005). Sharan (1980) proved in his research study that the procedure of collaboration and cooperation by connecting to each other in the same group has a great positive effect on both the educational acquisition for the students in addition to their cognitive leaning as well. Moreover, the self-evaluation maintenance model (Tesser and Campbell, 1982) evidenced that the students developed an attitude towards comparing their performance and critical thinking skills with other classmates in their cooperative group. This will positively promote the behaviour of the students.

One of the benefits of this strategy is the knowledge and the understanding that the students can gain in the journey of education. In this case, teachers will play a vital role towards providing the students with a framework via clear and direct training of social interface, allocating duties along with the essential sub-tasks when they are working as a group (Hartman, 2002). Additionally, CL confirms the significance of "communicative capacity" (Lovat, 2005), which has the ability to construct a strong connection between the members of the group depending on a trust established between them by inspiring each other. In the near future, this allows children to explore the thoughts and views for them and for their classmates as well (Lovat & Toomey, 2007). Large studies were released and accordingly revised the literature that is related to CL (Slavin, 2014; 2008, Johnson and Johnson, 2014; 2009). As a result of all these research studies, there is significant evidence about the positive learning outcomes for the students, which inspire the teachers in all schools to put the cooperative learning strategy into operation. Specific results came from Johnson & Johnson (1989) when they illustrated in their study the improvement of the students' attitudes towards professors and topic parts as well. Furthermore, students will have the capacity to reach a higher level of critical thinking skills along with understanding.

CL is considered to be one of the common methods that is connected to effective pedagogy, which had been ascertained in the research study by Mina and Miranda (2010) whereby they showed that involving the students in the cooperative learning strategy represents a robust predictor of a learner's academic achievement. Moreover, there is an important progressive relationship between the students' degree and his/her academic achievement with the positive contribution in cooperative learning strategy in the schoolrooms, which symbolises as a strong indication of the learner's performance on willingness valuation exams. According to Williams and Sheridan (2006) cooperation is indicated as a crucial component of "pedagogical quality." Therefore, teachers must think seriously about implementing the cooperative learning practices in their preparation for their classes in order for students to able to focus on these positive skills in the direction of progression.

The Influences of Cooperative Learning on Effective Studying of Learning

Science has become a part of our daily life. People nowadays are paying a lot of attention to developing scientific culture from its first stages where the researchers are involved in it, to the changing and growing stages where individuals can benefit from it (Gransard-Desmond, 2015). Therefore, science is important not only within school time but outside of it as well, where it can produce the potential to create an environment in which schoolchildren have the capacity to participate in different activities that are focused on the scientific field, linked to their endless interests (Polman and Hope, 2012). Science plays a vital part in several different countries where they use it to teach younger (preschool) children because of its importance. This enables children to gain an understanding of many concepts and ideas, which are related to scientific language. As a result of this, children will have a great opportunity to stimulate the growth of scientific notions on the way to scientific thinking (Andersson and Gullberg, 2012). According to a research study which had been clarified by Eshach (2006) and Sjøberg (2000) in which they evidenced that there is an interest and enjoyment, especially for young children in discovering, examining and studying their environments. This interest will help them to improve their abilities in the future along with developing their skills towards differentiation, discovering, documenting, and posing inquiries that are related to discussing science. (Lpfo¨98 revised 2010, p. 10).

There are various strong indications that cooperative learning has an active effect in developing and improving the gained skills and results for the students as well as increasing the value of the academic learning (Gillies, 2003; Johnson et al., 1981; Johnson & Johnson, 1994; Johnson et al., 2000b; Slavin, 1995, 1996). Through working as a team in order to accomplish shared aims (O'Leary and Griggs, 2010) and searching for clarifications, understanding, definitions and constructing a product by working in pairs or in small groups (Erlandson et al., 2010). An important result which had been collected by (Yildirim & Girgin, 2012), they substantiated that the teaching method which relies on working cooperatively and creating student-centered groups has an additional positive outcome on the learners' academic success compared to the teaching method which depends on following the teacher-centered method. In this case, schoolchildren will be able to maintain sustainability as well as improved performance particularly in their homework and tests (Hsiung, 2012). This strategy can also develop the harmony between the students in the schoolroom along with their self-esteem and it will be easier for the learners to connect with the topic (Li, 2012). One of the strategies that showed a significant progress in students' outcomes is the "Think-Pair-Share" strategy where the learners can make a positive advantage from thinking and working cooperatively in pairs then sharing the results which will help them to save time as well as explaining the findings to their peers. Furthermore, it will decrease the students' anxiety, particularly for the learners who are below the average level of success (Andrew and Alexandria, 2015). Another strategy of CL called Jigsaw, has the ability to enable all the students to become experts in a particular part of the topic within the group. Then each learner has to take the position of the teacher to clarify and explain his/her part of the lesson for other learners in the same group. As a result of this strategy, students will gain the benefits of varied skill, knowledge and an understanding of group work (Slavin et al., 2003) along with avoiding a lot of other difficulties for the learners that are working in a team (Doymus et al., 2010). Karacop and Doymus (2012) indicated the positive outcomes of increasing the academic achievement for the learners by using this strategy especially in chemistry classes. Problem-solving strategy is another successful strategy that has been proved to have a lot of advantages towards students' learning. A proven finding, which had been collected by Femi (2010) found the vital positive effect on learners' achievement at secondary level in science and practically in physics through using this strategy. The results from this examination indicated that all physics teachers must use CL strategy in their education process. Additionally, it is recommended that teachers of other subjects should implement this strategy in their programs. Cooperative learning strategy has been reinforced also by Tsay and Brady (2010) where they evidenced that there is a noticeable progression in the learners' academic results when they work cooperatively with peers or in small groups. This strategy will help them to obtain a higher score in their assessments and their final tests as well as gaining valued knowledge and experience. According to Saka (2010), using CL along with the guided discussion methods in different science classes afforded the learners with a higher level of educational achievement growth. Thereby gaining the opportunity for eliciting ideas as well as developing the students' qualified abilities in association with the skills that are related to the scientific procedures about active science education. This process of teaching should take place in the Faculty of Education before the real educating in classrooms where Angela and Rylee (2013) found that the level of information and knowledge for each teacher plays an important role in deciding the significant effective factors that will be implemented in the tutorial room and lead to developing the students' learning. Furthermore, training of the faculty members will translate the learning philosophy into an adequate process on a greater range in schools in addition to providing specialized development courses. Orlich et al., (2011) showed that qualified teachers have the ability to provide their learners with different types of teaching strategies with the purpose of developing their academic outcomes. They highlight that, "reflective teachers incorporate social aspects in their instructional planning. They cognitively make the necessary adjustments in their instruction so that all students have an opportunity for success." (Orlich et al., 2011, p 17).

When students work in cooperative teams, they increase their abilities in regards to improving their social competence. This will have a significant effect in improving their educational success as a result of receiving knowledge from other classmates (Bratt, 2008; Lafont et al., 2007; Thurston et al., 2010). Further studies have been implemented in order to find out the effects of using CL strategies on learners' achievement. Effandi & Zanaton (2007) revealed that this strategy has the ability to put the students in an active situation containing more ideas to share together with working and cooperating with each other on the road to completing educational tasks. Additionally, when the learners are working together, they have an opportunity to support other classmates with the aim of improving their personal knowledge and that of others (Johnson & Johnson, 2001; Jolliffe, 2007). Bobbette (2012) focused on increasing learner's appreciation of literature, together with strengthened societal abilities of working cooperatively through the CL approach. He proved that teachers can teach cooperation as a key skill through using specific methods that include age suitable resources addressed on the cooperation subject. Consequently, there will be a noticeable positive effect in enhancing and reinforcing cooperative learning strategies in the schoolroom. Several trained teachers have demonstrated students' development outcomes where they considered these findings as a result of implementing CL strategies in their classrooms (Sears and Pai, 2012). The opinion of several teachers has been surveyed and studied in search of the opportunity of a successful application for CL where the conclusion of this study specified the affirmative possibilities and expectations for the teachers when applying CL (Al Yaseen, 2011).

The Benefits of Using Cooperative Learning Strategies

According to Orlich et al., (2011), the main goal of teaching and learning is to assist and support all schoolchildren to be effective citizens in the future through increasing their positive abilities to gain important skills and talents. This can be achieved by supplying them with the important information, abilities, performances and helpful working practices to support them to acquire an occupation, which will help them to be a part of society's progression. Using the cooperative learning strategy can assist students to become motivated, take part in educational activities and raise their attention span, along with supporting them in acquiring the skills to view different aspects in the surrounding world through other individuals' perspectives (Ebrahim, 2011). Accordingly, the student's capability to empathize with others increases and they will be able to accept and help other students with different needs such as special education and guidance. Moreover, it offers the students lots of opportunities to obtain further information by learning what others are thinking, in addition to assisting them in developing leadership capabilities through respecting other classmates thoughts, accepting differences and debating with other students. Therefore, they will have the ability to communicate with others in a democratic society (Özer et al., 2013). Creating, working cooperatively and evaluating is considered as part of the essential factors required for 21st Century skills where learners can build higher-quality reasoning. By changing their individual effort into group work assists in producing the strategy of critical thinking and problem solving through collaboration and teamwork (Gradel and Edson, 2010). Furthermore, cooperative learning strategy has the ability to provide the students with a great experience by learning from each other's trial and error and the affiliation to a team (Dietz-Uhler and Lanter, 2011). As a consequence, students will be more active in participating in the particular subject in addition to turning the learning context into a student-centered context and this will become a regular instruction in the tutorial room (Schul, 2011). This might aid the students in minimizing their concern and the anxiety of making errors, which will lead to developing their confidence and self-efficacy. As a result of the students assured contributions in class, a healthy educational environment with diversity in the learning method where thoughts and notions can be stated along with developing understanding, imagination, awareness and decision-making talents will be produced (Andrew and Alexandria, 2015). Additionally, this strategy can help learners who are below the advanced level to obtain a higher level of positive skills that are related to critical thinking skills along with problem solving. It assists the teacher in immediately helping students who are suffering from difficulties with the learning process because he/she will more easily notice the learners who need support. (Özer Aytekin and Saban, 2013) The students who learn by using cooperative learning strategy show a positive attitude to progression in their science classes (Demirci, 2010). The author confirmed the effective use of this approach in improving the learners' attitude along with increasing their success. A lot of researchers had established this result in their study (Doymuş et al., 2009; Demir, 2008; Arslan et al., 2006; Doymuş and Şimşek, 2007; Doymuş et al., 2007).

REFERENCES

Abdalkhail, M., and Irani, Z. (2012). A study of influential factors on quality of education. *International journal of Humanities and applied sciences*, 1 (3), 94-97.

Adesoji, F. A. and Ibraheem, T. L. (2009). Effects of student team achievement division strategy and mathematics knowledge on learning outcomes in chemical kinetics. *The Journal of International Social Research*, 2(6): 1-11.

Adetunji, F. (2010). Effects of Problem-Solving and Cooperative Learning Strategies on Senior Secondary School Students' Achievement in Physics. *Journal of Theory and Practice in Education*, 6 (1):235-266. http://eku.comu.edu.tr/index/6/2/faadeoye.pdf

Adeyemi, S. B. (2002). Relative effects of cooperative and individualistic learning strategies on students' declarative and procedural knowledge in map work in Osun state. *Unpublished Ph. D Thesis*. Ibadan, Nigeria: University of Ibadan.

Akinbobola, A. O. (2006). Effects of cooperative and competitive learning strategies on academic performance of students in physics. *Journal of Research in Education*, 3(1): 1-5.

Al-Yaseen, W. (2011). Expectation of a Group of Primary School Teachers Trained on Cooperative Learning on the Possibility of Successful Implementations. *A Journal of Comparative and International Education*. 132(2).

Andersson, K. and Gullberg, A. (2012). What is science in preschool and what do teachers have to know to empower children? *Cult Stud of Sci Educ*, 9(2), 275-296.

Andrew, P. and Alexandria, L. (2015). An Exploratory Study on Using the Think-Pair-Share Cooperative Learning Strategy. *Journal of Mathematical Sciences*, 2(22-28).

Angela, H. and Rylee, A. (2013). Implementing cooperative learning in Australian primary schools: Generalist teachers' perspectives. *Issues in Educational Research*, 23(1).

Arslan, O., Bora, N. D., & Samancı, N. K. (2006). İşbirliğine dayalı öğrenme tekniklerinin 10. sınıf öğrencilerinin sinir sistemi konusunu öğrenmelerine etkisi [The effect of cooperative learning strategies on 10th grade pupils' achievement on nervous system]. *Eğitim Araştırmaları*, 6(23), 1-9.

Bobbette M. Morgan (2012). Teaching Cooperative Learning with Children's Literature. *National Forum of Teacher Education Journal*, 3(22).

Bratt, C. (2008). The jigsaw classroom under test: no effect on intergroup relations evident. *J Comm Appl Soc Psychol* 18:403–419.

Cammarata, L. & Tedick, D. J. (2007). Content-Based Language Teaching with Technology, National Educational Technology Standards for Students Curriculum Series: Foreign Language Units for All Proficiency Levels, 147-188.

Candas-Karababa, Z. (2009). Effects of Cooperative Learning on Prospective Teachers' Achievement and Social Interactions. *Ankara Universitesi Egitim Bilimleri Fakultesi Dergisi*, 36: 32-40.

Cheng, C.W. (2010). The effect of World-Wide-Web on learning achievement with different cooperative learning grouping methods. *Educators and Professional Development*, 12(4), 1-7.

Chin, C. & Osborne, J. (2008, March). Students' questions: a potential resource for teaching and learning science, *Studies in Science Education*, 44(1), 1-39.

Cohen, E. G. (1994). Restructuring in the classroom: Conditions for productive small groups. *Review of Educational Research*, 64, 1–35.

Cohen, E. G. (1994b). Restructuring the classroom: Conditions for productive small groups. *Review of Educational Research*, 64, 1-35.

Cooper, J & Mueck, R. (1990). Student Involvement in Learning: Cooperative Learning and College Instruction. *Journal on Excellence in College Teaching*, 1, 68-76.

Demir, K. (2008) Transformational leadership and collective efficacy: The moderating roles of collaborative culture and teachers' self-efficacy. *Eğitim Araştırmaları-Eurasian Journal of Educational Research, 33, 93-112.*

Demirci, C. (2010). Cooperative learning approach to teaching science. *Egitim Araştırmaları Eurasian Journal of Educational Research,* 40, 36-52.

Dietz-Uhler, B. and Lanter, J. (2011). Perceptions of Group-Led Online Discussions: The Benefits of Cooperative Learning. *Journal of Educational Technology Systems*, 40(4), 381-388.

Doymus, K. (2008) Teaching chemical bonding through jigsaw cooperative learning. *Res Sci Technol Educ* 26(1):47–57.

Doymuş, K. & Şimşek, Ü. (2007). Kimyasal bağların öğretilmesinde jigsaw tekniğinin etkisi ve bu teknik hakkında öğrenci görüşleri [The effect of the teaching of chemical bonds jigsaw techniques and pupil opinions about this technique] *Milli Eğitim Üç Aylık Eğitim ve Sosyal Bilimler Dergisi,* 35 (173), 231-244.

Doymuş, K., Şimşek, Ü., & Karaçöp, A. (2007). Genel kimya laboratuarı dersinde öğrencilerin akademik başarısına, laboratuar malzemelerini tanıma vekullanmasına işbirlikli ve geleneksel öğrenme yönteminin etkisi. [The effect of cooperative learning and traditional method on students' achievements, identifications and use of laboratory equipments in general chemistry laboratory course]. *Eğitim Araştırmaları-Eurasian Journal of Educational Research, 28, 31-43.*

Doymuş, K. Şimşek, Ü., & Karaçöp, A. (2009). The effects of computer animations and cooperative learning methods in micro, macro and symbolic level learning of states of matter. *Eğitim Araştırmaları-Eurasian Journal of Educational Research*, 36, 109-128.

Doymus K, Karacop A, & Simsek U (2010). Effects of jigsaw and animation techniques on students' understanding of concepts and subjects in electrochemistry. Educ Technol Res Develop 58(6), 671–691.

Dymond, S. K. & Bentz, J. L. (2006). Using Digital Videos to Enhance Teacher Preparation, *Teacher Education and Special Education*, 29(2), 98-112.

Ebrahim, A. (2011). The effects of cooperative learning strategies on elementary students' science achievement and social skills in Kuwait. *Int J of Sci and Math Educ*, 10(2), 293-314.

Effandi, Z. & Zanaton, I. (2007). Promoting cooperative learning in science and mathematics education: A Malaysian perspective. *Eurasia Journal of Mathematics, Science & Technology Education*, 3(1), 35–39.

Erlandson, B. E., Nelson, B. C., & Savenye, W. C. (2010). Collaboration modality, cognitive load, and science inquiry learning in virtual inquiry environments. *Education Technology, Research & Development*. https://doi:10.1007/s11423-010-9152-7.

Eshach, H. (2006). *Science literacy in primary schools and pre-schools*. Dordrecht, The Netherlands: Springer

Felder, R. (1996). Active-Inductive-Cooperative Learning: An Instructional Model for Chemistry? *J. Chem. Educ.*, 73(9), 832.

Felder, R. M. and Brent, R. (2005). Effective strategies for cooperative learning. *Journal of Cooperation & Collaboration in College Teaching 10 (2), 69-75.*

Femi A. (2010). Effective of Problem-Solving and Cooperative Learning Strategy on Senior Secondary School Students' Achievement in Physics. *Journal of Theory and Practice in Education*, 6 (1), 235-266.

Fitzgerald, G., Koury, K. & Mitchem, K. (2008). Research on computer-mediated instruction for students with high incidence disabilities, *Educational Computing Research*, 38(2), 201-233.

Gillies, R. (2003). Structuring cooperative group work in classrooms. *International Journal of Educational Research*, 39 (1-2), 35-49.

Gillies, R. M. (2004). The effects of cooperative leaming on junior high school students during small group learning. *Learning and Instruction* 14, 197-213.

Gillies, R. M. (2004). The effects of communication training on teachers' and students' verbal behaviors during cooperative learning. *International Journal of Educational Research*, 41(3), 257-279.

Gillies, R, M. (2006). Teachers' and students' verbal behaviours during cooperative and small-group learning. *The British Journal of Educational Psychology*, 76, 271–287.

Gradel, K. and Edson, A. (2010). Cooperative Learning: Smart Pedagogy and Tools for Online and Hybrid Courses. *Journal of Educational Technology Systems*, 39(2), 193-212.

Gransard-Desmond, J. (2015). Science educators: bridging the gap between the scientific community and society. *World Archaeology*, 47(2), 299-316.

Hackling, M., Peers, S. & Prain, V. (2007, September). Primary connections: Reforming science teaching in Avustralian primary schools, *Teaching Science*, 53(3), 12-16.

Haller, C. R., Gallagher, V. J., Weldon, T. L., & Felder, R. M. (2000). Dynamics of peer education in cooperative learning workgroups. *Journal of Engineering Education*, 89(3), 285–293.

Hartman, H. (2002). Scaffolding & Cooperative Learning. *Human Learning and Instruction* (pp.23-69). New York: City College of City University of New York.

Hénard, F. & Roseveare, D. (2012). Fostering Quality Teaching in Higher Education: Policies and Practices. *An IMHE Guide for Higher Education Institutions,* 7-11.

Hennessy D. & Evans, R. (2006). Small group learning in the community college classroom. *Commun Coll Enterp* 12(1):93–109.

Hourigan, R. (2006, Spring/Summer). The use of the case method to promote reflective thinking in music teacher education. *Applications of Research in Music Education*, 24(2), 33-44.

Hogan, K., Nastasi, B. K. & Pressley, M. (2000). Discourse patterns and collaborative scientific reasoning in peer and teacher-guided discussions. *Cognition and Instruction*, 17, 379–432.

Hsiung, C. M. (2010). Identification of dysfunctional cooperative learning teams based on students' academic achievement. *Journal of Engineering Education*, 99(1), 45–54.

Hsiung, C. (2012). The Effectiveness of Cooperative Learning. *Journal of Engineering Education*, 101(1), 119-137.

Huang, Y. M., Lin, Y. T., & Cheng, S. C. (2010). Effectiveness of a mobile plant learning system in a science curriculum in Taiwanese elementary education. *Computers & Education*, 54(1), 47-58.

Johnson, David, & Johnson, Roger W. (1989). *Cooperation and Competition: Theory and Research* Interaction Book Co, 7208 Cornelia Drive, Edina, MN 55435.

Johnson, D. & Johnson, R. (1989). Cooperation and competition: Theory and research. Medina, MN: Interaction Book Co.

Johnson, D. & Johnson, R. (1994). Learning together and alone: Cooperative, competitive and individualistic learning. Boston, MA: Allyn and Bacon.

Johnson, D., Johnson, R. & Smith (2000b). Constructive controversy: the educative power of intellectual conflict. *Change*, 32 (1), 28-38.

Johnson, D. & Johnson, R. (2014). Cooperative Learning in 21st Century. [Aprendizaje cooperativo en el siglo XXI]. *analesps*, 30(3).

Johnson, D., Maruyama, G., Johnson, R., Nelson, D., & Skon, L. (1981). Effects of cooperative, competitive and individualistic goal structures on achievement. *Psychological Bulletin*, 89, 47-62.

Johnson D.W. & Johnson, R.T. (1989). Leading the Cooperative School, Edina, MN: Interaction Book Company.

Johnson, D.W. & Johnson, R.T. (1999). Learning together and alone: Cooperative, competitive, and individualistic learning (5th ed.). Boston: Allyn and Bacon.

Johnson, D.W. & Johnson, R.T. (1999). Making cooperative learning work. *Theory into Practice*, 38(2), 67-74.

Johnson, D.W. & Johnson, R.T. (1999). 'Building community through cooperative learning'. *Theory into Practice* 38/2: 67–73.

Johnson D.W., Johnson R.T., & Smith K. (2007). The state of cooperative learning in postsecondary and professional settings. *Educ Psychol Rev* 19(1):15–29.

Johnson, D.W., Johnson, R.T. & Smith, K. (2007). The state of cooperative learning in postsecondary and professional settings. *Educational Psychology Review*, 19, 15-29.

Johnson. D. W., & Johnson, R. T. (2009). An educational psychology success story: Social interdependence theory and cooperative learning. *Educational Researcher*, 38, 365-379.

Johnson, R. T. & Johnson, D. W. (1994). An overview of cooperative learning. In J. Thousand, A. Villa & A. Nevin (Eds). *Creativity and Collaborative Learning*. Baltimore: Brookes Press.

Johnson, R. T., & Johnson, D. W. (2001). What is cooperative learning? Minneapolis, MN: Cooperative Learning Center, University of Minnesota. Retrieved from http://www.co-operation.org/pages/cl.html.

Jolliffe, W. (2007). *Cooperative learning in the classroom: Putting it into practice*. Thousand Oaks: Paul Chapman Publishing.

Jordan, D., & Le Metaias. J. (1997). Social skilling through cooperative learning. *Educational Research*, 39, 3-21.

Joyce, B., Weil, M., & Showers, B. (1992). *Models of teaching*. Boston, MA: Allyn and Bacon.

Karacop, A. and Doymus, K. (2012). Effects of Jigsaw Cooperative Learning and Animation Techniques on Students' Understanding of Chemical Bonding and Their Conceptions of the Particulate Nature of Matter. *Journal of Science Education and Technology*, 22(2), 186-203.

Kaufman, D.B., Felder, R.M., & Fuller, H. (2000). Accounting for individual effort in cooperative learning teams. *Journal of Engineering Education*, 89(2), 133–140.

Kim, J., Kim, M. and Svinicki, M. (2012). Situating Students' Motivation in Cooperative Learning Contexts: Proposing Different Levels of Goal Orientations. *The Journal of Experimental Education*, 80(4), 352-385.

Kolawole, E.B. (2008). Effects of competitive and cooperative learning strategies on academic performance of Nigerian students in mathematics. *Educational Research and Review*, 3(1): 33-37.

Lafont L, Proeres M, & Vallet C (2007) Cooperative group learning in a team game: role of verbal exchanges among peers. *Soc Psychol Educ* 10:93–113.

Li, W. (2012). Critical Analysis of Cooperative Learning in Chinese ELT Context. *JLTR*, 3(5).

Liang, L.L. & Gabel, D.L. (2005, August). Effectiveness of a constructivist approach to science instruction for prospective elementary teachers, *International Journal of Science Education*, 27(10), 1143-1162.

Lin, E. (2006). Cooperative Learning in the Science Classroom. *NSTA National Science Teachers Association.*

Lovat, T. (2005). Values education and teachers' work: A quality teaching perspective. Paper presented at the National Values Forum, National Museum, Canberra.

Lovat, T. & Toomey, R. (2007). Values education and quality teaching: the double helix effect. Terrigal: David Barlow Publishing.

Lpfo¨98, La¨roplan fo¨r fo¨rskolan, reviderad. (2010). [National curriculum for the preschool, revised 2010]. Skolverket: The Swedish National Agency for Education.

McLeish, K. (2009). *Attitude of Students Towards Cooperative Learning Methods at Knox Community College: A Descriptive Study*. Liberal Studies University, Jamaica.

Meng, J. (2010). Jigsaw Cooperative Learning in English Reading. *JLTR*, 1(4).

Mestre, J., & Cocking, R. R. (2002). Applying the science of learning to the education of prospective science teachers. In: *Learning Science and the Science of Learning: Science Educators' Essay Collection*, ed. R. W. Bybee, Arlington, VA: National Science Teachers Association Press.

Mina, T. & Miranda, B. (2010). A case study of cooperative learning and communication pedagogy: Does working in teams make a difference? *Journal of the Scholarship of Teaching and Learning*, 10(2), 78 – 89.

Ohland, M. W., Layton, R. A., Loughry, M. L., & Yuhasz, A. G. (2005). *Journal of Engineering Education*, 94(4), 319–326.

O'Leary, N. and Griggs, G. (2010). Researching the pieces of a puzzle: the use of a jigsaw learning approach in the delivery of undergraduate gymnastics. *J Furth High Educ* 34(1):73–81.

Orlich, D.C., Harder., R.J., Callahan, R.C., Trevisan, M.S., Brown, A.H. & Miller, D. (2011). Teaching Strategies: *A Guide to Effective Instruction*. 10th ed. United States of America: Wadsworth, Cengage Learning.

Özer Aytekin, K. and Saban, A. (2013). An evaluation of the use of the cooperative learning method in teaching Turkish at the 4th and 5th grade elementary classes. *International Journal of Academic Research*, 5(1), 84-92.

Pimmel, R. (2001). Cooperative learning instructional activities in a capstone design course. *Journal of Engineering Education*, 90(3), 413–422.

Polman, J., & Hope, J. (2012, March). *Citizen science journalism: A pathway to developing a scientifically literate and engaged public?* Poster session presented at the annual meeting of the National Association for Research in Science Teaching, Indianapolis, IN.

Prince, M. J. (2004). Does active learning work? A review of the research. *Journal of Engineering Education*, 93(3), 223–231.

Roon, R. J., Van Pilsum, J. F., Harris, I., Rosenberg, P., Johnson, R., Liaw, C., & Rosenthal, L. (1983). The experimental use of cooperative learning groups in a biochemistry laboratory course for first-year medical students, *Biochemical Education 11* (1), 12-15.

Saka, A. Z. (2010). Implementation of Cooperative Learning and Guided Discussion Methods in Science Teaching to Improve Professional Skills of Student Teachers. *Journal of Turkish Science Education*, 30-51.

Schaal, S. & Bogner, F. (2005). Human visual perception-learning at workstations, *Journal of Biology Education*, 40(1), 32-37.

Schul, J. (2011). Revisiting an Old Friend: The Practice and Promise of Cooperative Learning for the Twenty-First Century. *The Social Studies*, 102(2), 88-93.

Sears, D. and Pai, H. (2012). Effects of Cooperative Versus Individual Study on Learning and Motivation After Reward-Removal. *The Journal of Experimental Education*, 80(3), 246-262.

Sharan, S. (1980). Cooperative learning in small groups: Recent methods and effects on achievement, attitudes and ethnic relations. *Review of Educational Research*, 50, 241-271.

Shimazoe, J., and Aldrich, H. (2010). Group work can be gratifying: Understanding and overcoming resistance to cooperative learning. *College Teaching*, 58, 52-57.

Slavin, R.E. (1980). Cooperative Learning. *Review of Educational Research*, 50(2), 315 - 342.

Slavin, R. E. (1990). Learning together. *American School Board Journal*, 177, 22–23.

Slavin, R. E. (1990). *Cooperative Learning: Theory, Research, and Practice*. Allyn and Bacon, Toronto, ISBN 0-13-172594-7.

Slavin, R. E. (1995). *Cooperative learning: Theory, research and practice*. (2nd edn). Boston: Allyn and Bacon.

Slavin, R. (1996). Education for all. Exton, PA: Swets & Zeitlinger Publishers.

Slavin, R. E. (1996). Research on cooperative learning and achievement: What we know, what we need to know. *Contemporary Educational Psychology*, 21, 43–69.

Slavin, R. E., and R. Cooper. (1999). Improving intergroup relations: Lessons learned from cooperative learning progress. *Journal of Social Issues* 55 (4): 647–663.

Slavin, R., Hurley E., & Chamberlain A. (2003). Cooperative learning and achievement: theory and research. In: Weiner I and Reedheim D (eds) *Handbook of Psychology: educational psychology*. John Wiley & Sons: New Jersey, 191.

Slavin, R. (2008). Cooperative Learning, Success for All, and Evidence-based Reform in education. *Éducation et didactique*, 2(2), 149-157.

Slavin, R. (2014). Cooperative learning in elementary schools. *Education* 3-13, 43(1), 5-14.

Smith, K. A., Sheppard, S. D., Johnson, D. W., & Johnson, R. T. (2005). Pedagogies of engagement: Classroom-based practices. *Journal of Engineering Education*, 94(1), 87–101.

Souvignier, E. & Kronenberger, J. (2007). Cooperative learning in third graders' jigsaw groups for mathematics and science with and without training, *British Journal of Educational Psycholgy*, 77, 755-771.

Teng Z., Lu Z, & Yang T. (2015). Sports autonomy, cooperation, inquiry learning ability training pathway design research. *Journal of Chemical and Pharmaceutical Research*, 7(4):745-748. Available online www.jocpr.com

Terenzini, P. T. (2001). Collaborative learning vs. lecture/discussion: Students' reported learning gains. *Journal of Engineering Education*, 90(1), 123–130.

Tesser, A. & Campbell, J. (1982). A self-evaluation maintenance approach to school behavior. *Educational psychologist*, 17, 1-13.

Thurston A, Topping, K.J, Tolmie, A, Christie, D, Karagiannidou, E, & Murray, P. (2010). Cooperative learning in science: follow-up from primary to high school. *Int J Sci Educ* 32(4):501–522.

Tsay, M., & Brady, M. (2010). A case study of cooperative learning and communication pedagogy: Does working in teams make a difference? *Journal of the Scholarship of Teaching and Learning*, 10(2), 78 – 89.

Vygotsky, L.S. (1987). *Problems of general psychology including the volume thinking and speech*. New York: Plenum Press.

Williams, P. & Sheridan, S. (2006). Collaboration as one aspect of quality: A perspective of collaboration and pedagogical quality in educational settings. *Scandinavian Journal of Educational Research*, 50 (1), 83-93.

Willis, J. (2007). 'Cooperative learning is a brain turn on'. *Middle School Journal* 38/4: 4–13.

Yasemin, K., Kemal, D., Ataman, K. and Ümit, Ş. (2010). The Effects of Two Cooperative Learning Strategies on the Teaching and Learning of the Topics of Chemical Kinetics. *Journal of Turkish Science Education*, 7(2), 52-65.

Yildirim, B. & Girgin, S. (2012). The Effects of Cooperative Learning Method on the Achievements and Permanence of Knowledge on Genetics Unit Learned by the 8th Grade. *Elementary Education Online*, 11(4), 958-965. http://ilkogretim-online.org.tr

Zohar, A. & Dori Y. J. (2003). Higher order thinking skills and low-achieving students: Are they mutually exclusive? *The Journal of The Learning Sciences*, 12(2), 145–181.

CHAPTER 17

Learning Factors of Health Sciences

Rania S. Alayli

ABSTRACT

Learning Factors of Health Sciences, purpose is to explore the students' perceptions of factors influencing their learning of health sciences. The chapter provides future scholars with tools to promote students' learning of health science education, and to help the preparation of health work force being equipped with the needed knowledge and skills in this dynamic and rapid changing field. The chapter presents answer to the main research question: What factors are influencing students' learning of health sciences? Broad review of literature has been carried out to explore learning approaches used in the field of health sciences, particularly, critically review general issues of health sciences, and career opportunities. The sections of the chapter describe and discuss the approaches to learning, case-based learning (CBL), and critical thinking in health sciences education. The last section sheds light on the barriers to interactive strategies in health sciences education.

INTRODUCTION

Institutional requirements of health sciences education are increasing in the globe. In an attempt to meet the international standards, these institutions tend to put more efforts focusing on faculty and staff training to prepare them to improve education of health sciences (Tekian, Roberts, Batty, & Norcini, 2013). The World Health Organization announced that the need of professional workers in the health care fields is increasing to meet the demands of the changing criteria in the field (World Health Organization [WHO], 2013). It also confirmed that not only the quantity of the workers is needed, but also the competencies. Consequently, what is needed is a combination of innovative, collaborative and student-centered educational strategies to prepare a new generation of health care practitioners with a fundamental scientific knowledge and leadership abilities (Johnson, 2015). To keep them in practice, future health care practitioners need support during their learning journey. The scope of learning process will definitely determine their knowledge and skills (American

Hospital Association [AHA], 2013), identify their abilities to meet competencies, equip them with solid bases, encourage them to continue their education and enhance facing the challenges (Johnson & Johnson, 2016). The requirements to meet these challenges are not only the strategies of teaching that need to be modified, but also the willingness of students to learn, apply new concepts, acquire the required skills and competencies, and display the professional approaches and attitudes essential for the profession (Mello, Alves, & Lemos, 2014).

Due to the rapid growth and altering demands of the health care services, the United Arab Emirates (UAE) is experiencing fast and intense developments in the healthcare systems. UAE universities and colleges are offering health sciences education as fundamental requisites for students to pursue their studies in the health care systems prior to continue their specialized career path.

The purpose of this study is to understand the students' perceptions of factors influencing their learning of health sciences. The findings offer future scholars with tools to promote students' learning of health science education, and to help the preparation of health work force being equipped with the needed knowledge and skills in this dynamic and rapid changing field. The study uses qualitative descriptive interpretive approach to answer the research question:

What factors are influencing students' learning of health sciences?

Literature Review

Broad review of literature has been carried out to understand different learning approaches used in health sciences education. Factors and strategies that influence student's learning in the field have also been included. To adhere to the purpose of this study, only the literature related to general health sciences, and that is offered to prepare undergraduate students for future health related careers was considered.

The following sections reflect the approaches to learning, case-based learning (CBL), and critical thinking in health sciences education. The last section sheds light on the barriers to interactive strategies in health sciences education.

Approaches to Learning

Approaches to learning can be defined as the cognitive, affective and psychosocial behavior characteristics that offer consistent indicators of how learners learn (Mirghani et al., 2014). Therefore, they encompass educational means that assist students to improve their learning and teachers to improve their effective teaching. It is noteworthy that the student's traits, perceptions and the learning environment may change. Thus, the student's approaches to learning are not static and may change as well (Gijbels et al., 2014).

Learners may assume one of the two approaches to learning: surface or deep. Learners who are depending mostly on memorization rather than understanding to pass the assessment in order to finish the course of study are considered adopting the surface approach (Shah et al., 2016). However, those who are keen to understand, discuss, integrate and apply the knowledge are most likely adopting the deep approach. These learners are able to examine new knowledge and connect them to their own, trying to apply it into the needed and unfamiliar contexts (Dolamns et al., 2016).

In the field of health education, many factors may impact whether the student adopt the surface or deep approach to learning. For example, surface approach may be encouraged by inappropriate assessment measures especially if students have different expectations and educational backgrounds (Reitmanova, 2011). In this condition, barriers to learning may be developed (Mirghani et al., 2014). However, the deep learning approach is positively correlated with high academic achievers in the medical field (Kumar & Sethuraman, 2007). This occurs for the reason that learners tend to link contents, principles and evidences altogether to be able to critically analyze and evaluate the knowledge (Loyens et al., 2013).

Case Based Learning

What has been used in health education was the traditional lecture-based learning (McLean, 2016). Scholars in the field recognized the value of integration of clinical practice with the basic preparation of students in classrooms (Eisenstein et al., 2014). They recognized the need to stress the importance of the students' in-depth understanding of related clinical skills rather than just memorizing from books (Galvao et al., 2014). They identified the importance of including of actual clinical cases in the learning process. This is what has been called case-based learning (CBL) (McLean, 2016).

CBL is not a new concept, as it has been used in health education from the beginning of the 20th century (Li et al., 2014). Since that time CBL leads a paradigm shift in the process of learning in health sciences (Khan et al., 2015). The field of health care systems is in need of health practitioners who are demonstrating professional competence, problem solving and decision-making abilities (Smith & Coleman, 2008). It needs practitioners who are able to work efficiently within multi-disciplinary health care teams (Baker et al., 2007). To meet these requirements, case-based approach to learning has been developed with a shift to student–centered and practice-based learning strategies (Dickson, 2010).

For a deeper understanding, McLean (2016, p. 42) concluded that CBL is a "form of inquiry-based learning and fits on the continuum between structured and guided learning." In the same vein, Thistlewaite et al. (2012, p. e422) have identified that "the goal of CBL is to prepare students for clinical practice, through the use of authentic clinical cases. It links theory to practice, through the application of knowledge to the cases, using inquiry-based learning methods".

In CBL, students tend to work in groups and to apply critical thinking to help them reach a decision (Preeti et al., 2013), to solve problems and discover rules by constructing their own knowledge (Bate et al., 2014). This initiates their passion to learn (Khan et al., 2015) and supports the creation of self-directed and lifelong learner (Mansur et al., 2012) that promotes at the other end the patient's safety (Wittich et al., 2010).

On the other hand, the role of the teacher in CBL is the facilitation of students' learning rather than lecturing before them. Facilitators are specialized professionals who have relevant knowledge and skills to use to support students' learning to meet the expected outcomes (Goh, 2014).

Critical Thinking

For decades, psychologists and philosophers have considered critical thinking as an important issue in science education. Scholars have defined critical thinking as thinking and learning skills (Halpern, 1992), reasoning skills (Zohar & Tamir, 1993), and enhancing higher-order thinking skills (Mestre et al., 1992).

In the past, content knowledge was the focus of health science education. Sourced from a variety of theoretical and epistemological aspects (Brookfield, 2012), educators in the field are putting more efforts in favor of recognition of processes of critical thinking (Kahlke & White, 2013).

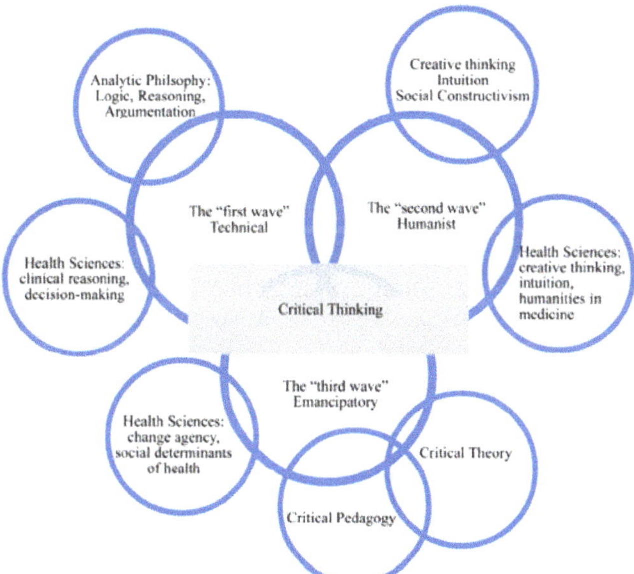

Figure 1: Three traditions in critical thinking. Reprinted from "Critical thinking in health sciences education: Considering three waves," by R. Kahlke & J. White, 2013, *Creative Education*, *4*, p. 22. Retrieved from http://www.scirp.org/journal/ce

In health sciences education, three approaches/ waves to critical thinking are highlighted: technical, humanist (Walters, 1994), and emancipatory (Kahlke & White, 2013). The technical approach/ wave to critical thinking is the prevailing approach in health sciences education (Morrall & Goodman, 2013). This approach considers critical thinking as a focus on instrumental tools that support the teaching of skills, techniques or procedures to enhance the clinical thinking processes (Kahlke & White, 2013) that is openly connected to clinical reasoning, judgment and problem solving (Gambrill, 2012). The second approach to critical thinking referred by Walter (1994) as the second wave and called by McLaren (1994) the humanist approach. It is mainly highlighting the creative and contextual processes to critical thinking (Walters, 1994). Creativity involves knowledge, motivation and intellectual abilities that lead to new scientific innovations (Sternberg, 1999). The third wave (McLaren, 1994) is called emancipatory approach (Kahlke & White, 2013). It reflects the knowledge as naturally constructed and considers the social deconstruction as its main notion. Yanchar et al. (2008) argued that this approach is applicable in health sciences education as it deals with the "identification and evaluation of ideas, particularly implicit assumptions and values, that guide the thinking, decisions, and practices of oneself and others" (p. 270). Figure 1 represents the "three waves" of critical thinking provided by Kahlke and White (2013).

Barriers to Interactive Strategies

Although an immense literature was concerned with the barriers to education, little was dedicated to health sciences. Barriers were mainly focused on students, faculty and theory and practice gap during the health sciences education. Student's barriers were mainly initiated by their expectations, attitudes and traditions (DaRosa et al., 2011). This creates a barrier to teacher's application of critical thinking and interactive teaching strategies (Twenge, 2009), as students prefer the traditional lectures methods over active learning. They consider lectures as easier to memorize, and help them pass their exams (Shell, 2001).

Although studies have shown barriers to effective teaching are the lack of time and formal preparation of faculty members (Voytovich et al., 2008), these findings were expressed by faculty themselves. Another barrier to health sciences education was found to be the faculty's attitude towards the institutional attempts to develop and improve their teaching skills (DaRosa et al., 2011). As some faculty underestimate the benefit of professional development value in enhancing their teaching skills (Skeff et al., 1997). Also, some faculty considers that what they have of clinical experience is enough to become an experienced faculty member (Friedland, 2002).

Another barrier to health sciences education was found to be the gap between theory and practice. Participants in Jahanpour et al. (2016) study reported that they need facilitators in the clinical training who are aware of what is being taught in the classroom and recommended the faculty themselves to be their mentors in clinical areas. As in some countries, faculty are mainly hold Master's or PhD degrees and are hesitant to be involved in clinical education. So, the role of clinical teaching is assigned to health care professionals who are either newly graduates or in distance of academic teaching (Jahanpour et al., 2016). This imposes negative affect on student's learning process and widens the gap between theory and practice.

METHODOLOGY

To fulfill the study aims, the researcher implemented the qualitative descriptive research method with the interpretive phenomenological analysis to explore how students perceive their learning experiences in health sciences.

Research Design

Semi-structured interviews were utilized as data collection tool to offer the researcher the opportunity to obtain in-depth understanding to answer the research question. Interviews were conducted through open ended and flexible conversation (Hanson et al., 2011) to highlight participants' concerns and experiences and to discover the reasons behind these experiences (Thorne et al., 2004). The interview questions are planned to help the researcher to obtain detailed explorations on the students' learning in health sciences. Also, room for explanation was given whenever new ideas were provoked throughout the interview sessions. The questions' content of the interview was obtained mainly from the literature and others have been added based on the researcher's experience. A primary pilot study was carried to find out whether the aimed theme was appropriate and covered.

The study was carried out in one of the colleges of health sciences in the UAE. Participants were students who receive the health sciences course. They are the individuals who are experienced and well-informed about the theme under exploration (Cresswell & Plano Clark, 2011). To obtain rich information from these students, purposeful sampling is considered the most suitable technique (Patton, 2002). However, all students who are enrolled in health sciences course in the college campus were invited to participate. The researcher reflected that the ten students who showed interest to participate in the study are of satisfactory sample size. As for a qualitative study, the sample size is acceptable once the participants are able to adequately answer the interview questions (Marshall, 1996). The interviews were conducted by the researcher over two sessions. Each session lasted 45-60 minutes. It was digitally recorded and transcribed verbatim. Ethical issues about the anonymity and confidentiality were also assured.

Analysis and Findings

Thematic framework analysis method was deemed to be appropriate for the analysis of the data collected. As it is commonly applied in research studies related to the health field and it is suitable to analyze data collected through interviews (Gale et al.,

2013). In this analysis, the researcher tends to report patterns of the data and then identify themes out of these patterns (Braun & Clarke, 2012). The steps of familiarization, coding, theme development, and interpretation were followed. It is noteworthy that the themes reflected all concerns expressed by participants during the interviews. Overall, four themes were developed and reported in the following sections. The students' responses are presented within quotes.

Collaborative Learning

In response to the questions on what students do to learn better. Many participants expressed that their ability to learn better is the presence of their peers. Others also reported that teamwork enhances their creativity. Some mentioned that their colleagues may remind them of ideas mentioned during the class and they were not aware of.

> "I prefer to study with my friends to prepare for the exams".
> "by studying and discussing issues with friends before exams".
> "Ask my friends if they want will study together".
> "I use different ways to study like studying with my friends".
> "study with others teaches us how to communicate with others".
> "group working will help solve problems and I learn how to deal with it".

On the other hand, some students feel that working in teams is time consuming and they prefer to study by their own. They reported that they may ask the teacher if they needed any support.

Others have expressed distraction when they study with others, however, they may offer help if they were asked to.

> "Because I like to work and do everything by my-self".
> "I prefer to study alone to have more knowledge about the content".
> "I am ready to help other students if they ask for help".
> "I will ask the teacher for explanation if I misunderstand anything".

Teacher's Character

When asked how teachers' role influences their learning, majority of participants pointed that the knowledge and experience the teacher has and the way s/he teaches has a lot of influence on their learning. Also, participants displayed that their learning is impacted by the teacher passion to the subject s/he teaches, they feel that this passion is transferred to them. Participants added that a knowledgeable and experienced teacher inspires them to be like her/ him in the future.

> "to give the right information to the student and to make sure that everyone understands".

> "the teacher helps me to do my job in trustful way and we know what we are doing".
> "by doing short quiz at the end of the class or at the beginning of the class will help me remember what we studied before".
> "by supporting students in their learning".
> "she inspires me and give me an idea how I will be in future".
> "Do their best in their job and give students the best learning in better way".
> "I like to be like her and acquire the knowledge she has".
> "to explain all the important points and help the students to achieve the outcomes of the lesson and answer their questions".
> "also use different ways for teaching".

Some participants also expressed that when the teacher is treating them with respect and responsible adults is enhancing their learning. Also, they displayed that the teacher's way of dealing with students in a fair manner is encouraging them to learn as they are not going to be penalized for others' mistakes.

> "To teaching with passion and treat all the students fairly".
> "supporting students and treat them as adult".
> "by showing respect and care about the students".
> "it is good to feel that we are college students and in school".

Student Personal Traits

This theme reflected the participants' perception of strategies they follow to learn better. Participants disclosed that stress and time management is a major concern for them. Responses to how they deal with stress and manage their time, students discussed that they tend to review and study the contents once it is discussed in the class and not at the time of assessments.

> "I feel always that I have no enough time".
> "sometimes I need to quit and change my major".
> "sometimes I cannot tolerate the stress and talk to the teacher for help".
> "I try to manage stress to learn about my mistakes and trying to avoid it"
> "like reading the content directly after the class picking the important key words".
> "I overcome stress by not postponing reading".
> "Also studying on time and catch up on what you have taken, summarize the main point and use different highlights and colors".

Participants reported that their study skill is an important factor that supports their learning.

> "skills like reading, writing it helps".
> "to improve my study skills and abilities to do what is needed".
> "I feel I am responsible about my grades because I can put goals and follow them".
> "short quiz for myself or solve quizzes online".
> "Reading related topics helps me learn more".

Majority of students have stated that their learning styles help them. Some said they learn better by doing or implementing a skill. Others stated that they are visual learners.

> "I learn better by applying what we learn on the real life".
> "watch a video or read something related to the course".
> "I like to do things I learn to feel like I can really do them".
> "I prefer to learn by notes and instructions to follow".

Supportive Environment

This theme is associated with responses on the questions of how the college influences their learning. The majority of the students feel that they have more positive learning experiences due to the provision of resources such as libraries and audiovisual aids. Some students mentioned that the spacious and well-ventilated classrooms make them feel comfortable during their classes.

> "providing books and library time help me to learn".
> "by using different ways such as: youtube, learning apps, books and internet".
> "If there is a visual thing like video to make the idea clear, activities."
> "Watch a video make me motivated to learn".
> "don't forget the big classroom is very comfy".
> "without proper air conditioning I cannot focus, because it is too hot".
> "I prefer not to come to college if there is no ventilation in the classroom".

Other students stated that short break times is not enough to eat and pray also has impacted their learning as they spent most of their day time at the college.

> "enough breaks to eat and pray is important so that we can concentrate".
> "sometimes I ask teachers for extra break time".
> "we buy food from vending machines to save time".

Some participants expressed that they preferred to have competitions during their learning. They stated that if competitions are planned and carried out by the college's management, will add more excitement and fun to the learning process and at the same time, support the learning environment.

> "competitions are also fun, and we learn from it".
> "activities and guest speakers make us change from the mood of studying".
> "I learn better with activities as they create excitement".

Discussions, Conclusion, and Recommendations

The findings of this study convey what students consider as factors associated with their learning of the health sciences. The study's participants proposed that their learning is deeply enhanced by their working in teams and collaboration with their peers during learning, as it supports them during the study for their assessments and making them aware of any catch-up misses. Lai (2011) has argued that an interaction among learners is able to produce expanded explanations and improve learning predominantly for students who are having low achievements. Moreover, a study carried out by Carmichael (2009) found that collaborative learning is able to produce better achievements and higher productivity levels. The interviews reflected that participants preferred collaborative learning to learn better, however, some participants communicated their comfort and better performance when they learn alone. As they consider learning in groups may distract them or waste their time.

Previously, Bandura (1986), in his social learning theory, recommended that the learner's behavior is greatly impacted by their exposure to role models. Also, was asserted by Lashley and Barron (2006), that role modeling plays an essential role in the student's learning process. Moreover, the study that was done by Bashir et al. (2014) has identified that most of their students' participants have taken their teachers as role models. This is consistent with the participant's perception in this study that their teacher is impacting and encouraging them learn to resemble them in the future.

To be able to respond to the student's needs, teachers require to be flexible, creative and responsible (Tulbure, 2011). Research studies have confirmed that teaching strategies have positive impact on student's achievement, their approach to learning and learning styles (Tulbure, 2011). Hightower et al. (2011) asserted that quality teachers are able to reveal their teaching skills, communication, commitment and abilities to create an environment that is conducive to learning. This has a positive influence on their understanding of learner's styles, needs and culture. This aspect in teacher's impact showed in this study participants' responses, as they stated that the methods of teacher's communication and dealing with them as professional and fair have positively affecting their learning process. Lai (2011) noted that to improve student learning, teachers require providing explanations whenever it is needed and provide opportunities to practice new skills. Although there are lot of evidences of teachers' impact on student's learning, there is lack of observations to proof this impact (Metzler & Woessmann, 2010).

Time management and stress is a major concern of students participated in this study. Also, it is evident in other studies that were done in the field. Zakaria (2016) has commented that time management is an essential element of the healthcare providers' work to avoid dissatisfaction and stress. To overcome stresses and feeling of being behind their peers, student nurses participants of Mirzaei et al. (2012) tend to spend most of their time on academic homework, assignments and projects. Thus, allowing less time to do other tasks of their daily living activities. On the other hand, Panchu et al. (2017) highlighted that health professional's personality plays an essential role in supporting them to cope with stress.

Students' personalities differ and so are their individual values and abilities in processing information (Chowdhury, 2006). Individual learner's learning style, organizing and academic skills and also the sense of responsibility have effects on their learning. This is what participants during the interviews have displayed. The result of a study carried out by Al-Naggar et al. (2015) on health sciences students has given evidences of the same. Researchers argued that the learner's openness and thoroughness are associated with positive academic achievements (Al-Naggar et al., 2015).

Learning styles varied among health professionals. The students in this study disclosed the preferences of more than one learning style. This is a similar finding to that of Daud et al. (2014). As their medical students' participants displayed multimodal style, and they preferred active learning strategies which are mostly suitable to fulfill all students learning styles. Also, another study of Nair and Lee (2016) has resulted in more than half of the undergraduate nursing students have diverging learning style. Barman et al. (2014) confirmed that students' academic performance can be improved if students themselves are aware of their learning styles and learning potentials.

The students' learning is not only influenced by teachers and learners themselves, learning is also associated by the educational environment, its quality and what it offers to learners (Hakimzadeh et al., 2014). The provision of teaching resources is an important aspect of students' learning (Anyango, 2012). These resources include, and not limited to, educational laboratories and libraries which provide stimulation to learning and improve students' engagement in the learning process (Meg, 2014). These aspects revealed in the students' responses in this study as they stated that libraries and audiovisual aids offer them better learning experiences.

Meg (2014) discussed that if the institutions where students learn provide uncomfortable setting such as classroom ventilation, maintenance, lighting and noise also has negative influence on students' learning. This is parallel to what participants in the current study specified as the spacious and ventilated classroom support their learning. As students will be disadvantaged in crowded classrooms that may hinder the teaching and learning processes as students feel difficulty to concentrate and cooperate (Chuma, 2012).

On the other hand, the themes above showed that participants require a sort of competitions to foster their learning. Mushtaq and Khan (2012) in their research study, "Factors Affecting Students' Academic Performance", pointed that students may perform better if extracurricular activities were planned within the curriculum. Moreover, the Centers for Disease Control and prevention confirmed that "extracurricular activities are associated with a variety of academic outcomes" (Centers for Disease Control and prevention [CDC], 2014, p. 8).

Participants mentioned the need of enough time to eat and pray as important part of their learning. This is evident in the work of Murphy (2007) when declared that reduced performance is associated with missing breakfasts. Moreover, MacLellan et al. (2008) found that students' low grades are linked to poor nutrition. Basch (2011) added that reduced amount of essential vitamins and minerals due to hunger is associated with higher absenteeism rates and failures among students.

Limitations to this study are mainly due to time and the sample population restrictions. The study was conducted in one of the colleges of health sciences that were convenient to the researcher. Also, in the campus where the study was conducted, the majority is female students as there are no males currently enrolled at the College. This aspect excluded the male students from being involved in the study who might have different perspectives.

In conclusion, the current study tries to explore factors affecting learning of health sciences. Its main goal is to use the findings to support future scholars with tools to promote students' learning of health science education, and to help the preparation of health work force being equipped with the needed knowledge and skills. To recommend for future studies, it is important to include observations of health sciences teachers to create an understanding on how the strategies they use impact student learning. Also, mixed methods of research studies are needed to investigate the students' opinion and attitudes towards factors affecting their learning of health sciences.

REFERENCES

Al-Naggar, R. A., Osman, M. T., Ismail, Z., Bobryshev, Y. V., Ali, M. S., & Menendez-Gonzalez, M. (2015). Relation between type of personality and academic performance among Malaysian health sciences students. *International Archives of Medicine Section: Psychiatry and Mental Health, 8*(182), 1-8.

American Hospital Association. (2013). *Developing an effective health care workforce planning model.* AHA. http://www.aha.org/content/13/13wpmwhitepaperfinal.pdf

Anyango, B. O. (2012). *Factors influencing girls performance in KCSE examination in mixed secondary schools* in Lower Nyokal division, MED project University of Nairobi.

Baker, C. M., McDaniel, A. M., Pesut, D. J., & Fisher, M. L. (2007). Learning skills profiles of master's students in nursing administration: assessing the impact of problem-based learning. *Nursing Education Perspectives, 28*(4), 190-195.

Bandura, A. (1986). *Social foundations of thought and action: A social cognitive theory.* Prentice-Hall.

Barman, A., Aziz, R. A., & Yusoff, Y. M. (2014). Learning style awareness and academic performance of students. *South East Asian Journal of Medical Education, 8* (1), 47-51. http://seajme.md.chula.ac.th/articleVol8No1/8_OR3_ArunodayaBarman.pdf

Basch, C. E. (2011). Healthier Students Are Better Learners: A Missing Link in Efforts to Close the Achievement Gap. *Journal of School Health, 81*(10), 593-598.

Bashir, S., Bajwa, M., & Rana, S. (2014). Teacher as a role model and its impact on the life of female students. *International Journal of Research Granthaalayah, 1*(1), 9-20.

Bate, E., Hommes, J., Duvivier, R., & Taylor, D. C. (2014). Problem-based learning (PBL): getting the most out of your students their roles and responsibilities: AMEE Guide No. 84. *Medical Teacher, 36*(1), 1-12.

Braun, V., & Clarke, V. (2012). Thematic analysis. In H. Cooper, P. M. Camic, D. L. Long, A. T. Panter, D. Rindskopf, & K. J. Sher (Eds.), *APA handbooks in psychology®. APA handbook of research methods in psychology, Vol. 2. Research designs: Quantitative, qualitative, neuropsychological, and biological*, p. 57-71.

Brookfield, S. D. (2012). *Teaching for critical thinking: Tools and techniques to help students question their assumptions*. Jossey-Bass.

Carmichael, J. (2009). Team-based learning enhances performance in introductory biology. *Journal of College Science Teaching, 38*(4), 54-61.

Centers for Disease Control and Prevention. (2014). *Health and academic achievement*. CDC. https://www.cdc.gov/healthyyouth/health_and_academics/pdf/health-academic-achievement.pdf

Chowdhury, M. (2006). Students' personality traits and academic performance: A five-factor model perspective. *College Quarterly*, *9*(3). http://collegequarterly.ca/2006-vol09-num03-summer/chowdhury.html

Chuma, P. C. (2012). *Challenges affecting teaching-learning in primary schools in kenya.* A case study of Central Division Mandera East District Executive Med Project, Moi University.

Cresswell, J. W., & Plano Clark, V. L. (2011). *Designing and conducting mixed method research* (2nd ed.). Sage

DaRosa, D. A., Skeff, K., Friedland, J. A., Coburn, M., Cox, S., Pollart, S., O'Connell, M., & Smith, S. (2011). Barriers to effective teaching. *Academic Medicine*, *86*(4), 1-7. http://www.pediatrics.emory.edu/resources/teaching/barriersbyd.pdf

Daud, S., Kashif, R., & Chaudhry, A. M. (2014). Learning styles of medical students. *South East Asian Journal of Medical Education*, *8*(1), 40-46. http://seajme.md.chula.ac.th/articleVol8No1/7_OR1_SeemaDuad.pdf

Dickson, C. A. W. (2010). Evaluating the student experience of inquiry-based learning: An educational initiative. *Practice and Evidence of Scholarship of Teaching and Learning in Higher Education*, 5(1), 33-45. http://eresearch.qmu.ac.uk/4562/1/4562.pdf

Dolmans, D. H. J. M., Loyens, S. M. M., Marcq, H., & Gijbels, D. (2016). Deep and surface learning in problem-based learning: A review of the literature. *Advances in Health Sciences Education*, *21*, 1087-1112. NCBI. https://www.ncbi.nlm.nih.gov/pmc/articles/PMC5119847/pdf/10459_2015_Article_9645.pdf

Eisenstein, A., Vaisman, L., Johnston-Cox, H., Gallan, A., Shaffer, K., Vaughan, D., O'Hara, C., & Joseph L. (2014). Integration of basic science and clinical medicine: the innovative approach of the cadaver biopsy project at the Boston University School of Medicine. *Academic Medicine*, *89*(1), 50-53.

Friedland, J. A. (2002). Social learning theory and the development of clinical performance. In J. C. Edwards, J. A. Friedland, & R. Bing-You (Eds.), *Residents' teaching skills* (pp.18-37). Springer.

Gale, N. K., Heath, G., Cameron, E., Rashid, S., & Redwood, S. (2013). Using the framework method for the analysis of qualitative data in multi-disciplinary health research. *BMC Medical Research Methodology*, *13*. https://bmcmedresmethodol.biomedcentral.com/articles/10.1186/1471-2288-13-117

Galvao, T. F., Silva, M. T., Neiva, C. S., Ribeiro, L. M., & Pereira, M. G. (2014). Problem based learning in pharmaceutical education: a systematic review and meta-analysis. *Scientific World Journal.* https://www.hindawi.com/journals/tswj/2014/578382/

Gambrill, E. (2012). *Critical thinking in clinical practice: Improving the quality of judgments and decisions* (3rd ed.). John Wiley & Sons, Inc.

Gijbels, D. (2008). Effectiveness of Problem-Based Learning. *National Research Council.* http://www7.nationalacademies.org/bose/PP_Commissioned_ Papers.htm.

Gijbels, D., Donche, V., Richardson, J. T. E., & Vermunt, J. D. (2014). Learning patterns in higher education. *Dimensions and Research Perspectives*. Routledge.

Goh, K. (2014). What good teachers do to promote effective student learning in a problem-based learning environment. *Australian Journal of Educational & Developmental Psychology*, *14*, 159-166. http://files.eric.ed.gov/fulltext/EJ1041678.pdf

Hakimzadeh, R., Ghodratil, A., Karamdost, N., Ghodrati, H., & Mirmosavi, J. (2014). Factors affecting the teaching-learning in nursing education. *Journal of Education*, 174-184.

Halpern, D.F. (1992). A cognitive approach to improving thinking skills in the sciences and mathematics. In D.F. Halpern (Ed.), *Enhancing Thinking Skills in the Sciences and Mathematics* (pp. 1-14). Erlbaum.

Hanson, J.L., Balmer, D.F., & Giardino, A.P. (2011). Qualitative research methods for medical educators. *Academic Pediatrics*, *11*, 375-386.

Hightower, A. M., Delgado, R. C., Lloyd, S. C., Wittenstein, R. Sellers, K., & Swansonl, C. B. (2011). *Improving student learning by supporting quality teaching: Key issues, effective strategies.* http://www.edweek.org/media/eperc_qualityteaching_12.11.pdf

Jahanpour, F., Azodi, P., Azodi, F., & Khansir, A. A. (2016). Barriers to practical learning in the field: a qualitative study of iranian nursing students' experiences. *Nursing and Midwifery Studies*, *5*(2), 1-3. https://www.ncbi.nlm.nih.gov/pmc/articles/PMC5002090/pdf/nms-05-02-26920.pdf

Johnson & Johnson. (2016). *Preparing Nurses for the Future.* http://www.jnj.com/caring/patient-stories/preparing-nurses-for-the-future

Johnson, S. (2015). *How has nursing changed and what does the future hold?* https://www.theguardian.com/healthcare-network/2015/mar/17/how-has-nursing-changed-and-what-does-the-future-hold

Kahlke, R., & White, J. (2013). Critical thinking in health sciences education: considering "three waves". *Creative Education*, *4*(12A), 21-29. http://dx.doi.org/10.4236/ce.2013.412A1004

Khan, M.A., Qamar, K., Khalid, S., Javed, H., Malik, M., Gondal, A., Zafar, H., Shoaib, A., Aman, R., Khan, A., Zain UlAbideem, A., Bilal, A., & Army, F. I. (2015). Comparison of case-based learning with conventional teaching students' perspective. *Pakistan Armed Forces Medical Journal, 65*(3), 415-419.

Kumar, L. R., & Sethuraman, K. R. (2007). Learning approaches in dental and medical students in AIMST: A comparison between deep and surface approaches. *The International Medical Education Conference*, Kuala Lumpur, Malaysia.

Lai, E. R. (2011). *Collaboration: A Literature Review.* http://images.pearsonassessments.com/images/tmrs/Collaboration-Review.pdf

Lashley, C., & Barron, P. (2006). The learning style preferences of hospitality and tourism students: Observations from an international and cross-cultural study. *International Journal of Hospitality Management, 25*(4), 552-569.

Li, S., Yu, B., & Yue, J. (2014). Case-oriented self-learning and review in pharmacology teaching. *The American Journal of the Medical Sciences, 348*(1), 52-56.

Loyens, S. M. M., Gijbels, D., Coertjens, L., & Cote´, D. (2013). Students' approaches to learning in problem-based learning: Taking into account students' behavior in the tutorial groups, self-study time, and different assessment aspects. *Studies in Educational Evaluation, 39*(1), 23-32.

MacLellan, D., Taylor, J., & Wood, K. (2008). Food intake and academic performance among adolescents. *Canadian Journal of Dietetic Practice and Research, 69*(3), 141-144.

Mansur, D. I., Kayastha, S. R., Makaju, R., & Dongol, M. (2012). Problem based learning in medical education. *Kathmandu University medical journal, 10*(40), 78-82.

Marshall, M. N. (1996). Sampling for qualitative research. *Family Practice, 13*(6), 522-525. https://47-269-203-spr2010.wiki.uml.edu/file/view/Research_I_20090916221539453.pdf/116402723/Research_I_20090916221539453.pdf

McLaren, P. (1994). Foreword: Critical thinking as a political project. In K. S. Walters (Eds.), *Re-thinking reason: New perspectives on critical thinking* (pp. ix-xv). State University of New York Press.

Mclean, S. F. (2016). Case-based learning and its application in medical and health-care fields: A review of worldwide literature. *Journal of Medical education and curricular development, 3*, 39-49.

Meg, C. A. (2014). *Influence of school environmental factors on teaching-learning process.* http://eap.uonbi.ac.ke/sites/default/files/cees/education/eap/FINAL%20REPORT.pdf

Mello, C. C. B., Alves, R. O., & Lemos, S. M. A. L. (2014). *Methods of health education and training: literature review, 16*(6). http://www.scielo.br/scielo.php?pid=S1516-18462014000602015&script=sci_arttext&tlng=en

Mestre, J.P., Dufresne, R.J., Gerace, W.J., Hardiman, P.T., & Tougher, J.S. (1992). Enhancing Higher Order Thinking Skills in Physics. In D.F. Halpern (Eds.), *Enhancing thinking skills in the sciences and mathematics* (pp. 77-94). Erlbaum.

Metzler, J., & Woessmann, L. (2010). The impact of teacher subject knowledge on student achievement: Evidence from within-teacher within-student variation. *IZA DP No. 4999* http://ftp.iza.org/dp4999.pdf

Mirghani, H. M., Ezimokhai, M., Shaban, S., & Berkel H. J. M. van. (2014). Superficial and deep learning approaches among medical students in an interdisciplinary integrated curriculum. *Education for Health, 27*(1), 10-14. http://www.educationforhealth.net/temp/EducHealth27110-3132922_084209.pdf

Mirzaei, T., Oskouie, F., & Rafii, F. (2012). Nursing students' time management, reducing stress and gaining satisfaction: A grounded theory study. *Nursing and Health Sciences, 14*, 46-51. https://www.researchgate.net/publication/221794272_Nursing_students'_time_management_reducing_stress_and_gaining_satisfaction_A_grounded_theory_study

Morrall, P., & Goodman, B. (2013). Critical thinking, nurse education and universities: Some thoughts on current issues and implications for nursing practice. *Nurse Education Today, 33*(9), 935-937. http://www.nurseeducationtoday.com/article/S0260-6917(12)00384-X/fulltext

Murphy, J. M. (2007). Breakfast and learning: An updated review. *Current Nutrition & Food Science, 3*, 3-36.

Mushtaq, I., & Khan, S. N (2012). Factors affecting students' academic performance. *Global Journal of Management and Business Research, 12*(9), 17-22. http://www.dl.icdst.org/pdfs/files/3deebc68c9747ac3dbc60045c5ad9993.pdf

Nair, M. A., & Lee, P. (2016). Exploration of the learning style among undergraduate nursing students from an Indian perspective. *IOSR Journal of Nursing and Health Science, 5*(5), 1-4. http://www.iosrjournals.org/iosr-jnhs/papers/vol5-issue5/Version-1/A0505010104.pdf

Panchu, P., Bahuleyan, B., & Vijayan, V. (2017). n exploration into the inter relationship between personality and metacognitive awareness of I year medical students. *International Journal of Health Sciences & Research, 7*(2), 132-136. http://www.ijhsr.org/IJHSR_Vol.7_Issue.2_Feb2017/21.pdf

Patton, M.Q. (2002). *Qualitative evaluation methods.* Sage Publications

Preeti, B., Ashish, A., & Shriram, G. (2013). Problem based learning (pbl) An effective approach to improve learning outcomes in medical teaching. *Journal of Clinical and Diagnostic Research, 7*(12), 2896-2897.

Reitmanova S. (2011). Cross-cultural undergraduate medical education in North America: Theoretical concepts and educational approaches. *Teaching and learning in medicine, 23*, 197-203.

Shah, D. K., Yadav, R. L., Sharma, D. Yadav, P. K., Sapkota, N. K., Jha, R. K., & Islam M. N. (2016). Learning approach among health sciences students in a medical college in Nepal: A cross-sectional study. *Advances in Medical Education and Practice, 7*, 137-143. https://www.ncbi.nlm.nih.gov/pmc/articles/PMC4786058/pdf/amep-7-137.pdf

Shell, R. (2001). Perceived barriers to teaching for critical thinking by BSN nursing faculty. *Nursing Health Care Perspectives, 22*, 286-291.

Skeff, K. M., Stratos, G. A., Mygdal, W., DeWitt, T. A., Manfred, L., Quirk, M., Roberts, K., Greenberg, L., & Bland, C. (1997). Faculty development: A resource for clinical teachers. *Journal of general internal medicine, 12*(2), S56-S63.

Smith, L., & Coleman, V. (2008). Student nurse transition from traditional to problem-based learning. *Learning in Health & Social Care, 7*(2), 114-123.

Sternberg, R. J. (1999). *Handbook of Creativity*. Cambridge University Press.

Tekian, A. Roberts, T., Batty, H. P., & Norcini, J. (2013). Preparing leaders in health professions education. *Medical Teachers, 36*, 269-271. https://www.researchgate.net/publication/258101875_Preparing_leaders_in_health_professions_education

Thistlewaite, J. E., Davies, D., Ekeocha, S., Kidd, J.M., MacDougall, C., Matthews, P., Purkis, J., & Clay, D. (2012). The effectiveness of case-based learning in health professional education. *Medical Teacher, 34*, E421-E444. https://doi: 10.3109/0142159X.2012.680939.

Thorne, S., Kirkham, S.R., & O'Flynn-Magee, K. (2004). The analytical challenge in interpretive description. *International Journal of Qualitative Methods World, 3*. http://www.ualberta.ca/~iiqm/ backissues/3-1/html/thorniest.html

Tulbure, C. (2011). Learning styles, teaching strategies and academic achievement in higher education: A cross-sectional investigation. *Social and Behavioral Sciences, 3*, 398-402. http://ac.elscdn.com/S1877042812001590/1-s2.0-S1877042812001590-main.pdf?_tid=a916fdf0-56d5-11e7-b89f00000aacb35d&acdnat=1498086312_f36dcab60adf77ac69fa865ae05b6544

Twenge, J.M. (2009). Generational changes and their impact in the classroom: Teaching Generation Me. *Medical Education, 43*(5), 398-405.

Voytovich, A., Longo, W. E., Kozol, R. A., & Chandawarkar, R. Y. (2008). Improving surgical residents' performance on written assessments of cultural competency. *Journal of surgical education, 65*, 263-269.

Walters, K. S. (1994). Introduction: Beyond logicism in critical thinking. In K. S. Walters (Eds.), *Re-thinking reason: New perspectives in critical thinking* (pp. 1-22). State University of New York Press.

WHO (2013) Transforming and scaling up health professionals' education and training. *World Health Organization Guidelines.* http://apps.who.int/iris/bitstream/10665/93635/1/9789241506502_eng.pdf

Wittich, C. M., Lopez-Jimenez, F., Decker, L. K., Szostek, J. H., Mandrekar, J. N., Morgenthaler, T. I., & Beckman, T. J. (2010). Measuring faculty reflection on adverse patient events: Development and initial validation of a case-based learning system. *Journal of General Internal Medicine, 26*(3), 293-298.

Yanchar, S. C., Slife, B. D., & Warne, R. (2008). Critical thinking in disciplinary practice. *Review of General Psychology, 12*, 265-281.

Zakaria, A. M. (2016). Effectiveness of Learning Module on Time Management Ability and Delegation Skills for Head Nurses. *IOSR Journal of Nursing and Health Science, 5*(2), 31-40. http://www.iosrjournals.org/iosr-jnhs/papers/vol5-issue2/Version-1/F05213140.pdf

Zohar, A., & Tamir, P. (1993). Incorporating critical thinking into a regular high school biology curriculum. *School Science and Mathematics, 93*(3), 136-140.

PART 6

CHAPTER 18

Web 2.0 Applications into Mathematics Instruction

Nada Albarbari

ABSTRACT

Technology provides an extensive range of possibilities for teachers in establishing a collaborative, interactive mathematics class. The worldwide practice of reforming mathematics instruction has emanated from findings regarding college students and others entering the job market, who, according to recent studies, lack the competency to analyze situations despite their adequate knowledge of the subject matter. This chapter aims to provide comprehensive description and discourse on teacher implementation of contemporary techniques that integrate mathematics and technology as essential to scientific and economic development. Web 2.0 technology can be applied in the classroom in various ways and forms, three of which are addressed for the purposes of this study.

INTRODUCTION

Educators of the twenty-first century strive to equip students with skills to succeed in a rapidly transforming world highly reliant on technology (Lee & Ge 2010). Over and above their typical responsibilities, teachers are challenged to implement contemporary techniques integrating technologies, and the conventional practice of displaying a lesson or activity onscreen in the classroom no longer suffices. Unless technology is used as an essential daily classroom tool to advance the learning process, the integration is considered incomplete (Cuhadar & Kuzu 2010).

The significant revolution in Web 2.0 tools and applications has undoubtedly facilitated their use in class like never before. Therefore, studies on the benefits and drawbacks of transforming classwork activities into computer tasks as well as teacher and student perceptions of this change dominate the field of education research (Malhiwsky 2010; Sistek-Chandler 2012). Learning is no longer restricted to the traditional forms of educational institutions; Internet cloud computing and related services have transformed it into a pervasive phenomenon that might happen anywhere at any time (Johnson, Adams & Cummins 2012). This has made people become cognisant of ways to employ knowledge and communication to enrich their learning experiences.

Recent studies in the education field have emphasised the many different ways Web 2.0 and social media can be used to enhance learning practices and positively influence both students and teachers (Heibergert & Loken 2011; Cain & Policastri 2011; Kelm 2011). Web 2.0 applications grant users the opportunity to communicate, access, broadcast, and collaboratively create in a safe and user-friendly setting (Bates 2011). The meaning of Web 2.0 as a collaborative interaction and communication tool in the classroom is grounded in Vygotsky's social learning theory (1978), through which he elucidated the significant role of social interaction and language in cognitive development and meaning making.

According to Ollis (2011), Web 2.0 applications are excellent tools to fulfil the academic and emotional requirements and obligations of learners in the twenty-first century and can provide a complete experience of learning within a group and via social interaction. Thongmak (2013) has affirmed that Web 2.0 has created innovative forms of social interaction that prove the positive impact of collaboration and gradually enhance students' learning outcomes, enrich their shared retention of knowledge, and develop their analytical and reasoning skills.

Web 2.0 technologies can provide an extensive range of possibilities to teachers establishing a collaborative mathematics class. The worldwide practice of reforming mathematics instruction has emanated from findings regarding college students and others entering the job market, who, according to recent studies, lack the competency to analyse situations despite their sometimes ample knowledge of a subject (Carbonneau, Marley & Selig 2013).

Since the adoption of electronic devices in schools, mathematicians have worked on developing technological solutions to close the gaps between research and findings in constructivist approaches to learning (Aqel 2011). First among targeted research areas is the incorporation of writing activities across different curriculum subjects, including mathematics (Hendry et al. 2011). The ability to verbalise maths problems is as crucial as the mastery of procedural calculations, allowing students to not only solve word problems but also conveniently articulate their thoughts and expand their comprehension of topics and therefore take part in higher cognitive tasks and discussions (Morris 2006; Millard, Oaks & Sanders 2002; Steen 2007).

The National Council of Teachers of Mathematics (NCTM) has published *Principles and Standards for School Mathematics* (2000), which points out that technology is one of the six main six principles of mathematics and emphasises its potential to provide a broad range of alternative solutions and approaches. Bates (2011) has also stated the importance of creating opportunities for alternative solutions, offering that teachers should motivate their students to look for unconventional ways to solve problems rather than simply following instructions. Mason and Rennie (2010) and Zgheib and Dabbagh (2012) have also stated that teachers must encourage students to discuss collaboratively and analyse alternative solutions to decide whether to accept or reject them. This research is designed to investigate the capability of Web 2.0 communication applications to stimulate learners' reasoning and develop their problem-solving skills by providing opportunities for constructive writing and professional dialogue (Bicer, Capraro & Capraro 2013; Johnston 2012; Cooper 2012; NCTM 2000).

THEORETICAL FRAMEWORK

To develop a theoretical framework for understanding the effectiveness of integrating writing activities in mathematics instruction using Web 2.0 as a means of communication towards a common goal, I have drawn on three learning theories: the social constructivist learning theory, the communities of practice theory, and the self-determination theory.

Social Constructivist Learning Theory

The majority of collaborative learning theories are based on Vygotsky's theory of social constructivism, which considers society the source of human intelligence and describes learning as a social process generated by individuals and knowledge as an inevitable consequence of social interaction. Hence, individuals obtain knowledge when joining knowledge communities (Vygotsky 1978). Vygotsky has agreed with the theories of most of the early constructivists, like Piaget, and their assumptions regarding learning mechanisms, but he has further recognised the fundamental roles of social context and language on the cognitive development of children.

Vygotsky believes that learning occurs in two stages: the first is the acquisition of knowledge as a result of interaction with a community, and the second is the mental processing and accommodation of this knowledge. One imperative dimension of Vygotsky's theorem is the zone of proximal development (ZPD), which he developed as an extension of Piaget's theory of cognitive development, which defined children as 'lone scientists' and asserted that children possess all the required learning mechanisms and need no adult guidance. Vygotsky's ZPD is the 'distance between the actual developmental level as determined by independent problem-solving and the level of potential development as determined through problem-solving under adult guidance or in collaboration with more capable peers' (Vygotsky 1978, p. 86).

Vygotsky views collaborative learning as one very effective way of improving skills and developing cognition, where less capable students benefit from more competent peers within the ZPD. Hence, the ZPD expresses a level of assignments in which learners need support and scaffolding from a more knowledgeable person to achieve and master a subject. The expressions 'ZPD' and 'scaffolding' are now used interchangeably in the literature despite the fact that Vygotsky has never used them so. Rather, Wood et al. (1976) were the firsts to adopt these terms in education. The zone is the range in which students are ready to acquire knowledge with the assistance of social interaction and collaboration, and scaffolding is the approach to developing this intellectual knowledge (Gunawardena et al. 2009).

Vygotsky's theory of the ZPD is a convenient foundation to establish an online collaborative learning setting in which the teacher assists and all class members contribute. The outcomes of online collaboration are not only the direct outputs of applications but mostly the skills students learn from taking responsibility for their learning (Buzzetto-More 2010; Koohang, Riley & Smith 2009).

Self-Determination Theory

Self-determination theory (SDT) represents an extensive framework for the study of internal and external forces driving human motivation. SDT originated with Edward L. Deci and Richard M. Ryan, and then underwent several ameliorations by many scholars worldwide. SDT focuses on the social conditions that promote self-determined actions and the consequences of inner resources in developing different types of motivation, each type conveyed by its particular regulatory style (Visser 2010; Johanston 2012).

Ryan and Deci (2000) have organised the main four types of motivation along a continuum according to each type's level of self-regulation and have used the term 'internalisation' to describe the process of moving from one type of motivation to another within the continuum (Deci et al. 1991). The internalisation process is not necessarily linear and has been found to be beneficial to students' intellectual development and practical progression (Deci et al. 1991; Ryan & Deci 2000). It can be developed and enhanced along with students' social and personal development (Reeve & Jang 2006), persistence, and creativity (Hahn & Oishi 2006) with the aid of three psychological needs; these are competence, autonomy, and relatedness, three qualities that accompany the self-determination theory (Ryan & Deci 2000).

Competence is an indicator of an individual's confidence in his or her capability to accomplish a duty, and autonomy is an indicator of confidence in one's capability to make an appropriate choice independently (Ryan & Deci 2000; Johanston 2012). A large body of literature has supported the role of psychology in promoting intrinsic motivation (Hahn & Oishi 2006; Ryan & Deci 2000; Reeve & Sickenius 1994; Deci et al. 1991; Reeve 2005; Reeve & Jang 2006; Visser 2010; Johanston 2012). SDT suggests that dissatisfaction with any of these psychological elements within a community will cause an undesired reaction in the learning environment and in learners' inner enthusiasm and tendencies to move towards self-regulated motivation (Hahn & Oishi 2006).

Mathematics Communication

Communication is a fundamental component of the learning process, especially in mathematics. The NCTM's communication standards, issued in 1989, 1991, 1995, and 2000, have highlighted the significance of mathematical fluency in communication. Hiebert et al. (1998) have described mathematical communication as the ability to clearly and reasonably express mathematical ideas verbally or in writing along with the capacity to contribute to the discussion by sharing ideas and listening to and evaluating others' ideas. Hiebert et al. (1998) have also stressed that this fluency in communication stimulates learners' capability to gain insight into their knowledge and their peers' knowledge and formulate concepts and ideas. Pugalee (2004) has found in his exploratory study on ninth grade students that writing out problem-solving approaches has a positive effect in cultivating young learners' metacognitive structures.

This part of the literature review focuses on writing and discussion as effective types of class communication to enhance students' abilities to explain and comprehend their thoughts and doubts. Bicer, Capraro, and Capraro (2013) have described mathematical communication as the act of consolidating a variety of cognitive and social activities to boost critical thinking and logical reasoning. Steele (2001) has referred to the association of writing and discussion in mathematics as the employment of a sociocultural classroom environment with the intention of formulating mathematical structure based on cognitive reasoning.

CONSTRUCTIVIST LEARNING AND INTERACTION

The classroom environment is influential in education as an imperative context in which actual learning takes place based on productive interactions between students and instructors (Beavers et al. 2015). The social constructivist learning theory suggests that individuals generate knowledge using their cognitive development through social interaction, relating prior knowledge to new experiences. Constructivism refers to the learning process and focuses on students as active participants with teachers and other students (Buzzetto-More 2010; Beavers et al. 2015). Vygotsky claims that social constructivism and individual cognition occur in social contexts where more knowledgeable peers and adults structure information and transfer this information via language (Cuhadar & Kuzu 2010).

Social constructivism affirms the significance of social contexts and the collaborative quality of knowledge acquisition and defines learning as integrating individual knowledge with experiences and interaction outcomes, in contrast with the individual adaptation of knowledge in cognitive constructivism (Buzzetto-More 2010). In social constructivism, intellectual functions of differing levels of difficulty are related to the social development of individuals and to cooperation with knowledgeable peers or adults (Cuhadar & Kuzu 2010).

Vygotsky (1978) has proposed that scaffolding occurs in the social context when learners acquire assistance from instructors, adults, or more experienced peers. Storch (2007) has suggested that the move from the student-centred classroom is an evident trend in today's world. In constructivist education, the teacher is a facilitator who performs planning and provides support for students. Pedagogical instruction is now more transactional, where students accept more responsibility for their learning, such as assimilating and accommodating learned knowledge and developing the necessary language to achieve this (Beavers et al. 2015; Kenney, Shoffner & Norris 2014).

WRITING FOR LEARNING AND FOR PROBLEM SOLVING

Principles and Standards for School Mathematics (NCTM 2000) has suggested reinforcing written and verbal mathematical dialogue as teachers' responsibilities shift from traditional to constructivist practices. Starting in the 1970s, researchers called for the incorporation of writing in mathematics curricula and investigated the benefits of this approach (Emig 1977). In 1992, Countryman released the revolutionary publication *Writing to Learn Mathematics*, which was considered one of the most influential in providing a comprehensive description of several approaches for integrating writing into mathematics (Kenney, Shoffner & Norris 2014; Beavers et al. 2015). Countryman (1992) suggested that writing helps students develop their sense of reasoning, analyse their abilities and qualities, find gaps in their understanding, realise misconceptions, and reflect critically on their learning (Wilson & Clarke 2004).

Students are expected to give comprehensive and coherent explanations while analysing maths problems. However, many students can achieve correct answers in maths but lack the ability to justify their methods. According to Beavers et al. (2015), having students translate their mathematical comprehension into writing develops their literacy, provides insights into their understandings, and stimulates their metacognition to meet higher-order thinking questions and tasks, like evaluation and creation (Beer, Capraro & Capraro 2013).

Studies about this approach have found many benefits aside from improving reasoning skills. Kenney, Shoffner, and Norris (2014) have described writing as a reproductive expressive procedure that encourages students to develop knowledge and abilities, directs them to distinguish what they know, and motivates them to explore what they don't. Adams (2010) has noted that not only is students' cognition in mathematics positively affected by writing activities, so are their communication skills, capability to reflect on both work and abilities, and motivation for teamwork (Capraro, Capraro & Rupley 2011). This section covers the two primary types of writing activities used to perform this study.

Mathematical Journals

Educators have exploited several approaches to motivate students to use writing in mathematics. Journals seem to dominate that field of research. Kostos and Shin (2010) have suggested that pedagogic research does not provide a clear definition for 'journal writing'. Instead, mathematicians customarily define it according to the nature of its practice and intended outcome. Johnston (2012) believes different definitions agree that journal writing involves a personalised, informative, and purposive document that educates students by encouraging them to think critically and analytically about their work in progress.

Journal writing is an opportunity for students to present and organise acquired knowledge in their very own ways. It is an exceptional method to reflect on learning with summaries, questions, comments, and feedback (Kostos & Shin 2010). Students should be given the opportunity to observe and reflect on communication skills and participate in active learning situations (Leonard 2008). According to Carter (2009), students should be explicit with mathematical discourse and be able to interpret other people's mathematical logic, and the use of maths journaling is an efficient way to meet both demands. Johnston (2012) and Craig (2011) have asserted that students do not currently write within mathematics instruction to document their learning procedures from their perspectives, supported by their doubts and certainties, and that the literary quality of this writing should not affect the quality of the content.

A large body of literature has supported the use of journaling in mathematics as an efficient technique to enhance students' reasoning and critical thinking skills. Research in this field has mostly based this instructional approach on the constructivist theory of learning and has focused on its connection to Vygotsky's sociocultural learning theory (1978). Vygotsky's theories strongly emphasise the significance of social interaction, language, and scaffolding in the development of cognition and linking prior knowledge to acquired knowledge. Vygotsky (1986) has described language as a tangible form of thinking, the means by which one can make meaning (Beavers et al. 2015; Kenney, Shoffner & Norris 2014).

Nowadays, mathematics educators recognise the importance of showing solution steps and strategies, as students are expected to be able to justify their answers and explain their reasoning. Leonard (2008) has mentioned that maths journals help students explain their thoughts to the teacher adequately and, more importantly, make sense of those thoughts. Cooper (2012) has indicated that teachers can encourage students to be more persuasive when explaining their understanding and more precise when evaluating their learning through journaling in mathematics.

Mathematics Technology

New technologies have been implemented in maths classrooms since the arrival of Web 2.0. Early computer-assisted instruction showed the many advantages of different software but also highlighted a lack of human interaction. New technology has opened many doors for classroom inclusion, as applications like chat rooms, forums, and blogs can easily be used to implement writing activities, since they match young learners' technological interests and do not require formal writing (Cooper 2012).

Writing in maths using technology involves deep understanding and requires no immediate response. Rather, students depend on their communication and writing skills to gain insights into their learning, whereas teachers are responsible for creating tasks that ensure student engagement. Sunstein et al. (2012) have conducted a study to investigate the effects of writing in mathematics on 86 high school students from two different schools in two completely different areas and cultural environments. The project was based on having students collaborate to solve mathematical problems via email. The two groups of students were supposed to only exchange their mathematical knowledge and information, but after a short while, they started to share their life experiences and varied lifestyles, connecting those elements to the mathematical topics at hand. Sunstein et al. (2012) concluded that their students gained more benefits from technology than planned or anticipated, as aside from enriching their mathematical information, students got the chance to cultivate and develop communication and language skills.

Morris (2006) has stated that for many years, students solved mathematical problems without a true understanding of their processes or reasoning, and while that was sometimes acceptable, the idea behind letting students clarify their procedures or thoughts is to help them tackle higher-level thinking tasks and questions. This can easily be connected with the aid of technological media because of students' familiarities with these applications. Writing to peer audiences and expecting feedback and interaction inspire students to elucidate concepts, develop their awareness, and simplify ideas within an authentic context (Chen et al. 2013).

Web 2.0 Technologies in Education

Education has witnessed a significant transformation in the past several decades, from accentuating the duties of teachers to pointing out the crucial role of learning. Recent studies have shifted towards facilitating instruments and approaches by which students have more control of their own learning processes (Zakaria 2013). Discussion of the efficient use of Web 2.0 applications to raise students' engagement in their learning has comprised a large share of these studies. Brown and Adler (2008) have stated that the power of Web 2.0 tools lies in their capability to boost the efficiency of class discussion by providing suitable collaboration platforms.

Additionally, the massive amount of research studies on the topic of Web 2.0 has led to the recognition of an explicit and inclusive definition of 'Web 2.0' itself (Fuchs 2010). According to Zakaria (2013), Web 2.0 technologies might have different definitions depending on context. However, all definitions agree unanimously that Web 2.0 applications are characterised by their power to provide effective collaboration mediums and reinforce the involvement of users, unlike earlier forms of technology, wherein users were mostly passive receptors. Tim O'Reilly (2005) initially created the term 'Web 2.0' and used it to describe the change in the information technology world that brought the Internet to users as a platform for creation and personalisation by means of accessible publishing tools. Web 2.0 technologies comprise any Internet tools and features that allow the user to be a social producer. O'Reilly (2005) believes that Web 2.0 applications gained popularity as their active users' imitated social interactions from offline.

Web 2.0 technology offers various intriguing ways to implement more effective collaborative learning. Clark et al. (2009) have claimed that the use of Web 2.0 tools in the classroom not only promotes constructive learning but also widens the scope of traditional teaching and learning approaches and fills gaps between theories and practice. Rethlefsen, Piorun, and Prince (2009) have found that knowledge exchange within classroom discourse works more effectively with the use of Web 2.0 technology. While Mason and Rennie (2007) have stated that the appropriate use of different Web 2.0 tools consolidates group work conception and cultivates students' collaboration abilities, they have also noted that shared working spaces and group communication excite young people and therefore should motivate them to learn.

With the rapid growth of Web 2.0 tools and applications, the expectations of schools and college graduates experienced with the use of this technology, some of whom have even improved it, have increased as well. Albion (2008) and Hazari and North (2009) have stated that students are expected to master technological skills while studying other subjects as long as the use of technology forms an essential part of their learning routine. Early use of Web 2.0 technology in education focused on using each tool for the explicit purpose it was designed for: presentation, reading, or writing. In no time, educators started to call for the direct use of this technology to enhance students' motivation and engagement. Kemker, Barron, and Harmes (2007) have found that students' motivation to solve problems considerably increases when they use technology in reading, writing, and data analysis. Similarly, Jakubowicz (2011) has shown that the use of Web 2.0 technology comes first among teachers' options to raise students' motivation for reading and writing.

In this context, Web 2.0 technologies also function to enhance many essential skills for a productive and motivated classroom. For instance, Web 2.0 can help teachers differentiate their instruction in class and afford equal learning opportunities to students with different learning styles (Jakubowicz 2011). It also encourages students' interest in topics and enhances cognitive thinking and problem solving, along with communication and contextual learning abilities (Barlow 2008; Greenhow, Robelia & Huges 2009; Gooding 2008).

WEB 2.0 TECHNOLOGY IN THE CLASSROOM

Web 2.0 technology can be applied in the classroom in various ways and forms, three of which are addressed for the purposes of this study.

Electronic Journals

Electronic journals allow for the documentation of personal viewpoints and reflection on certain topics. They can be an open space for learners to add input or a private forum for expression of thoughts to be seen only by the instructor (Jakubowicz 2011). Some students hesitate to speak their minds in front of the class and find writing safer, and journals give them enough time to formulate their opinions to avoid embarrassment. As such, electronic journals are a way for students who can't respond spontaneously in the classroom to take part in the class community and collaborative problem-solving teams.

Discussion Boards

Discussions boards are virtual spaces where many learners can contribute to the same online conversation. Online classrooms influence learning processes positively when managed and directed in a way that allows all participants to take place. Xia, Fielder, and Siragusa (2013) believe that discussion boards are the best ways to build productive communities of practice because of their ability to promote peer interaction and enrich collaborative learning. Employing discussion boards in instruction can provide the scaffolding needed to enhance active learning. Miyazoe and Anderson (2010) have argued that discussion boards reinforce the building of a constructive learning environment through cooperative knowledge and the immediate feedback students most often receive from their teachers and peers.

According to Khoshneshin (2011), class online social interaction under teacher supervision encourages productive teamwork and collaborative problem solving and grants students the chance to be in charge of their learning by building their conceptual understanding through debates and reflection. Xia, Fielder, and Siragusa (2013) have found that continued participation on class discussion boards has an evident positive effect on teamwork quality and students' final grades.

Social Networking

Buzzetto-More (2012) has defined social networks as computer-assisted communities that facilitate complex connections between users through applications that enable communication, presentation, storage, and interaction. Chen and Bryer (2012) have addressed the use of social networking as a pedagogical element of Vygotsky's social constructivism theory (1978), as Vygotsky built his theory on the belief that learning is a social activity and that effective learning happens through engagement in a collaborative problem-solving environment organised by the teacher.

The use of social networking as a platform for learning challenges teachers to create interesting concept-related tasks to stimulate students' desire to contribute and reflect on their learning practices. The use of social networks enriches learning by making it more student centred and accessible at all times and by providing a suitable environment for constructive, reflective, collaborative, informative, and personalised activities (Al-Kathiri 2014; Fogg et al. 2011; Zaidieh 2012). At the beginning of the e-learning revolution in education, many educators raised the lack of human interaction as a great obstacle towards supporting this innovation (Firpo & Ractham 2011). Social networks have resolved this issue and many others successfully and have also increased teachers' interest in e-learning, especially through closed and nonthreatening social networks (Zaidieh 2012).

MOTIVATION IN MATHEMATICS

With the aim of establishing a link between the integration of Web 2.0 writing applications and students' motivation to learn mathematics, it is important to briefly define motivation. Wæge (2010) has defined it as the capability to control and direct actions, mentioning that this capability is regulated by demands and objectives. With their self-determination theory, Ryan and Deci (2000) have argued that humans are motivated by external and internal factors, and they have designated three essential psychological needs that may impact students' intrinsic motivation to learn: autonomy, competence, and social belonging or relatedness. The need for competence describes the desire to take on challenging tasks and execute them successfully to attain mastery, whereas the need for autonomy describes the desire to make decisions when carrying out a task and to experience freedom (Niemiec & Ryan 2009).

Wæge (2010) has stated that students' need for competence inspires them to solve challenging mathematical problems and feel accomplishment at their mastery of skills, while the need for autonomy is associated with a willingness to engage in several tasks while feeling free to make meaningful choices. Johnston (2012) believes teachers should attempt to prompt students' intrinsic motivation by providing challenging tasks, opportunities for decision making, and systems of rewards. Students' motivation to learn mathematics can be influenced by a constructive teaching approach, collaborative work, and student-centered teaching. Under supervision of the teacher, students can develop a sense of freedom and mastery (Wæge 2010; Niemiec & Ryan 2009; Kahn et al. 2013).

Self-determination theory describes motivation as a range starting with intrinsic motivation, moving toward extrinsic motivation, and ending with amotivation, which, according to Deci and Ryan (2000), is the absence of any interest in a task. Green-Demers et al. (2006) have suggested four main reasons for amotivation, these being low self-confidence, effort assumptions, a history of academic achievement, and the nature of the task undertaken. A wide range of studies in mathematics motivation have agreed that the consideration of diverse learning approaches, along with dynamic collaboration, teacher influence, and autonomy support, is a significant factor in motivating student learning (Maulana et al. 2011; Kim, Murayama et al. 2013; Kahn et al. 2013; Wæge 2009).

References

Al-Kathiri, F., (2014). Beyond the Classroom Walls: Edmodo in Saudi Secondary School EFL Instruction, Attitudes and Challenges. *English Language Teaching*, *8*(1), 189-204.

Badawy, A.H.A., (2012). Students Perceptions of the Effectiveness of Discussion Boards: What can we get from our students for a freebie point. *International Journal of Advanced Computer Science and Applications, (3)*9. 136-144.

Barlow, T. (2008). Web 2.0: Creating a classroom without walls. *Teaching Science, 54*(1), 46-48.

Beavers, A., Fox, B.L., Young, J., Bellows, E.M. and Kahn, L., (2015). Integrating Writing in the Middle-Level Mathematics Classroom: An Action Research Study. *MLET: The Journal of Middle Level Education in Texas,2*(1), p.4.

Brendefur, J. & Frykholm, J., (2000). Promoting mathematical communication in the classroom: Two preservice teachers' conceptions and practices. *Journal of Mathematics Teacher Education, 3*(2), pp.125-153.

Brodahl, C., Hadjerrouit, S. & Hansen, N.K., (2011). Collaborative writing with Web 2.0 technologies: education students' perceptions.

Broeck, A., Vansteenkiste, M., Witte, H., Soenens, B. & Lens, W., (2010). Capturing autonomy, competence, and relatedness at work: Construction and initial validation of the Work-related Basic Need Satisfaction scale. *Journal of Occupational and Organizational Psychology, 83*(4), 981-1002.

Brown, J. S., & Adler, R. P. (2008). Minds on fire: Open education, the long tail, and learning

Buzzetto-More, N. (2012). Understanding social media. In C. Cheal, J. Coughlin, & S. Moore (Eds.), *Transformation in teaching: Social media strategies in higher education* (pp. 1-18).

Carbonneau, K.J., Marley, S.C. & Selig, J.P., (2013). A meta-analysis of the efficacy of teaching mathematics with concrete manipulatives. *Journal of Educational Psychology, 105*(2), 380-400.

Chen, Y.C., Hand, B. & McDowell, L.E., (2013). The Effects of Writing-to-Learn Activities on Elementary Students' Conceptual Understanding: Learning About Force and Motion Through Writing to Older Peers. *Science Education,97*(5), 745-771.

Christiansen, C., (2010). *Creating Classroom Communities of practice: Students as practioners of content.* (Doctoral dissertation, The Evergreen State College).

Clark, W., Logan, K., Luckin, R., Mee, A., & Oliver, M. (2009). Beyond Web 2.0: Mapping the classroom. *International Journal of Technology in Teaching and Learning, 4*(2), 134-147.

Cooper, A. (2012). Today's technologies enhance writing in mathematics. *Clearing House: A Journal of Educational Strategies, Issues and Ideas, 85*(2), 80-85.

Craig, T.S. (2011). Categorization and analysis of explanatory writing in mathematics. *International Journal of Mathematical Education in Science and Technology, 42*(7), 867-878.

Cuhadar, C. & Kuzu, A. (2010). Improving Interaction through Blogs in a Constructivist Learning Environment. *Turkish Online Journal of Distance Education, 11*(1), 134-161.

Davier, A.A. & Halpin, P.F. (2013). Collaborative problem solving and the assessment of cognitive skills: Psychometric considerations. *ETS Research Report Series, 2013*(2), pp. i-36.

Dick, T. P., & Hollebrands, K. F. (2011). Focus in high school mathematics: Technology to support reasoning and sense making. NCTM.

Fuchs, C., (2010). Web 2.0, prosumption, and surveillance. *Surveillance & Society, 8*(3), 288-309.

Galligan, L., Hobohm, C. & Loch, B., 2012. Tablet technology to facilitate improved interaction and communication with students studying mathematics at a distance. *Journal of Computers in Mathematics and Science Teaching, 31*(4), 363-385.

Gooding, J. (2008). Web 2.0: A vehicle for transforming education. *International Journal of Information and Communication Technology Education, 4*(2), 44-53.

Greenhow, C. (2011). Web 2.0 and classroom research: What path should we take? *Educational Researcher, 38*(4), 246-259.

Gunawardena, C.N., Hermans, M.B., Sanchez, D., Richmond, C., Bohley, M. and Tuttle, R., (2009). A theoretical framework for building online communities of practice with social networking tools. *Educational Media International, 46*(1), 3-16.

Henning, J.E., McKeny, T., Foley, G.D. & Balong, M., (2012). Mathematics discussions by design: creating opportunities for purposeful participation. *Journal of Mathematics Teacher Education, 15*(6), 453-479.

Jakubowicz, J. (2011). *Using Web 2.0 Technology in the Classroom: Blogging for Motivation* (Doctoral dissertation, St. John Fisher College).

Jewett, P. and MacPhee, D. (2012). Adding collaborative peer coaching to our teaching identities. *The Reading Teacher, 66*(2), 105-110.

Kenney, R., Shoffner, M. and Norris, D. (2014). Reflecting on the Use of Writing to Promote Mathematical Learning: An Examination of Preservice Mathematics Teachers' Perspectives. *The Teacher Educator, 49*(1), 28-43.

Lasker, R.D. and Weiss, E.S. (2003). Broadening participation in community problem solving: a multidisciplinary model to support collaborative practice and research. *Journal of Urban Health, 80*(1), 14-47.

Lee, B., & Ge, S. (2010). Personalisation and sociability of open knowledge management based on social tagging. *Online Information Review, 34*(4), 618-625.

Liang, S., (2014). College mathematics classroom for pre-service teachers: developing students' ability of communication that promotes deeper learning. *Scientific Journal of Pure and Applied Sciences, 3*(1), 21-25.

Malhiwsky, D.R., (2010). Student Achievement Using Web 2.0 Technologies: A Mixed Methods Study. *ProQuest LLC.*

Malhiwsky, D.R., 2010. Student Achievement Using Web 2.0 Technologies: A Mixed Methods Study. *ProQuest LLC.*

Maulana, R., Opdenakker, M. C., den Brok, P., & Bosker, R. (2011). Teacher-student interpersonal relationships in Indonesia: Profiles and importance to student motivation. *Asia Pacific Journal of Education, 31*(1), 33-49.

Murphy, J., & Lebans, R. (2008). Unexpected outcomes: Web 2.0 in the secondary school classroom. *International Journal of Technology in Teaching and Learning, 4*(2), 134-147.

Nenthien, S. and Loima, J., (2016). Teachers' Motivating Methods to Support Thai Ninth Grade Students' Levels of Motivation and Learning in Mathematics Classrooms. *Journal of Education and Learning, 5*(2), 250.

Ollis, J. C. (2011), Web 2.0 at A Non-Traditional Charter School: A Mixed Methods Study. A Thesis, Master of Science, the Florida State University College of Education

Pugalee, D. K. (2004). A comparison of verbal and written descriptions of students' problem-solving processes. *Educational Studies in Mathematics, 55,*27-47.

Roicki, J., (2008). *Effects of discussion and writing on student understanding of mathematics concepts* (Doctoral dissertation, University of Central Florida Orlando, Florida).

Rollett, H., Lux, M., Strohmaier, M., Dösingerm, G., & Tochtermann, K. (2007) The Web 2.0 way of learning with technologies. *International Journal of Learning Technology*, 3(1), 87-107.

Rosen, Y. and Tager, M., (2013). Computer-based assessment of collaborative problem-solving skills: Human-to-agent versus human-to-human approach. *Research & Innovation Network, Pearson Education.*

Sistek-Chandler, C. (2012). Connecting the digital dots with social media and web 2.0 technologies. *Journal of Research in Innovative Teaching* 5(1), 78-87. *Publication of National University, 78*

Stein, M.K. and Smith, M., (2011). *5 Practices for Orchestrating Productive Mathematics Discussions.* National Council of Teachers of Mathematics. 1906 Association Drive, 20191-1502.

Steinbring, H. (2005). Analyzing mathematical teaching-learning situations *Studies in Mathematics, 59*, 313-324.

Tatsis & Koleza, (2008) Social and socio-mathematical norms in collaborative problem-solving. *European Journal of Teacher Education, 31*(1), 89-100.

Visser, C., (2010). Self-determination theory meets solution-focused change: Autonomy, competence, and relatedness support in action. *InterAction-The Journal of Solution Focus in Organisations, 2*(1), 7-26.

Vygotsky, L. S. (1987). Thinking and speech. In Reiber, R. W. & Carton, A. S. (Eds.), *The collected works of L. S. Vygotsky.* Plenum Press, 39-243.

Wenger, E., (2008). Communities of practice: a brief introduction. 2006.*Available from the Internet at: http://www. ewenger. com/theory/index. htm (cited 5/22/2008) b17.*

Xia, C., Fielder, J. & Siragusa, L., (2013). Achieving better peer interaction in online discussion forums: A reflective practitioner case study. *Issues in Educational Research, 23*(1), 97-113.

Yackel, E., & Cobb, P. (1996). Sociomathematical norms, argumentation, and autonomy in mathematics. *Journal for Research in Mathematics Education, 27*(4), 458-477.

Zaidieh, A.J.Y., (2012). The use of social networking in education: challenges and opportunities. *World of Computer Science and Information Technology Journal (WCSIT), 2*(1), 18-21.

Zakaria, M.H., (2013). E-learning 2.0 experiences within higher education: theorising students' and teachers' experiences in Web 2.0 learning. PhD thesis, Queensland University of Technology. https://eprints.qut.edu.au/61958/

CHAPTER 19

Science Handheld Technology Instruction

Nagib Balfakih & Sufian Forawi

ABSTRACT

The rapid movement toward the use of palm and handheld devices for communication makes it likely that these particular technologies will play an important role in advancing teacher and student science experiences in the near future. The major purpose of the study is to investigate the effectiveness of the use of sensor probeware technology and guided inquiry in teaching and learning of sciences. To achieve this goal, six in-service science teachers and their 38 students participated in this study during a summer camp in Al Ain, United Arab Emirates. Science teachers were trained as part of this study to use the sensor probeware technology and guided-inquiry instruction to investigate water quality. Then each group of students investigated water quality at different locations. Qualitative as well as quantitative data were collected from teachers and students. Results indicated that students and teachers had benefited from the use of sensor probeware technology (t- test 13.784 p<.000). Participants showed great interest in conducting science guided-inquiry activities and using the sensor probes.

Keywords: Handheld Technologies, Sensor Probe, In-service Science Teachers.

INTRODUCTION

The national standards of technology, as well as, the standards of science and other various subject areas provide accountability measures to integrate technology with teaching and learning. Teacher education programs need to be responsive to new directions of performance education as suggested by global initiatives such as the Interstate New Teacher Assessment and Support Consortium (INTASC), a resource that outlines the knowledge, dispositions, and performances deemed essentials for teachers (Hodges & Bam, 2018; Squire, et. al., 2007). As an example, in using new probeware technologies, such as handled and sensor devices, the assumption is that we can replace one medium for another, keeping the benefits of traditional laboratory measuring instruction, while adding a host of new conveniences. The past experiences with innovative technologies would suggest one technology cannot be so easily swapped for another. The introduction of a new tool into human activity often changes that activity in ways unanticipated and sometimes profound. The rapid movement toward all types of palm and handheld devices for communication makes it likely that these particular technologies will play an important role in advancing students' experiences in science as well as teachers' scientific knowledge and instruction. The handheld technology selected and the ways it is implemented influences the way mathematics is taught and learnt, which in turn generates positive effects in the mathematics education (Tan & tan, 2015). Transformation of these experiences into teaching and learning has been dismal worldwide. Particularly, there isn't much research on the implications of using these new sensor probeware technologies with sciences in the United Arab Emirates (UAE).

The integration of technology in teaching is a major goal in the UAE educational curricula. Today, technology has changed the face of the world; however, its uses in education are still limited. It offers the opportunity for students to communicate and interact with multi-media learning resources and simulated environments (Holzinger, 2005). In addition, young students are naturally attracted to new technology and are more motivated to learn new concepts when they are integrated with technology. Some studies showed that the use of technology in teaching increases the students' curiosity and encourages problem and constructivist learning (Bennett, et al., 2009; Forawi & Liang, 2005; Sharples, 2005; Sharples, et. al., 2002).

With public schools, the situation is completely different; UAE has good commitment from the principals as well as teachers and all of the stakeholders of education to work on and support students' progress and attainment by applying the most recent and modern methods of technology in classrooms, however, as it is highly believed that even the use of computers was not fully exploited as a teaching and learning method in all schools of the UAE. It has been announced at the beginning of the year 2011 that five initiatives for information technology in the schools of Abu Dhabi and that is really promising of a good future for our schools in the region.

More importantly, UAE has many future plans to well equip all public and private schools with all kinds of modern technology to keep the education system at the top of all countries and to meet the explosion of information revolution in the whole world (World Economic Forum, 2017). This realizes that using technology in education can add various media and different delivery methods that allow better learn ability that suites all learners' needs and interests. Many training workshops have been held in different areas especially in the government sector to train teachers of how to use the latest technological teaching tools as we all know that interactive teaching can be more effective than lecturing.

Universities need to take responsibility in preparing students for lifelong learning as well as the care that can manage and support such essential educational systems. While engineering and science degrees traditionally do not emphasize the importance of lifelong learning skills new programs of study are now being introduced, often using electronic portfolios to support engagement with learning objectives and reflection. As the literature shows, portfolios are often used in conjunction with assessment. However, their roles in improving learning and developing reflection are important areas note emphasizing. As indicated by similar UAE studies in this chapter, use of multimedia and electronic portfolio has opened new ways to incorporate such powerful tools to enhance learning at grade and university levels and therefore support self-learning and reflection. Despite the challenges of motivating students to develop the electronic portfolios and accommodating time spend in creating them within courses and schoolwork (Heinrich, Bhattacharya, & Rayudu, 2007), these tools can still be effective to use in education in developing countries. In particular, students' experiences in developing the 'programatic' e-portfolio at the United Arab Emirates University are recognized and they promise improvement in student achievement at higher education in the country.

Educational institutions need to change their organizational structure and processes if they want to succeed in adapting and utilizing available technology in the educational field as they are the basic requirement for integrating learning technologies (Tan & Tan, 2015). It is found that to incorporate new technology will be almost impossible with same traditional structure of educational organizations.

Technology concerns knowledge not merely artifacts. To transfer it effectively requires prepared minds on the part of the receivers and some measure of shared cognitive frameworks. The culture has to change as well to make any necessary changes a reality. It also requires coordinated policies on investment, education and training, employment, the economy and development. Thus, technology is the need of today and the future, and the application of this technology means a leap towards success. Consequently, an attention should be paid to introduce technology to our students at an early age in schools to have a shortcut from present into future progress. Finally, it is so important to mention that the UAE is keen to create equal educational chances for our students to get their full rights just like other students in other countries (Shaw, 2002).

There are some trials to use technology, basically computers, in classrooms focusing mainly on data display. The use of technology could be extended to let the students themselves construct their learning anywhere and anytime. This could be achieved by the use of mobile learning. This application could be enhanced with e-learning (Forawi & Liang, 2005; Tatar, et. al., 2003).

The shift towards more authentic assessment in science in the UAE grading system requires a shift in the way students learn and conduct activities. The learning will change from teacher-centered to student-centered. The students will actively participate and construct for themselves meaningful knowledge and experiences. It is believed that the use of new technologies such as handhelds and mobile devices could achieve this goal (Holzinger, et. al., 2005; Motschnig-Pitrik, et. al., 2002).

Mobile learning (ML), also known as m-learning, is a new way to access learning content using mobiles. Mobile learning supports, with the help of mobile devices, continuous access to the learning process. This can be done using devices like your phone, laptop or tablet. You can learn wherever and whenever you want! (Pedro, Barbosa, & Santos, 2018). With the advent of mobile learning, educational systems are changing ML is enhanced by the use of constructivist teaching methods, such as guided-inquiry. Students work in small groups to test hypotheses and investigate projects. All students are required to select a problem that relates to their community. From this perspective, all the activities that students do are central to constructivist learning. Students use scientific inquiry and problem-solving methods which are in the heart of constructivist learning approach (Newell, et. al., 1972; Holzinger, 2005). Students explore with the guidance of their teachers and the help of other group members. MLearning in science is conducted through a question, use of a local problem, and dealing with an attractive device by students to independently achieve their goals. Roschelle (2003) reports that both students and teachers respond favorably to handheld applications; however, their impact in the classroom has still to be investigated.

Recent theories of learning focus on the nature of cognition as a way of thinking, and interacting with a community setting (Squire, et. al., 2007; Wenger, 1998). It is seen in the physical context and how the person is interacting with the learning environment (National Research Council. 2000; Sternberg, et. al., 2005). For learning and teaching, wireless handheld devices (WHDs) support social interactivity, are contextually sensitive, facilitate cognition distributed between people and tools or contexts, and provide individualized scaffolding (Klopfer, et. al., 2003). When used in conjunction with constructivist learning principles (Salomon, 1993) and guidelines for differentiating instruction (Tatar, et. al., 2003), handhelds have the potential to change both what and how we teach. This study, therefore, aimed to incorporate handheld technologies into guided inquiry teaching to address a UAE community problem to enhance teachers' and students' learning experiences.

RESEARCH METHODOLOGY

Statement of the Problem and Questions

Science standards emphasize the use of inquiry as a teaching method, and for students to be active participants. In addition to that, the integration of technology in science assessment is essential for improving students' science learning and process skills. Therefore, this study has two aims, first, to enhance teachers' and students' scientific knowledge and skills in using sensor probe technologies. Second, is to investigate the use of guided-inquiry skills regarding incorporation of sensor probe technology in science teaching and learning.

In-service teachers are required to incorporate technology into science teaching. This study focused on investigating teachers' and students' science knowledge and skills of the use of the new palm and sensor probe technologies. This rationale is based on the importance of documenting results on the effectiveness of such technologies with teachers and students, especially in the UAE. The major questions investigated were:

1. What can in- service teachers gain from a workshop on the use of palm, sensor probes, and scientific guided inquiry?

2. How do in-service teachers view the effectiveness of the use of probeware technologies and scientific guided inquiry?

3. What, if any, is the impact of probeware technologies and scientific guided inquiry on K-12 students?

Based on these major questions, the researchers used a mixed method approach to collect quantitative and qualitative data in this study. Surveys, test, interviews and observations were used to gather data which will be reported and analyzed at the end of the study.

Sample

Six in-service science teachers and 38 high school students were selected randomly from 100 teachers and 1500 students who participated in a one-month summer camp. Each teacher had more than seven years in teaching. The teachers' majors were in general science, chemistry, physics, and geology. None of the teachers had used handheld technology in their science teaching before. All teachers attended eight hours of workshop. Each one of them supervised a group of about six students. The total number of the students was 38; 8 students in 9th grade, 15 students in 10th grade and15 students in 11th grade. The 9th and 10th grade students were combined in a group for the purpose of the analysis because the topic of water quality is presented in the 11th grade curriculum in the UAE.

Instruments

Several assessment instruments used: (1) Project Participant Goal Survey, (2) Science Guided-inquiry Survey, and (3) Water Quality test.

The first and second instruments were only used with in-service science teachers. However, the third instrument was used with K-12 students. Follow up interviews were conducted with a convenient partial sample of teachers.

Quantitative and qualitative data was collected from participating in-service science teachers as well as K-12 students.

Procedures

The workshop was conducted for in-service science teachers one week prior to the summer day camp, in how to use handheld technology to test water quality in five criteria: pH, calcium, fluoride, nitrate, and ammonia. Then at the beginning of the summer camp, participating students were given the water quality pretest. Then, they were instructed by teachers on how to use sensor probeware, applied to water quality activities using guided-inquiry approach in which students were asked particular questions to investigate, collect, and test water samples. The students had the chance to collect the water samples from different locations, homes, the Gulf, Ayn Alfayda spring, bottled water, etc, for a two-week period. In the third week the students wrote a full report including introduction, data collection, data analysis, discussion, and conclusion. In the fourth week students presented their work to the camp's participants and displayed it on posters for visitors. In the last day of the summer camp a posttest of the water quality test was given to the students while the two instruments, the Project Participant Goal Survey and the Science Guided-inquiry Survey, were administered to teachers. A follow up interview was given to students two weeks after competing science camp.

Results and Discussions

Quantitative and qualitative data collected from the various instruments were described and coded in an appropriate format. Data, then, was analyzed using qualitative and statistical methods. The data collected answered the research questions. The two sections, Teacher Results and Student Results below presented and discussed main results of the study.

Teacher Results

All teachers had indicated that handheld technology enhanced their use of guided inquiry science instruction. It allowed students to better comprehend science materials. None of the participating teachers had prior experience with the sensor probe technology. As part of the teachers' demographic data, Table 1 indicated that two of teachers represented each of the three major levels, elementary, middle and high school. Three teachers had background in general sciences while the other three had background in chemistry, geology, and physics one each. One teacher had teaching experience of less than two years, while the other five teachers had more than seven years of teaching experience. Four teachers had bachelor's degrees in science, one had a Bachelor of Education, and one had a diploma in education. These teacher demographics help us to understand the context of the teacher sample and subsequently the result of the study in general. The teacher sample is very small; hence, it may be considered as a case study regarding the UAE context. However, teachers' impact on students, in this study, is measured quantitatively as presented in the student section.

These results support previous research on the importance of science and technology in advancing teaching and learning (e.g. (Klopfer, et., al., 2003; Motschnig-Pitrik, et, al., 2002; Sternberg, et. al., 2005). Also, the new addition to the research is that this study was conducted with teachers and students in the United Arab Emirates where it was the first of its type in the country. The results explored questions and issues with regard to science instruction and technology in the UAE. In particular, teachers found the workshop on how to teach investigative water quality topics with guided inquiry to be beneficial and worth teaching during the day camp experience.

Table 1: Teachers' Survey Responses

#	**Items**	**SA%**	**A%**	**N%**	**DA%**	**SD%**
1	I believe the workshop was a ositive experience	66.7	33.3			
2	Sensor probe technologies were easy to use	66.7	33.3			
3	My interest in the subject increased as a result of taking the workshop	66.7	33.3			
4	I was able to better understand concepts taught using sensor probe technology	50.0	50.0			
5	I increased my professional networking	66.7	16.7	16.7		
6	Instructors used sensor probe technology effectively to engage participants	83.3	16.7			
7	Instructors demonstrated sensor probe technology effectively	83.3	16.7			
8	Instructors were knowledgeable of workshop's content	83.3	16.7			
9	Instructors clearly outlined workshop's objectives and materials	83.3	16.7			
10	Guided-Inquiry activities were connected to workshop's objectives	66.7	33.3			
11	I am familiar with science guided-inquiry instruction	50.0	16.7			
12	Activities increased my skills in using guided-inquiry	66.7	33.3			
13	Instructors demonstrated well guided-inquiry instruction	66.7	33.3			
14	Use of sensor probe facilitated use of guided inquiry	83.3	16.7			
SA= Strongly Agree, A= Agree, N= Neutral, DA= Disagree, SD= Strongly Disagree						

The results also showed how it is effective to introduce these new technologies in an informal setting, ex. science day camp. While teachers were teaching at all the three different level of schooling, they all have indicated ease and effectiveness in dealing with the new technology and the guided inquiry approach to teach the water quality theme, which is taught at the high school level in the UAE. Most of previous research has been done on a particular science field, e.g. biology, or chemistry or physics (Tatar, et. al., 2003), however, this study had teachers who teach general science and ecology as well.

Student Results

The student sample revealed several results based on the students' participation in the summer day camp water quality activities, their pre and post test responses to the content-based water quality test developed by the researchers using UAE curriculum framework on the subject, and the follow up interview two weeks after completing the summer camp. These results explored and answered the third major study's question below:

> 3- What, if any, is the impact of probeware technologies and scientific guided inquiry with K-12 students?

A group comparison was conducted using independent samples t-tests to determine if there was a significant difference in average of the obtained 20 item scores between pre and post test. Grade levels were also compared with the sample students for the same score. For calculation purpose, the average of the 20 item scores were added together as dependent variable, overall (pre & post), and grade (9th &10th, 11th) were used as independent variables in the t-test. Two major results from student test are:

- Students' performance has improved significantly from pretest to post test using 2 tailed t-test, $t = 13.784$ $p <.000$. (See Table 3). This indicated that the instruction provided during the summer day camp, using sensor probe technologies and guided inquiry, has influenced change of participating students' water quality knowledge as measured by the water quality test.
- The statistical results indicated that there was no significant difference between pre and post test, between combined 9th &10th and 11th students in average item score. This result showed that, despite the improvement of the overall water quality performance of participating students, the grade level, being 9th &10th or 11th had no impact on result.

This is considered a pioneer study in which UAE students were participated. A major criterion for recruiting the students was their interest in science and the attempt to increase their scientific knowledge. Use of open-ended guided inquiry activities, as well as the new sensor probeware seemed to significantly influence the results on this study. As reported in the mobile learning literature review, the use of such sensors advances learning allows for greater mobility and preserves environment by not disturbing it by collection of data, but rather making the measures on the spot. Additional studies are recommended to investigate the variable examined in this study and others with students to shed light on use of sensor probeware and inquiry instruction in developing countries.

Table 2: Paired Samples Statistics

		Mean	N	Std. Deviation	Std. Error Mean
Pair 1	Pos	13.7105	38	2.07830	.33715
	Pre	6.5789	38	2.60593	.42274

		Mean	N	Std. Deviation	Std. Error Mean
Pair 1	Pos	13.7105	38	2.07830	.33715
	Pre	6.5789	38	2.60593	.42274

Students who attended the summer camp were 9th, 10th, and 11th grade students. The study's convenience sample included only 10th and 11th grade students. The students' main reason for attending was to have informal educational and leisure instruction to stay away, as possible, from the very hot weather of the UAE. The temperature can reach 45 degrees during the summer months. The students, as well as the organizers, did not plan to have topics and activities related to sciences at the camp. Rather, they preferred having topics related to arts. The researchers convinced them that volunteer teachers will have training in water quality testing as part of the study and that the students will take part in using the new sensor probe technologies. So, they agreed to include our science instruction and consequently allow us to collect data. To report on students' science background, especially in water quality testing, only 15% of the participating students had ideas about water quality before joining the summer camp. Those students were 11th grade students whom their major was science. The water quality topic is a project assigned by the Ministry of Education to 11th grade science major students. However, previously students used only calcium and chloride tests in those instances. In addition, the instruments used were old and were handled only by the teachers.

The rest of the 11th grade participating students had no formalized knowledge and experience in water quality topics. But they had covered related topics and used some simple measures, such as acidity test using pH strip, and determining nitrate using very complicated instrument, that they could not identify, which was done by the teacher. On the other hand, participating 10th grade students knew only how to determine the acidity of a solution using litmus paper. To sum up here, all participating students knew little about the water quality testing prior to taking part in the study.

At the beginning and before handling the sensor probes and handhelds, the students were assigned to groups of about 3 to 4 students and asked to collect data regarding the water quality standard. All groups had their reports one day before the workshops started. Then the teachers introduced the students to the new technology and taught them how to use it to measure temperature. Then the next step was the $_{P}H$ measurement. At this point, the students became excited because they had the opportunity to test variety of water samples. They tested drinking water, mineral water, and distilled water. They found out that the water samples had different $_{P}H$. Then the teacher trained them to use other sensors for calcium, chloride, ammonia, and nitrate. Students were asked to bring in a water sample of choice the next day. Students were very excited to bring water samples from many places that even the teachers themselves did not expect. Samples were brought from home, mineral water, swimming pool, distilled water, well, water their camels and goats drink, water that had used by them in washing face, spring water, water from the gulf, and even a sample from the Dead Sea. All groups were self motivated and collected data cooperatively and with enthusiasm and provided a report at the end of the session. Students had learned a lot from this experience. They understood why they should not use well or spring water to wash clothes. They found that their swimming pool water had a large amount of chlorine. They felt that their drinking water is very valuable. They indicated that if they wasted water in the past because of ignorance, today their attitude is changed as a result of the experience they had during the summer camp to not waste the valuable water of the community.

The news about this experience was presented in the camp site's newsletter. Other locations contacted camp supervisors asking them to have this water quality testing and technology probes as part of their program. The good thing about this camp was that even though more than 50% of the students were not science majors they expressed their enjoyment and interaction with science activities and materials. Overall filed notes and interviews showed how the experience was enjoyed by teachers and students alike.

Participating students responded to a follow up four-question interview two weeks after completing the summer science camp. Students' responses were translated from Arabic to English and checked for commonalities and main perceptions on how effective was use of the sensor probe technologies in impacted their learning were presented below.

What benefits have you gained from using the handheld sensor technologies and conducting the water quality tests?

Most students have mentioned that they gained new knowledge and skills related to technology and water quality tests. In addition to that they gained new information about the water that they use and drink. They equally enjoyed working with their teachers and peers in an informal setting—meaning the guided-inquiry approach. This was a good indication that students were able to use guided-inquiry instruction effectively.

> *"This is the first time, for me, to use the handheld and sensors. I think they are very good and can help me to understand science better. Our teacher had made it fun and exciting experience for us in using them."*

Did the project increase your awareness about water quality and its value?

All the students mentioned that the project enabled them to be introduced to water quality issues. Students were allowed to search the internet for additional related information and standard measures of water. This gave them a lot of information about the water that they drink and how to maintain its quality. They said a lot of their previous behavior will change such as wasting water and adding chemicals to the water that change its quality. They will rather preserve water, use it wisely, and communicate that to other family members and friends. It is worth mentioning that the UAE is considered a country which has drinking water shortages.

> *"I will definitely change my attitude regarding use of water and not waste it, as I used too."*

Do you think you can contribute to keeping our environment safe?

All of them said yes. They gave very nice suggestions such as not using excessive fertilizers, not throwing chemicals in water, treating factories waste properly, and not allowing ships to spill crude oil in the Gulf. They stated that other individuals in their surroundings should have similar awareness to make a greater impact. They suggested that the UAE should have tough legislated policies to preserve water and the environment.

Has any of the water quality testing results surprised you?

Students were not that surprised. Most results were expected, except the accuracy by which they tested samples repeatedly. This is one very evident advantage of the use of sensor probe testing in which one reading with two decimal places is displayed. Unlike students arguing about what color indicated what level of measure as in the traditional devises such as paper strips. The only surprises they had were that they found our drinking water to have levels of chemicals such as chloride and calcium.

To discuss the students' overall responses to the questions above, science teaching by using handheld technology provided the students the freedom to answer essential question during summer classes, which was: what water in your region fulfill the water quality standards? Students collected samples from different locations and found this experience a very interesting and useful one. Some of the students expressed their excitement for learning science through this strategy, guided inquiry. They indicated a wish for all science concepts to be taught this way- meaning using guided-inquiry approach. They enjoyed the freedom of investigation, independency in carrying on the hands-on activities, and the integration of technology with science and society. This study had a particular focus as indicated in the introduction however such qualitative responses opened a door for more understanding of issues related to the study. The water quality test indicated that students scored in the posttest significantly higher than the pretest. Another observation was that participating students showed interest and satisfaction of the probeware activities. While students were excited to use these new sensor probe technologies, they were equally enthusiastic in conducting on site guided-inquiry activities as well as science process investigations that they were part of identifying their sites and materials.

Implications and Conclusion

The following are a few possible implications related to the present study. The percentage of use of mobile for communication in the UAE is the highest in the world. However, the use of handheld technology in this study was introduced the first time ever in the UAE. No doubt this project will contribute to close the gap in science education and technology in the UAE, especially, if the application of this device finds its way to schools. Results of this study will help the researchers to establish a framework to train in-service as well as pre-service science teachers to effectively use handheld technology. The researchers' future plan is to integrate handheld technology in school science projects and have an interactive web site to allow greater participation. In addition, handheld technology can be connected to a computer and students will be able to manipulate many variables and make predication and inferences.

- The study was conducted by gathering data on the use of new technologies in science teaching. Palm and handheld devices have been used by many students and teachers in educational and non-educational contexts. However, they have not been used in schools in the UAE. This study attempted to introduce use of palm handheld and sensor probes to enhance science teaching and learning in the country.

- The use of the equipment is essential to gather data on the effective use of palm handhelds and sensor probes in this region. Students, as well as teachers, collected data from the field without harming or disturbing the environment, by using easy, effective, and accurate scientific instruments.

- This study is considered a pioneer in investigating the effectiveness of use of these technologies in this country. Consequently, the study results are expected to provide insights into the appropriate use of palm handheld and sensor probe technologies with K-12 students.

- Finally, the equipment improved participants' scientific inquiry knowledge and skills by demonstrating sample of relevant hands-on science activities. Participants benefit from integrating the use of inquiry science approaches and technology to impact K-12 student learning.

- The study was conducted on a small number of students in one Educational Zone. Additional studies are needed to be able to generalize the findings.

A study by Omer (2014) that interviewed some UAE teachers indicated main responses such as: 'It is wonderful that we can use different kinds of technology to help our students learn in an effective and a fun way.' 'Using Computers adds a lot to education as it is continuation of the lifestyle of the present generation.' As all know most children use mobiles, computers,...etc. in their homes for hours and they like using it especially at school so as teachers, we have to take advantages of this situation and make computers available for them for the sake of their learning. Some teachers implied that their school keeps on forgetting the importance of providing us with special training to enable us as teachers to use technology in the right way and to work on improving our skills and abilities in teaching our students. These added to the need and challenges of use of technology related devises in UAE context (WEF 2017). In particular, training and professional development (Forawi, 2015) is a crucial factor as teachers always need adequate training in any new technologies, such as the science sensor probeware.

In conclusion, while this was a pioneer study that explored the use of sensor probeware and guided inquiry instruction with UAE teachers and students, there were few studies found which related to science sensor probeware that support the aforementioned results. The literature review, however, suggested the need for addressing how the experiences gained by such devices contribute to science teaching and learning. Therefore, this study has provided valuable information on this regard. Further investigation should be done with similar samples. Participants in this study, teachers and students, have developed interest and understanding of guided-inquiry activities and sensor probe technologies. The outcome was valuable as well as it relates to informal science camp experience. More studies will be recommended to further support the findings of this study. Teachers should be followed up during the year to investigate the usefulness of their summer experience with sensor probeware technologies and use of guided inquiry with other students.

REFERENCES

Bennett, K. R.& Cunningham, A. C. (2009). Teaching Formative Assessment Strategies to Preservice Teachers: Exploring the Use of Handheld Computing to Facilitate the Action Research Process. *Journal of Computer in Teacher Education*, 25, 3, 99-105.

Brooks, J. G. & Brooks, M. G. (1993). *In search of understanding: The case for constructivist classrooms.* Association for Supervision and Curriculum Development.

Forawi, S. A. & Liang, X. (2005). Science electronic portfolios: Developing and validating the scoring rubric. *Journal of Science Education,* 2, 6, 97-99.

Hodges, T & Baum, A. (2018). *Handbook of Research on Field-Based Teacher Education.* Eds. IGI Global Publisher.

Holzinger, A. (2005). Multimedia Basics, Volume 2: Learning. *Cognitive Fundamentals of multimedia Information Systems.*

Holzinger, A. & Motschnik-Pitrik, R. (2005). Considering the Human in Multimedia: Learner-Centered Design (LCD) & Person-Centered e-learning (PCeL). In: Mittermeir, R. T. (Ed.) Innovative Concepts for Teaching Informatics. 102-112.

Klopfer, E. Squire, K. & Jenkins, H. (2003). *Augmented reality simulations on handheld computers.* Paper presented at the 2003 American Educational Research Association Conference.

Motschnig-Pitrik, R. & Holzinger, A. (2002). Student-Centered Teaching Meets New Media: Concept and Case Study. *IEEE Journal of Educational Technology & Society.* 5, 4, 160-172.

National Research Council. (2000). *How people learn: Brain, mind, experience, and school* (Expanded ed.). National Academy Press.

Newell, A. & Simon, H. (1972). *Human Problem Solving.* Prentice-Hall.

Pedro, L., Barbosa, C. & Santos, C. (2018). A critical review of mobile learning integration in formal educational contexts. *International Journal of Educational Technology in Higher Education,*15(10), https://educationaltechnologyjournal.springeropen.com/articles/10.1186/s41239-018-0091-4

Roschelle. J. (2003). Keynote paper: unlocking the learning value of wireless mobile devices. *Journal of Computer Assisted Learning.* 19, 260–272.

Salomon. G. (1993). *Distributed cognitions: Psychological and educational considerations.* Cambridge University Press.

Sharples, M. (2005). The design of personal mobile technologies for lifelong learning. *Computers & Education*, 34, 177-193.

Sharples, M., Corlett, D. & Westmancott, O. (2002). The Design and Implementation of a Mobile Learning Resource. *Personal and Ubiquitous Computing*, 6, 3, 220-234.

Squire, K. & Jan, M. (2007). Mad City Mystery: Developing Scientific Argumentation Skills with a Place-Based Augmented Reality Game on Handheld Computers, *Journal of Science Education and Technology, 16, 1, 5-29. Feb.*

Sternberg, R. & Preiss, D. (2005). (Eds.). *Intelligence and technology: The impact of tools on the nature and development of human abilities*. Lawrence Erlbaum Associates.

Tan, C. & Tan, P. (2015). Effects of the handheld technology instructional approach on performances of students of different achievement levels. *Computers & Education, 82, 306-314.*

Tatar, D., Roschelle, J., Vahey, P. & Penuel, W. (2003). Handhelds Go to School: Lessons Learned. *Computer,* 36, 9, 30-37.

Wenger, E. (1998). *Communities of practice: Learning, meaning, and identity*. Cambridge University Press.

World Economic Forum (WEF 2017). How digital technology is transforming Dubai. https://www.weforum.org/agenda/2017/05/how-digital-technology-is-transforming-dubai/ .

PART 7

CHAPTER 20

Language of Tertiary Level Science Instruction

Nesrin Tantawy

ABSTRACT

As a result of the supremacy of English that globally sweeps across the higher education landscape, English is seen to have played a crucial role in scientific literacy in its fundamental sense which refers to coaching students on how to synthesize scientific texts, and its derived sense or students' being well-founded and knowledgeable in science. It is widely believed that in order to attain scientific literacy, learners should be proficient in the language through which science is delivered. Science is distinguished from other epistemologies because of the consistent need for experiential standards, rational justifications, sound arguments, and plausible analysis and reasoning.

INTRODUCTION

As a result of the supremacy of English that globally sweeps across the higher education landscape, English is seen to have played a crucial role in scientific literacy in its fundamental sense which refers to coaching students on how to synthesize scientific texts, and its derived sense or students' being well-founded and knowledgeable in science. It is widely believed that in order to attain scientific literacy, learners should be proficient in the language through which science is delivered. Science is distinguished from other epistemologies because of the consistent need for experiential standards, rational justifications, sound arguments, and plausible analysis and reasoning.

Language is universally considered as a rudimentary vehicle of globalization, a term described by Doiz et al. (2013a) as a 'ubiquitous concept' that is inextricably intertwined with the multifaceted process of academic internationalization which is in turn defined as "the policies and practices undertaken by academic systems and institutions-and even individuals to cope with the global academic environment" (p. xvii). The galloping phenomena of globalization, internationalization or the "the implementation of specific measures to tackle the global context" (Doiz et al., 2013b, p. 1407), and modernization have elected the global English language as today's lingua franca, a commonly used language for inter-communicational functions among non-native speakers (Canagarajah, 2006; Phillipson, 2008; Weber, 2011; Dahan, 2013; Taguchi, 2014; Dahan, 2014; Chapple, 2015).

English Language Supremacy in UAE

As stated by Hopkyns (2014), English has not been pervasively used across the United Arab Emirates (UAE) until the oil discovery followed by the dramatic economic, social and infrastructural growth in the late 1950s. Since Emirati Nationals make up almost 20% of the population residing in the UAE (Findlow, 2006; Dahan, 2014), expatriates are being imported from around the world to work in the fields of medicine, construction, education, business, and retail work, the situation that has led to the current demographic and linguistic diversity. The drive to communicate has impelled such multilingual and multicultural communities to adopt English as the lingua franca while Arabic, the official language, has become the third most spoken language in the UAE (Hundley, 2010, as cited in Raddawi & Meslem, 2015). Accordingly, English has become deeply ingrained in the educational system of the country (Al-Issa & Dahan, 2011; Moore-Jones, 2015; Boyle, 2011; Dahan, 2014). Findlow (2008) emphasizes the centrality of higher education as a primary element in UAE's pursuit of an international identification with modernism, liberalism, power, and equality; regarding English as the language of science and academia, the UAE government has contributed to endowing the English language with legislative recognition through promoting it as the medium of instruction in UAE higher institutions (Belhiah & Elhami, 2015; Al-Bakri, 2013; Moore-Jones, 2015).

As put by Alhamimi (2015), English is entrenched in a set of cultural, political, economic and even social practices; hence, endorsing an English-medium curriculum in order to guarantee access to a specific professional, cultural or economic status is equivalent to applying profound changes to domestic, religious, and social affiliations. Recently, universities across the Arab Gulf, where English is not the first language, have been progressively offering more English-medium degrees as local students are supposed to be taught solely in English by expatriate teachers as a means of coping with the global status of English (Chuang, 2015, p. 63; Roche et al., 2015). Conforming to the fact that English plays a leading role in the global market, the UAE finds no choice other than preparing its workforce to function in the world economy through achieving a proper command of English (Troudi, 2009).

English is the medium of instruction in most science subjects in almost all levels of education; such scientific courses are active areas of knowledge in which almost all new discoveries and relevant information are offered in English (Ismail et al., 2011). However, some studies contend that EMI courses might have negative impact on the overall learning of the science-content subject (Chuang, 2015). Troudi (2009) questions the effectiveness of EMI in the UAE and the Arab world saying that students are more eligible to excel in academic subjects when are taught in their mother tongue as the most familiar language to them. Moreover, learning is believed to be a means through which students engage in learning activities, communicate, and interact in order to exchange knowledge; for sound learning, students should then be well-acquainted with the medium of instruction (Shahzad et al., 2013). Despite the fact that scientific knowledge is broadly offered in English, EMI "presents academic and social challenges in science education at university undergraduate level" as claimed by Alhamami (2015, p. 105).

More to the point, language policy of all academic institutions is often foisted by a governing body of authority which determines learners' academic performance in all content subjects including science. Teachers and learners, as the two main stakeholders in this process, are equally influenced by such policies and practices, yet their opinions and attitudes are "rarely considered, and usually excluded from this vital decision" (Alenezi, 2010, p. 2; Belhiah & Elhami, 2014; Alhamimi, 2015).

English-Medium Instruction (EMI) in the Global Eye

A medium of instruction represents the language by means of which teaching "non-language academic/ content subjects", e.g. science, happens (Lo & Lo, 2013, p. 47). In step with the escalating demand for globalization, English as the unsurpassed global language has become the chief foreign language used as a means of instruction at European and worldwide universities (Doiz et al., 2013a). Teaching English is sweeping across the educational system of many countries around the world; in such countries, English has no official status but bears symbolic supremacy as a global language that determines intellectual and even social ranks in the international arena; the need for internationalization of the educational systems has compelled a number of countries to make English proficiency, which is the measurable performance on a real-world task that integrates a number of sub-skills (Nunan, 2012), a national policy target. English-medium education represents academic programs being delivered in English as a medium of instruction in order to build students' English professional skills, develop their knowledge of diverse academic disciplines, and prepare them to be actively involved in the international community. In this context, English is seen as a tool not a subject; mastery of the English language is considered a by-product of obtaining content, academic knowledge. As a vehicle for attaining modernization and scientific development, EMI is used in several countries as an internationalization strategy in the higher-education context (Taguchi, 2014).

EMI at Tertiary Level

The social worth of higher education cannot be overlooked; there said to be a significant correlation between higher education, economic development, and family and cultural values. Among the intended benefits of higher education is the professional mobility, fine life quality, and advanced social status and knowledge of world affair (Doyle & Tagg, 2014). In view of the fact that 'Englishization' and globalization are inseparable in many contexts (Marsh, 2006), internationalization is claimed to be corresponding to English-medium higher education (Phillipson, 2008). It has been evident in research that the dissemination of the English hegemony is entrenched in successful economic, social, technological advancement, and international communication (Shahzad et al., 2013; Fung & Yip, 2014; Vu & Burns, 2014; Huang, 2015). Education policy makers, thus, felt the compelling urge to produce an English-proficient labour force to compete in the global market (Troudi, 2009). Consequently, English has developed from its status as a foreign or second language to the language of academic disciplines at tertiary level (Nadeem, 2012; Ebad, 2014; Chapple, 2015; Moore-Jones, 2015; Belhiah & Elhami, 2015). As emphasized by Doiz et al. (2013a), the adoption of university study programmes being executed in a foreign language is one of the most substantial outcomes of internationalization. English has also become an influential means by which people are to be included or excluded from further studies, social and occupational positions (Hopkyns, 2014).

EMI in Science Education

As a result of the supremacy of English that globally sweeps across the higher education landscape, English is seen to have played a crucial role in scientific literacy in its fundamental sense, coaching students on how to synthesize scientific texts, which would in turn pour into its derived sense, being well-founded and knowledgeable in science (Fung & Yip,, 2014). Many scholars believe that in order to attain scientific literacy, learners should be proficient in the language through which science is delivered. Remarkably, there is a growing consensus that English is the universal language of science and research, a phenomenon that is to be viewed as a by-product of the dominance of English as a global language (Amin, 2009; Troudi & Jendli, 2011; Fung & Yip 2014). Not to mention the macro-level educational policy that endorses employing a foreign language as the language of academia and science; not only do policy makers intrinsically associate academic success and development with the English language proficiency (Syed, 2003; Moore-Jones, 2015), but they also assume that inadequate English language proficiency is a negative indicator of the students' performance in science assessment (Maerten-Rivera et al., 2010). In international scientific activities, the use of English is inevitable; scientists intend to use English, at a cost of surplus effort and time, in order to be recognized for being internationally qualified and to secure their status "in the global scientific knowledge web" (Huang, 2005, p. 393).

The growing diversity of the student population in today's classrooms places challenges for teachers in terms of enabling students of diversified languages and cultures to gain academic content-knowledge. While literacy development is demanding for all students, the challenge is immense for those learners who need to develop language proficiency alongside academic knowledge. Thus, teaching practices that support content learning and promote English language proficiency simultaneously are endorsed (Lee & Fradd, 2001). It is widely agreed by researchers that science pedagogy and English language development do not have to be treated as discrete forms of instruction; science students are not obliged to wait till they become English proficient users before they engage in scientific reasoning and gain knowledge of complex scientific content. As a matter of fact, science is believed to afford a highly communicative learning environment of hands-on activities that reduces the language burden for students. Besides, the integration of science and English pedagogy can foster higher-order thinking (Zwiep & Straits, 2013).

On one hand, science students usually lag behind when they are not familiar with the academic register of science; the instructional discourse in science pedagogy is grounded in rhetorical and grammatical features of the English language that are not well-known to students. Science teachers, on the other hand, are not inclined to explain correlations between the language patterns and genres and the meaning of scientific concepts they represent. In order for learners to internalize the scientific register they get exposed to in teacher talk, science teachers are required to mediate between the learners' present linguistic repertoire and science discourse. ESL learners who learn English-medium science need to be scaffolded as they turn into bilinguals in two senses: they acquire a second language, and they learn to use the scientific language which is unlike the everyday discourse (Huang, 2005; Amin, 2009). As a scaffolding strategy, teachers need to integrate language and science learning objectives so as to meet the literacy needs of acquiring the linguistic-scientific register by diverse learners, a notion that aligns with and endows special prominence to Content-based Learning approaches (CBI) (Stevens et al., 2009; Torres-Gúzman & Howes, 2009).

Success in science is said to be an influential gatekeeper to potential academic attainment (Ash et al., 2009). Science education in the Arab world, including UAE, takes place in a complex multilingual context where English as a medium of instruction dominates the science domain. The dispute on whether science should be taught in Arabic or in English is said to mirror the tension between the need for accessing science literacy in an international language and the ambition of buttressing a competitive community of Arab scientists conducting research in Arabic (Amin, 2009). English language learners are usually the target population of any change in science educational policies (Maerten-Rivera et al., 2010); according to Boujaoude and Dagher (2009), the quality of teaching methods and science curricula are the two main issues that encounter ESL science students in the Arab region.

Nature of Science and Linguistic-Scientific Literacy

The nature of science refers to the epistemology of science or science as an approach to gaining the knowledge, as well as the values and beliefs intrinsic to developing scientific knowledge. In turn, science can be defined as 'argument' for its inimitable patterns of making hypotheses about problems and then establishing connections through argumentation. Science is distinguished from other epistemologies through employing experiential standards, rational justifications, sound arguments, and plausible analysis and reasoning (Yore et al., 2003; Yore, 2012).

Scientific discourse implies utilizing distinctive linguistic features that differ from everyday language use; it implements a range of genres, each of which has particular linguistic aspects that help communicate scientific knowledge and reasoning, e.g. exploration, explanation, analysis, making inferences, decomposition and classification, drawing analogies, and reporting. Comparably, teacher-student and student-student interactions in the science classroom require practicing a variety of language functions such as requesting, making observations, informing, seeking clarifications, and concluding. In addition to the formal, abstract, and precise nature of the scientific register, scientists also use investigative, interpretive, and figurative language features while constructing and communicating scientific knowledge (Amin, 2009). Hence, a conceptual understanding of science extends beyond the basic acquisition of a body of scientific facts; it has to do with practicing knowledge production through using the linguistic aspects pertinent to experimentation, presentation, negotiation of scientific interpretations (Yore et al., 2003; Goldberg et al., 2009). Generally speaking, science provides ESL learners with the input necessary for developing scientific and linguistic literacy (Lee, 2005).

While linguistic literacy is defined as "the ability to consciously access one's own linguistic knowledge and to view language from various perspectives" (Ravid & Tolchinksy, 2002, p. 418), scientific literacy implies one's functional understanding of the theories, principles, concepts and processes of science and one's apprehension of the intricate relationship between science and culture (DeBoer, 2000). Norris and Phillips (2003) founded a persuasive claim about scientific literacy that is based on analysis of both language and nature of science; scientific literacy is represented in two senses: the fundamental sense that revolves around being a conventional scientifically-literate learner who possesses the basic linguistic abilities of speaking, reading, and writing science; and the derived sense that involves the learner's knowledge of science curriculum. A similar scheme is proposed by Prain (2004) as he posits that science literacy, which is the coveted outcome of science learning, is not only focused on "technical concepts and terminology", but also on "communicating science concepts and applications" through language (p. 33).

Scientific knowledge is co-constructed during social interactions that take place through language; language is a mediating tool and a semantic system for constructing scientific concepts and for developing language (Torres-Gúzman & Howes, 2009). The interrelation between language and science learning is best conceptualized in Vygotsky's Sociocultural theories of thinking and learning which consider language as being central to the regulation of thoughts (Amin, 2009). From a Sociocultural perspective, attaining scientific literacy indicates the call for looking at the dynamics of student-teacher interactions and use of language in everyday classroom-discourse; the language of classroom is believed to be resourceful for students' spoken as well as written scientific discourse (Reveles, 2009). By and large, science forms an optimal source of comprehensible language input, positive affective contexts of interest, and occasions for learners to engage in authentic communicative interactions. Thinking and reasoning processes relevant to science learning enhance the concurrent development of scientific concepts and language (Stevens et al., 2009).

Owing to the fact that scientific knowledge is communicated through language, students need to be linguistically and scientifically literate (Webb, 2010). From a wider perspective, speaking, listening, writing, and reading skills are highly valued within the scientific community as students construct new understandings of scientific concepts, access information, and inform people about science. Additionally, ESL students who do not possess the basic linguistic elements or unacquainted with their use are expected to face difficulties in reasoning scientific concepts in EMI contexts (Yore et al., 2003).

Constructivism, Sociocultural, and Student-Centred Learning Theories

Constructivism as a theory is founded on observing and studying how people learn with its major construct lies with active learning. Active learning is a process in which the teacher engages in an active discourse with the learners with the intention of facilitating and regulating their learning. In the constructivist model, the educator assumes a distinctive role of assisting learners in constructing knowledge; rather than relying solely on the teacher's knowledge or textbooks, students interact actively with the teacher as a mediator whose role is to scaffold students by bringing them closer to the content (Brandon & All, 2010). In the same vein, Social Constructivism proposes the notion that knowledge is a socially and culturally-constructed human product; meaning is created in the individual's mind through social activities and interactions with others and environment. Social constructivist research draws attention to the motivational and affective dimensions of literacy as well as the role of family members, teachers, and peers in mediating learning and their impact on classroom pedagogical dynamics (Kalina & Powell, 2009).

In an English-medium setting, English is not only a target language but also a medium of education; consequently, there should be some strategies for "building linguistic bridges between learner language and the target register" (Gibbons 2003, p. 267) in order to help students to construct new content knowledge; in teacher-student interactions, there is a substantial linguistic and conceptual gap that needs to be mediated by the use of tools such as language and cultural practices (Lantolf, 2000). Mediation, the teachers' linguistic choices in content-based classrooms, is a basic construct of Vygotsky's sociocultural theory. Similarly, Bransford et al. (2000) concludes that teachers should develop awareness of the theories of knowledge that inform the subject matter instruction and knowledge of how learners' cultural convictions and individual characteristics can shape learning. Scaffolding and contingency are two constructs that are deeply rooted in mediation; scaffolding is used by educators in order to depict the aided performance in relation to teacher-student interaction; constructivist teachers generate chances for teacher-steered and peer scaffolding to enhance meaning formation and fill gaps of knowledge

(Stalmeijer, 2015). Contingency refers to how an educator gauges the amount, type and quality of assistance needed by learners; the distance between the teacher's and the learner's talk reflects the level of scaffolding being offered within the classroom setting (Gibbons, 2003).

Constructivism has been considered as the domineering paradigm to inform science education in schools and colleges (Fensham, 2004); Constructivist pedagogy in science education fosters student-centeredness through valuing the importance of exploring the learner's existing knowledge, building on it, and clarifying misconceptions that might impede acquisition of target knowledge (Taber, 2010). Bächtold (2013) asserts that science ought to be instructed in any way that is likely to involve the active participation of learners in their own learning process which is proven to enhance comprehension, application, and retention of scientific content-knowledge. The relationship between learners and content is at the heart of student-centred learning. Constructivism is believed to align with various student-centred classroom practices; Constructivism describes content as being the means to knowledge rather than the end-product (Weimer & Weimar, 2002).

As Harris and Cullen (2010) put it, learner-centred paradigms appreciate learners' individual differences and, thus, place learning, not knowledge, as the foundation of decision-making procedures; learners' backgrounds should be utilized to enlighten the teaching-learning process. In addition, Paulo Freire (2003, as cited in Harris & Cullen, 2010) puts forward another educational philosophy that promotes student-centred learning and draws a spotlight on the concept of 'conscientization' which alludes to the students' "awareness of the sociocultural reality of their lives and their ability to take action to transform that reality (p. 43); the notion of 'conscientization' is believed to interpret much of the students' attitudes towards adopting EMI.

Teachers' Cognition and Beliefs

The study of teachers' beliefs illuminates how teachers conceptualize their classroom practices. In the spirit of constructivism, there said to be interconnections between a teacher's personal and professional life which have provoked interest in investigating the concept of teachers' beliefs regarding applying EMI principles (van Huizen et al., 2005). According to Mansour (2009), teachers' beliefs, though difficult to investigate, is a prominent psychological dimension of teachers' education that is believed to dictate their classroom instructional behaviours, planning, and decisions; he also defines beliefs as a stimulating construct of a teacher's professional knowledge that is shaped by context, content, and personality. Contextual restraints including a bureaucratic administration that imposes particular rules related to institutional international ranking, e.g. EMI policy, are substantial contributors to such disparity (Cornbleth, 2001). To put in a nut shell, any educational system being promoted in a society is inextricably connected to the cultural, political, and social contexts within which it operates; teaching-learning process cannot be fully investigated unless such limiting factors are taken into consideration.

In a similar manner, Borg (2003) uses the term 'teacher cognition' to point out teachers' knowledge, beliefs, and thoughts and their impact on teachers' career. A teacher's capability of making active instructional decisions is influenced to a great extent by their own learning experiences through schooling and professional development. Borg summarizes the situation saying that "teacher cognitions and practices are mutually informing, with contextual factors playing an important role in determining the extent to which teachers are able to implement instruction congruent with their cognitions" (p. 81). Accordingly, Tsai (2002) disputes that ample research bodies have shown evidence that teachers' beliefs about teaching and learning science as well as the nature of science is rooted in their personal school science experience including medium of science instruction.

Content-Based Instruction (CBI)

For ESL students to thrive in an English-speaking academic context, they need to be academically as well as functionally literate; ESL learners are required to employ English to access, comprehend, and analyse relationships among a wide range of content concepts. ESL students join college to go beyond simply learning English; they need to acquire pronounced expertise in various disciplines which is seen achievable only by becoming familiar with the linguistic register of these content-areas through carefully planned content-based courses (Kasper, 2000). The recent pervasiveness of CBI in foreign or second language, tertiary-level classrooms is in response to the status of English as an international academic language and as a medium of instruction, the changing demographics of students, and global communication needs (Crandall & Kaufman, 2003; Stoller, 2008).

As described by Stoller (2008) "CBI is an umbrella term referring to instructional approaches that make a dual, though not necessarily equal, commitment to language and content-learning objectives" (p. 59). CBI entails a complete integration of both language and content learning; learning language through studying subject matter is believed to help learners attain language proficiency (Hung & Hai, 2016). Content, in CBI, is a non-language, traditional school-subject; content-based approaches support purposeful use of language in which language is the medium for content learning, and content is the resource for language learning (Kasper, 2000). CBI seems to support synergistic, not sequential, proficiency in both content and language which is made possible through adopting an academic environment where learners are exposed to meaningful content-related discourse communicated in the second language. Rodgers (2014) asserts that the separation between subject-matter and language instructions reflects a more conventional, and decontextualized way of language and content teaching which would lead to a superficial level of understanding on the part of the learners.

If well implemented, CBI is claimed to enable college ESL students to develop sophisticated literacy skills which would be of much importance in their future life (Weimer & Weimar, 2002; Kasper, 2000). In content courses, students are supposed to think critically in order to direct questions as well as discuss, synthesize and evaluate information. Moreover, CBI is said to develop and refine a variety of English academic skills which should prepare college students to perform efficiently and attend to their needs inside and outside the language classroom, amongst which are reading, listening and taking notes, academic writing, and oral communications (Crandall & Kaufman, 2003). According to Vygotsky's sociocultural approach to second language acquisition, communicative competence is naturally acquired while learning about specific disciplines, e.g. science, as students need to negotiate language as well as content forms with their peers or with the teacher; teacher-student interactions in a content-based classroom enhance learners' language proficiency because such interactions offer opportunities for developing new academic register being delivered in the second language (Gibbons, 2003).

According to Babbitt and Mlynarczyk (2000), there are three approaches to CBI, namely theme-based courses, sheltered courses, and adjunct courses. While theme-based courses, referred to as 'unidisciplinary,' self-contained, content-based ESL courses (Kasper, 2000), are designed around content from other disciplines being delivered by language experts, sheltered courses are content courses that incorporate significant language and academic skills being taught by content specialists; in both classes, ESL students only are included. On the other hand, adjunct courses associate a language with a content course and might comprise non-ESL students. In the context of the current research, the implemented EMI is seen to be based in the sheltered-courses approach as it is conveyed by a science specialist; it, yet, lacks two basic aspects which are incorporating portions of language academic skills and being coordinated with language instructors.

Key Studies on EMI Worldwide

Being a very lucrative market, international education has been promoted in many universities worldwide through increasingly offering English-medium courses; English as a medium of instruction has been playing an eminent role as a means by which governments pursue identification with international programs and thus increase the number of international students, facilitate staff and student mobility, and rise on the global ranking grid (Kirkpatrick, 2014). World ranking or the institutional world-class position that depends on the number of foreign students and faculty as well as the number of scholarly-cited publications has become an imperative (Salomone, 2015). English, as a symbol of power and wealth, has prevailed as the language of global communication and education, and mastering the English language should pave the students' way for educational attainment and career progression and, therefore, serve the governments' strategic plans of accomplishing educational modernization, economic globalization, and technological innovation (Nunan, 2003).

Asian and European Contexts

Numerous studies probed into the issue of implementing EMI as an educational policy all over the world. In Hong Kong, a study conducted by Yip and Tsang (2006) yielded results that English-medium instruction (EMI) students encountered more learning problems in learning science than their Chinese-medium instruction (CMI) peers, a situation that was ascribed to the nature of subject matter that requires familiarity with scientific terminology and abstract thinking, and thus demands higher language proficiency level. On the other hand, high achievers showed preference for EMI which was attributed to their capability of mastering scientific language as well as content and their awareness of the requirements for global communication. Another study administered by Evans and Morrison (2011) examined the challenges of EMI from first-year university students' perspectives and how they overcame such obstacles; the results revealed that Chinese students' primary goal was developing their English language skills in order to secure their places in the professional English-oriented context despite the language-related problems they experienced in their freshman year due to the transition to EMI. Among these factors are difficulties in understanding and using discipline-based technical and academic vocabulary, grasping the content of lectures, and acquiring academic writing style. In their meta-analytical study, Lo and Lo (2013) concluded that while EMI students performed better in their second language proficiency at the expense of their mother tongue, they struggled in content subjects compared to their CMI counterparts which might imply the failure of Hong Kong educational policy of achieving additive bilingualism.

In the same vein, Lei and Hu (2014) conducted a study to investigate whether Chinese undergraduate students' English proficiency level and anxiety toward English language learning (in-class) and use (out-of-class) were influenced by implementing English-medium instruction (EMI). Adopting EMI was expected to improve Chinese college-students' English proficiency in a way that would facilitate their access to advanced knowledge available only in English and thus equip the students with whatever it takes to place them on the global society educational and academic map. Unlike many other studies that tackled the issue of EMI elsewhere around the world, Lei and Hu (2014) found that the English-medium program (EM) had almost no positive impact on the students' English proficiency, English language learning, or English language use compared to the Chinese-medium program (CM). Moreover, EMI was seen to have increased the study burden placed on the students which comes in line with what Hay-hoe et al. (2011) assert that teaching content subjects in English would only diminish the students' understanding. However, the students held positive attitudes towards EMI which was interpreted by the researchers as related to the students' educational background and their individual abilities to benefit from the program in improving their English proficiency; the latter finding resonates with what Bolton and Tong (2002), Hu (2009), and Muthanna & Miao (2015) mentioned regarding the Chinese students' eagerness towards taking further studies in English.

University students' perception was put in the limelight in Taiwan by many scholars, among which are Wu (2006), Chang (2010), Yeh (2014), and Huang (2015); most of the students in all four studies held positive attitudes towards EMI policy and perceived it as beneficial in enhancing both English language skills, especially listening and oral communication, and content-knowledge as well as augmenting students' employability and further academic studies chances; such findings are found to align with those of Ismail et al. (2011) after investigating Malaysian university students' inclination towards EMI in science and Mathematics streams. Regarding their learning issues, students imputed them to their own inadequate English competency level which in turn raised their affective obstacles and to their professors' lack of sufficient English proficiency; as a result, students developed their own cognitive scaffolding strategies, e.g. note-taking, attentiveness, peer support, and extra reading. An additional sociocultural dimension was added to the feasibility of implementing EMI through the study of Shahzad et al. (2013) on Pakistani students; English-medium schooling background, convenient teaching and learning strategies, and encouraging home environment are among the significant contextual factors that can motivate students to progress academically.

In like manner, Kagwesage (2012) organized a similar study to examine higher-education students' views of EMI policy and the challenges that might have followed. Students in Rwanda showed no different attitudes than those in the above-mentioned studies; university students were quite aware of the new demand for globalization so that they showed great willingness to work hard on refining their English language abilities. However, apart from comprehension problems caused by their low proficiency level, students mentioned other obstacles placed by their teachers that impeded their understanding of content instructions, e.g. teachers' speed and pronunciation.

Lecturers' perspective on EMI is addressed by Başıbek et al. (2014) and Werther et al. (2014) in Turkey and Denmark respectively. Turkish lecturers' attitudes disclosed a degree of disparity as they were highly motivated to implement EMI rather than the Turkish medium of instruction for its widely perceived benefits of promoting learners' academic and social future lives; however, they were realistic about their students' ability to get deeper understanding of content knowledge through mother tongue. The study had implications not only on the students' readiness for EMI but also the teachers' preparedness to adopt a foreign language as the sole medium of instruction. Danish instructors' questionnaire and interview results also reflected their acknowledgement of the importance of EMI in enticing foreign students and staff; nevertheless, their attitudes towards teaching subject matter through English were inconsistent. Some thought about how EMI would positively impact their language expertise, while others were discouraged by their fear of exposing their linguistic meagreness which, according to participants, should be resolved by offering professional development to faculty members. A third study by Guarda and Helm (2015) concerning lecturers' teaching practices after shifting to English as a partial medium of instruction was done in Italy; this study managed to establish a link between language mastery and methodological approaches on one hand, and challenges posed to both students and instructors on the other hand. Instructional language shift to a language other than mother tongue, predominantly at advanced levels, is said to have affected the instructors' pedagogical practices and teaching skills; thus, a call for an equivalent shift in teaching methodology that focuses on student centeredness and deals with anticipated cognitive problems is perceived essential. The impact of EMI on students' medical register and competence was explored by Hoekje (2011) who concludes that with regard to the clinical nature of medical field, language influences the practice of medicine in many ways among which is discourse, register, grammar, and pronunciation in the medical students' interactions with the heterogeneous patient-populations; English language skills might affect the students' future patients', not to mention employers', perceptions of their adeptness. Similarly, Olmstead-Wang (2011) highlights the fact that adequate language skills are fundamental to the presence in international medical communities, e.g. conferences and publications.

Arab World Context

Language barriers in medical education were tackled from students' and staff members' perspectives by Sabbour et al. (2010); the study results indicated that while most of the students did not regard EMI as an impediment to learning, almost half of them resorted to translating medical terms to Arabic. For faculty members, less than 50% thought that EMI forms an obstacle to students' learning in their first year only. Moreover, there was an overall rejection to the idea of Arabization of medical education in Egypt.

Conversely, in Saudi Arabia, Alhamami (2015) conducted a research study on science instructors who admitted the negative influence of EMI which posed academic and social challenges on science undergraduate students; they also disclosed their preference of using Arabic, the students' mother tongue, as the pedagogical medium which is against the institutional policy. Ellili-Cherif and Alkhateeb's study (2015) of the effects of implementing EMI policy in the Qatari higher education context presented the students' ambivalent impression of the language shift to Arabic medium instruction, after ten years of adopting EMI, in three universities; students pointed to favouring Arabic medium instruction, despite their appreciation of the high status English occupies and the peril of studying content subjects in Arabic might pose on their employment and study prospects. Omani students' attitudes resembled those of Saudi and Qatari students; according to Abdel-Jawad and Abu Radwan (2011) and Al-Bakri (2013), Omani students acknowledged the global role of English in education, yet they had cognitive and affective concerns regarding learning subject matter in a foreign language.

UAE Context

Similar studies were administered in UAE so as to explore stances held by both university students and lecturers, for playing a crucial role in informing policy makers of the learners' needs, towards EMI (Hopkyns, 2014). In Findlow's (2006) piloted study, for example, 50% of the participants preferred EMI over AMI for academic career goals. In his small-scale research on teachers' opinions, Troudi (as cited in Wachob, 2009) highly recommended a systematic use of Arabic in teaching sciences at the tertiary level as the only possible alternative that might resolve potential problems including the double burden of learning a new concept through a foreign language. Troudi and Jendli (2011) investigated Emirati university students' perspectives on EMI policy and the results were relatively informative as it added contextual and sociocultural elements to the students' language preference. The nature of previous schooling experience, learners' English competence, parental background and viewpoint of English-medium instruction had a say in the participants' attitudes. Emirati students proved to have a realistic view of the debate on instructional language of content subjects; they associated English with employability and Arabic with religion, identity and culture discourses. Belhiah and Elhami (2014) surveyed and interviewed 600 participants who made it clear that EMI had considerable positive effect on their linguistic skills which would grant them a wider employment pool; nevertheless, disadvantages of such policy for students with low linguistic proficiency were asserted by both student and teacher participants. A question about the capability of those students who possess a poor command of academic knowledge as a result of applying EMI policy was raised by both parties.

Conclusion

Due to the very unique status of the UAE as a leading country in the Gulf region and its inimitable demographic context that have led to the emergence of English as a lingua franca at all social levels (Boyle, 2011), bilingual education, though seems logical for many scholars (Raddawi & Meslem, 2015; Troudi & Jendli, 2011; Ellili-Cherif & Alkhateeb, 2015), is not the best solution for the present educational situation in the UAE. Employing the bilingual model through utilizing the Arabic and English languages in science instruction manipulates students affectively, the situation that is perceived by many as adding insult to the injury instead of healing it; students who do not speak Arabic feel insecure and undesirably stressed in a classroom setting where explanations are delivered in an unfamiliar language.

A more compatible alternative to bilingual education is a well-implemented content-based instruction through which long-term goals of EMI would be attained; CBI aims to enhance students' motivation, interest, and positive attributions in order to help them acquire content knowledge and language skills simultaneously. Considering such affective variables as students' emotional reactions plays a critical role in granting students' better learning opportunities. The approach of offering 'sheltered courses' or 'integrated instruction' that include considerable language skills and are taught by content specialists to ESL learners is seen to be efficient for equipping science learners with the required English academic skills and content knowledge (Verma, 2008).

Revisiting the current language educational policy, which is defined as "mechanism used to create de facto language practices in educational institutes, especially in centralized educational system" (Shohamy, 2014), is the first step towards resolving the consequential issues related to EMI. Language proficiency, personal attitude, and effective teaching approaches are three pillar pillars on which EMI rests (Werther et al., 2014); therefore, integrating objectives of content and language learning unveils many challenges for policy makers, program planners, curriculum designers, materials writers, teacher educators, teacher supervisors, and learners. Implementing CBI also may have implications on teacher recruitment, qualifications- including target language proficiency, certification, training, and assessment. Teachers' attitude toward CBI is another inextricable obstacle; undervaluing language experts' role in the language-content teachers' partnership negatively impacts both language teachers and learners to whom these courses are presented in the first place (van Wyk, 2014).

However, a further investigation of CBI assessment process is recommended; since it is hard to isolate content learning from language learning in the assessment procedure, teachers encounter the predicament of deciding whether students' failure to demonstrate knowledge is caused by linguistic obstacles or a deficiency in understanding the content materials (Stoller, 2008). In addition, using the learners-as-researchers approach in which students are actively involved as co-researchers would grant insightful understanding of problematic classroom behaviours. Besides, taking into account the impact of the language of science instruction on classroom dynamics would inform policy makers.

REFERENCES

Abdel-Jawad, H. R., & Abu Radwan, A. S. (2011). The status of English in institutions of higher education in Oman: Sultan Qaboos university as a model. In A. Al-Issa, & L. Dahan (Eds.). *Global English and Arabic: Issues of language, culture, and identity* (pp. 123-151). New York: Peter Lang.

Al-Bakri, S. (2013). Problematizing English medium instruction in Oman. *International Journal of Bilingual & Multilingual Teachers of English, 1*(2), 55-69.

Alenezi, A.A. (2010). Students' language attitude towards using code-switching as a medium of instruction in the college of health sciences: An exploratory study. *ARECLS, 7*, 1-22.

Alhamami, M. (2015). Teaching science subjects in Arabic: Arab university scientists' perspectives. *Language Learning in Higher Education, 5* (1),105-123.

Al-Issa, A., & Dahan, L.S. (Eds.). (2011). *Global English and Arabic: Issues of language, culture and identity*. New York: Peter Lang.

Amin, T. (2009). Language of instruction and science education in the Arab region: Towards a situated research agenda. In S. Boujaoude, & Z.R. Dagher (Eds.). *The world of science education: Handbook of research in the Arab states* (pp.61-82). United States: Sense Publishers.

Ash, D., Tellez, K., & Crain, R. (2009). The importance of objects in talking science: The special case of English language learners. In K.R. Bruna, & K. Gomez (Eds.). *The work of language in multicultural classrooms: Talking science, writing science* (pp. 269-288). United States: Routledge.

Babbitt, M., & Mlynarczyk, R. W. (2000). Keys to successful content-based programs: Administrative perspectives. In L.F. Kasper (Ed.). *Content-based college ESL instruction* (pp. 26-47). United States: Lawrence Erlbaum Associates.

Bächtold, M. (2013). What do students "construct" according to constructivism in science education? *Research in Science Education, 43*(6), 2477-2496.

Başıbek, N., Dolmacı, M., Cengiz, B.C., Bür, B., Dilek, Y., & Kara, B. (2014). Lecturers' perceptions of English medium instruction at engineering departments of higher education: A study on partial English medium instruction at some state universities in Turkey. *Procedia Social and Behavioral Sciences, 116*, 1819–1825. https://doi: 10.1016/j.sbspro.2014.01.477

Belhiah, H., & Elhami, M. (2014). English as a medium of instruction in the Gulf: When students and teachers speak. *Language Policy, 14*(1), 3-23.

Bolton, K., & Tong, Q. (2002). Introduction: Interdisciplinary perspectives on English in China. *World Englishes, 21*(2), 177-180.

Borg, S. (2003). Teacher cognition in language teaching: A review of research on what language teachers think, know, believe, and do. *Language Teaching, 36* (2), 81–109.

Boujaoude, S., & Dagher, Z.R. (Eds.). (2009). *The world of science education: Handbook of research in the Arab states.* United States: Sense Publishers.

Boyle, R. (2011). Patterns of change in English as a lingua franca in the UAE. *International Journal of Applied Linguistics, 21*(2), 143-161.

Brandon, A.F., & All, A.C. (2010). Constructivism theory analysis and application to curricula. *Nursing Education Perspectives, 31*(2), 89-92.

Bransford, J.D., Brown, A. L., & Cocking, R. R. (Eds.). (2000). *How people learn: Brain, mind, experience, and school.* Washington, D.C.: National Academy Press.

Canagarajah, A.S. (2006). Negotiating the local in English as a lingua franca. *Annual Review of Applied Linguistics, 26*, 197-218. https://doi.org/10.1017/S0267190506000109

Chang, Y.Y. (2010). English-medium instruction for subject courses in tertiary education: Reactions from Taiwanese undergraduate students. *Taiwan International ESP Journal, 2*(1), 55-84.

Chapple, J. (2015). Teaching in English is not necessarily the teaching of English. *International Education Studies, 8*(3), 1-13.

Chuang, Y. (2015). An EMI pedagogy that facilitates students' learning. *English Language Teaching, 8* (12), 63-73.

Cornbleth, C. (2001). Climates of constraint/restraint of teachers and teaching. In W. B. Stanley (Ed.). *Critical issues in social studies research for the 21st century* (pp. 73-95). Greenwich, CT: Information Age Publishing.

Crandall, J., & Kaufman, D. (2003). *Content-based instruction in higher education settings.* Alexandria, VA: Teachers of English to Speakers of Other Languages, Incorporated (TESOL).

Dahan, L.S. (2013). Global English and Arabic: Which is the protagonist in a globalized setting? *Arab World English Journal, 4*(3), 45-51.

Dahan, L.S. (2014). The legacy of Arabic: Cultural heritage versus global English and globalization in the Arabian Gulf. *International Journal of Liberal Arts and Social Science, 2* (8), 112-120.

DeBoer, G.E. (2000). Scientific literacy: Another look at its historical and contemporary meanings and its relationship to science education reform. *Journal of Research in Science Teaching, 37* (6), 582–601.

Doiz, A. & Lasagabaster, D., & Sierra, J. (Eds.). (2013a). *English-medium instruction at universities: Global challenges*. United Kingdom: Multilingual Matters.

Doiz, A., Lasagabaster, D., & Sierra, J. (2013b). Globalisation, internationalisation, multilingualism and linguistic strains in higher education. *Studies in Higher Education, 38*(9), 1407-1421.

Doyle, T., & Tagg, J. (2014). *Helping students learn in a learner-centered environment: A guide to facilitating learning in higher education*. United States: Stylus Publishing (VA).

Ebad, R. (2014). The role and impact of English as a language and a medium of instruction in Saudi higher education institutions: Students-instructors perspective. *Studies in English Language Teaching, 2*(2), 140-148. DOI:10.22158/selt. v2n2p140

Ellili-Cherif, M., & Alkhateeb, H. (2015). College students' attitude toward the medium of instruction: Arabic versus English dilemma. *Universal Journal of Educational Research, 3*(3), 207–213.

Evans, S., & Morrison, B. (2011). Meeting the challenges of English-medium higher education: The first-year experience in Hong Kong. *English for Specific Purposes, 30* (3), 198–208.

Fensham, P.J. (2004). *The evolution of science education as a field of research: Defining an identity*. United States: Kluwer Academic Publishers.

Findlow, S. (2006). Higher education and linguistic dualism in the Arab Gulf. *British Journal of Sociology of Education, 27*(1), 19-36.

Findlow, S. (2008). Islam, modernity and education in the Arab States. *Intercultural Education, 19*(4), 337-352.

Fung, D., & Yip, V. (2014). The effects of the medium of instruction in certificate-level physics on achievement and motivation to learn. *Journal of Research in Science Teaching, 51*(10), 1219–1245.

Gibbons, P. (2003). Mediating language learning: Teacher interactions with ESL students in a content-based classroom. *TESOL Quarterly, 37*(2), 247-273.

Goldberg, J. S., Welsh, K. M., & Enyedy, N. (2009). Negotiating participation in a bilingual middle school science classroom: An examination of one successful teacher's language practices. In K.R. Bruna, & K. Gomez (Eds.). *The work of language in multicultural classrooms: Talking science, writing science* (pp. 133-154). United States: Routledge.

Guarda, M., & Helm, F. (2017). "I have discovered new teaching pathways": The link between language shift and teaching practice. *International Journal of Bilingual Education and Bilingualism, 20*(7), 897–913.

Harris, M., & Cullen, R. (2010). *Leading the Learner-Centered campus: An administrator's framework for improving student learning outcomes*. United Kingdom: Wiley, John & Sons.

Hoekje, B. (2011). International medical graduates in U.S. higher education: An overview of issues for ESP and applied linguistics. In B. Hoekje, & S. Tipton (Eds.). *English language and the medical profession: Instructing and assessing the communication skills of international physicians* (pp. 3-20). Netherlands: Emerald Group Publishing.

Hopkyns, S. (2014). The effect of global English on culture and identity in the UAE: a double-edged sword. *Learning and Teaching in Higher Education: Gulf Perspectives, 11*(2), 1-20

Hu, G. W. (2009). The craze for English-medium education in China: Driving forces and looming consequences. *English Today, 25*(4), 47–54.

Huang, D.F. (2015). Exploring and assessing effectiveness of English medium instruction courses: The students' perspectives. *Procedia Social and Behavioral Sciences, 173*, 71–78. doi: 10.1016/j.sbspro.2015.02.033

Huang, H.-J. (2005). Listening to the language of constructing science knowledge. *International Journal of Science and Mathematics Education, 4*(3), 391–415.

Hung, B. P., & Hai, T. T. (2016). Teachers' and students' attitudes towards the implementation of content-based instruction in higher education in Ho Chi Minh City. *English Language Teaching, 9*(5), 106-118.

Ismail, W.R., Mustafa, Z., Muda, N., Abidin, N.Z., Isa, Z., Zakaria, A.M., Suradi, N.R.M., Mamat, N.J.Z., Nazar, R.M., Ali, Z.M., Rafee, N.M., Majid, N., Jaaman, S.H., Darus, M., Ahmad, R.R., Shahabuddin, F.A., Rambely, A.S., Din, U.K.S., Hashim, I., Azlan, M.I. (2011). Students' inclination towards English language as medium of instruction in the teaching of science and mathematics. *Procedia Social and Behavioral Sciences, 18*, 353–360.

Kagwesage, A. M. (2012). Higher education students' reflections on learning in times of academic language shift. *International Journal for the Scholarship of Teaching and Learning, 6* (2), 1-14.

Kalina, C., & Powell, K. C. (2009). Cognitive and social constructivism: Developing tools for an effective classroom. *Education, 130*(2), 241-250.

Kasper, L.F. (2000). *Content-based college ESL instruction.* United States: Lawrence Erlbaum Associates.

Kirkpatrick, A. (2014). The language (s) of HE: EMI and/or ELF and/or multilingualism? *The Asian Journal of Applied Linguistics, 1*(1), 4-15.

Lantolf, J.P. (Ed.). (2000). *Sociocultural theory and Second language learning* (2nd ed.). Oxford, U.K.: Oxford University Press.

Lee, O. (2005). Science education with English language learners: Synthesis and research agenda. *Review of Educational Research, 75* (4), 491–530.

Lee, O., & Fradd, S.H. (2001). Instructional congruence to promote science learning and literacy development for linguistically diverse students. In D. R. Lavoie, & W. Roth (Eds.). *Models of science teacher preparation* (pp. 109–126). Springer Science + Business Media.

Lei, J., & Hu, G. (2014). Is English-medium instruction effective in improving Chinese undergraduate students' English competence? *International Review of Applied Linguistics in Language Teaching, 52*(2), 99-125.

Lo, Y.Y., & Lo, E.S.C. (2013). A meta-analysis of the effectiveness of English-medium education in Hong Kong. *Review of Educational Research, 84*(1), 47–73.

Maerten-Rivera, J., Myers, N., Lee, O., & Penfield, R. (2010). Student and school predictors of high-stakes assessment in science. *Science Education, 94*(6), 937-962.

Mansour, N. (2009). Science teachers' beliefs and practices: Issues, implications and research agenda. *International Journal of Environmental and Science Education, 4*(1), 25-48.

Marsh, D. (2006). English as a medium of instruction in the new global linguistic order: Global characteristics, local Consequences. *Proceedings of The Second Annual Conference for Middle East Teachers of Science, Mathematics and Computing*, 29-38.

Moore-Jones, P. J. (2015). Linguistic imposition: The policies and perils of English as a medium of instruction in the United Arab Emirates. *Journal of ELT and Applied Linguistics (JELTAL), 3*(1), 63-73.

Muthanna, A., & Miao, P. (2015). Chinese students' attitudes towards the use of English-medium instruction into the curriculum courses: A case study of a national key university in Beijing. *Journal of Education and Training Studies, 3*(5), 59-69.

Nadeem, M. (2012). Urlish: A code Switching/code mixing pedagogical approach in teacher education. *Journal of Research and Reflection in Education (JRRE), 6*(2), 154-162.

Norris, S.P., & Phillips, L.M. (2003). How literacy in its fundamental sense is central to scientific literacy. *Science Education, 87*(2), 224–240.

Nunan, D. (2003). The impact of English as a global language on educational policies and practices in the Asia-Pacific region. *TESOL Quarterly, 37*(4), 589-613.

Olmstead-Wang, S. (2011). Developing curriculum and strategies for a Chinese-language medical university in Taiwan adopting English as a medium of instruction. In B. Hoekje, & S. Tipton (Eds.). *English language and the medical profession: Instructing and assessing the communication skills of international physicians* (pp. 149-171). Netherlands: Emerald Group Publishing.

Phillipson, R. (2008). Lingua franca or lingua frankensteinia? English in European integration and globalisation 1. *World Englishes*, *27*(2), 250–267.

Prain V. (2004) The role of language in science learning and literacy. In C. S. Wallace, B. B. Hand, & V. Prain (Eds.). *Writing and learning in the science classroom*. Springer Netherlands.

Raddawi, R., & Meslem, D. (2015). Loss of Arabic in the UAE: Is bilingual education the solution? *International Journal of Bilingual & Multilingual Teachers of English*, *3*(2), 85-94.

Ravid, D., & Tolchinsky, L. (2002). Developing linguistic literacy: A comprehensive model. *Journal of Child Language*, *29*(02), 417-447.

Reveles, J. M. (2009). Academic identity and science literacy. In K.R. Bruna, & K. Gomez (Eds.). *The work of language in multicultural classrooms: Talking science, writing science* (pp. 193-218). United States: Routledge.

Roche, T., Sinha, Y.K., & Denman, C. (2015). Unravelling failure: belief and performance in English for academic purposes programs in Oman. In R. Al-Mahrooqi, & C. Denman (Eds.). *Issues in English education in the Arab world* (pp. 37-59). Cambridge Scholars Publishing.

Rodgers, D. M. (2014). Making the case for content-based instruction. *Italica*, *91*(1), 16-28.

Sabbour, S.M., Dewedar, S.A., & Kandil, S.K. (2010). Language barriers in medical education and attitudes towards Arabization of medicine: student and staff perspectives. *Eastern Mediterranean Health Journal*, *16*(12), 1263-1271.

Salomone, R. (2015). The rise of global English: Challenges for English-medium instruction and language rights. *Language Problems and Language Planning*, *39*(3), 245–268.

Shahzad, M.N., Sajjad, S., Ahmed, M.A., & Asghar, Z. (2013). The role of "radical change" in medium of instruction and its impact on learning. *Journal of Language Teaching and Research*, *4*(1), 36-44.

Shohamy, E. (2014). *Language policy: Hidden agendas and new approaches*. London: Routledge.

Stalmeijer, R.E. (2015). When I say cognitive apprenticeship. *Medical Education*, *49*(4), 355–356.

Stevens, L.P., Jefferies, J., Brisk, M. E., & Kaczmarek, S. (2009). Linguistics and science learning for diverse population: An agenda for teacher education. In K.R. Bruna, & K. Gomez (Eds). *The work of language in multicultural classrooms: Talking science, writing science* (pp. 291-316). United States: Routledge.

Stoller, F.L. (2008). Content-Based instruction. In N. Deusen-Scholl, & N. H. Hornberger (Eds). *Encyclopedia of language and education* (pp. 59-70.). Springer Science + Business Media.

Syed, Z. (2003). The sociocultural context of English language teaching in the Gulf. *TESOL Quarterly*, *37*(2), 337-341.

Taber, K.S. (2010). Paying lip-service to research? The adoption of a constructivist perspective to inform science teaching in the English curriculum context. *The Curriculum Journal, 21*(1), 25–45.

Taguchi, N. (2014). English-medium education in the global society. *International Review of Applied Linguistics in Language Teaching, 52*(2), 89-98.

Torres-Gúzman, M., & Howes, E. V. (2009). Experimenting in teams and tongues: Team teaching a bilingual science education course. In K.R. Bruna, & K. Gomez (Eds.). *The work of language in multicultural classrooms: Talking science, writing science* (pp. 317-340). United States: Routledge.

Troudi, S. (2009). The effects of English as a medium of instruction on Arabic as a language of science and academia. In P. Wachob (Ed.). *Power in the EFL classroom: Critical pedagogy in the Middle East* (pp. 199-216). Cambridge Scholars Publishing.

Troudi, S., & Jendli, A. (2011). Emirati students' experiences of English as a medium of instruction. In A. Al-Issa, & L. Dahan (Eds.). *Global English and Arabic: Issues of language, culture, and identity* (pp. 23-48). New York: Peter Lang.

Tsai, C.C. (2002). Nested epistemologies: Science teachers' beliefs of teaching, learning and science. *International Journal of Science Education, 24*(8), 771–783.

van Huizen, P., van Oers, B., & Wubbels, T. (2005). A Vygotskian perspective on teacher education. *Journal of Curriculum Studies, 37*(3), 267–290.

van Wyk, A. (2014). English-medium education in a multilingual setting: A case in South Africa. *International Review of Applied Linguistics in Language Teaching, 52*(2), 205–220.

Verma, G. (2008). Using sheltered instruction to teach English language learners. *Science Scope, 32*(3), 56.

Vu, N. T., & Burns, A. (2014). English as a medium of instruction: Challenges for Vietnamese tertiary lecturers. *The Journal of Asia TEFL, 11*(3), 1-31.

Webb, P. (2010). Science education and literacy: Imperatives for the developed and developing world. *Science, 328*(5977), 448–450.

Weber, A.S. (2011). Politics of English in the Arabian Gulf. *Proceedings of FLTAL: 1st International Conference on Foreign Language Teaching and Applied Linguistics,* 60-66.

Weimer, M., & Weimar, M. (2002). *Learner-centered teaching: Five key changes to practice.* San Francisco: Jossey-Bass Inc.

Werther, C., Denver, L., Jensen, C., & Mees, I.M. (2014). Using English as a medium of instruction at university level in Denmark: The lecturer's perspective. *Journal of Multilingual and Multicultural Development, 35*(5), 443–462.

Wu, W.-S. (2006). Students' attitudes toward EMI: Using Chung Hua University as an example. *Journal of Education and Foreign Language and Literature, 4*, 67–84.

Yeh, C. (2014). Taiwanese students' experiences and attitudes towards English-medium courses in tertiary education. *RELC Journal, 45*(3), 305–319.

Yip, D. Y., & Tsang, W. K. (2007). Evaluation of the effects of the medium of instruction on science learning of Hong Kong secondary students: Students' self-concept in science. *International Journal of Science and Mathematics Education, 5*(3), 393-413.

Yore, L., Bisanz, G.L., & Hand, B.M. (2003). Examining the literacy component of science literacy: 25 years of language arts and science research. *International Journal of Science Education, 25*(6), 689–725.

Yore, L.D. (2012). Science literacy for all: More than a slogan, logo, or rally flag! In K. Tan, & M. Kim (Eds.). *Issues and challenges in science education research* (pp. 5-23). Springer Netherlands.

Zwiep, S.G., & Straits, W.J. (2013). Inquiry science: The gateway to English language proficiency. *Journal of Science Teacher Education, 24*(8), 1315–1331.

CHAPTER 21

Language of School Level Science Instruction

Fatema Al Awadi

ABSTRACT

This chapter presents a critical analysis of a textbook that is part of the UAE Ministry of Education, Madares Al Ghad (MAG) system, that is aimed to teach science curriculum in English medium. The research was conducted on a textbook sample from a collection of textbooks that are used in grades one to nine in the MAG systems. These textbooks were prepared in terms of how the second language is integrated with teaching another subject that was normally taught in Arabic in most of the government schools. A qualitative research method was considered as a methodology of the study through considering the science textbook analysis as an artifact. Three themes were identified which are content, structure and objectives, English skills, and the strategies. The study outcomes demonstrated the importance of language proficiency in understanding scientific concepts. It also suggested some pedagogical implications for teachers. It recommended further consideration to the use of IBL in teaching science using the target language. It also recommended focusing on issues facing teachers when teaching another subject using the second language as in the UAE.

INTRODUCTION

Science is considered to be an essential subject to be taught in both UAE federal and private schools. Thus, the language used in teaching science has been controversial as it can be taught either in Arabic, which is the first language in the UAE, or English for foreign language learners. As a result, if that particular subject is taught in English, it requires not only teaching the science content but also teaching the language needed to comprehend the context entirely. This will include a combination of teaching and learning both subjects at the same time which teachers of foreign language learners learning science must be aware of. In addition, the students must have good foundations in the English skills to be able to perform the scientific classroom tasks.

This study will examine the science textbooks that were used in the MAG system and how the English language is used.

This chapter is going to deal with the aspects of teaching science through the English language in government schools. Dealing with an example of a previous science curriculum taught in English as part of the previous MAG system curriculums. Since the Ministry of Education has canceled the MAG system in the UAE, it would be an excellent opportunity to examine the previous curriculum taught in English, which is now transformed into Arabic. Taking into consideration the analysis of the English science curriculum by reviewing its content as well, whether it has been effective in teaching students through the implementation of English as a foreign language. A further objective is to discuss the usefulness of the content in terms of the methods, strategies, objectives, assessments, available resources and outcomes. An additional aim of this research is identifying the shortfalls in the English science curriculum, try to provide recommendations and suggest some pedagogical implications that would serve teaching science in the second language.

What motivated me to write this study is the continuous discussion about whether teaching science in English needs a lot of effort than teaching in the first language. Another point is the transformation from the MAG system teaching math and science in English into teaching them in Arabic. Keeping in mind that most private schools in the UAE consider teaching science in English, which is also preferred by most parents and stakeholders. Another aspect noted when observing teaching science in MAG schools needs good understanding and ability to use the language by both teacher and students to understand the scientific concepts in the curriculum. In addition, supporting learning science concepts requires integrating language teaching strategies with experimenting science notions.

A research question was stated to help me narrow down the ideas and find conclusions for this research, and the question is as follows:

> How can the use of the English language facilitate teaching and learning science in government schools?

Related Literature

English as a language can be applied in different subjects since it is considered as a way of communication between the teacher and students. It has different usages and applications depending on the taught subject, and what mainly required to be covered (Harmer, 2012). Therefore, it requires a mutual understanding between both teachers and students to facilitate the learning of the subject (Arslan & Akbarov, 2012). Hence, the following section will provide different related pieces of literature that I used to base my research and ideas upon.

English for Specific Purposes (ESP)

With English being a language of meaning delivery to students, it is meant to provide comprehensible input on the content taken (Buri, 2012). Krashen's concept leads to shape students understanding of the concepts acquired through the use of the second language, which is used purposefully within a clear context comprehended by both teacher and students (Buri, 2012; Lightbown & Spada, 2013). Teachers teaching dual educations such as English, along with another subject are faced with some challenges that they must be aware of. The first challenge is to balance between language learning and subject needs, also to consider students' first language and cultural background (Banuelos, 2008; Slavin, 2014). Moore (2007, p. 320) stated that "Language and discourses are codependent and rely on each other for developing understanding and scientific knowledge within sociocultural contexts". However, if teachers viewed themselves as only content teachers or only language teachers, this will negatively impact on the students' learning the concept and acquiring the language (Tan, 2011; Moore, 2007). Accordingly, teachers must improve their language proficiency in order to cope and become able to teach the concepts accurately to the students, they also must develop an awareness of important key concepts in teaching other subjects (Tan, 2011). Similarly, teachers who are considered as experts in using both the language and the subject content effectively, are considered having teaching abilities with high interaction and inquiry-based activities that can facilitate learning. This require effective and continuous studies and training to enhance these abilities (Lee et al., 2009; Tan, 2011).

English Language Teaching Through Content (Science Teaching in English)

Furthermore, it has been discussed that the teacher should ease the learning process by making the meaning more comprehensible as input for learning, as mentioned by Krashen (Lightbown & Spada, 2013). Consequently, teaching science consists of teaching concepts, instructions, and experiments that need good language proficiency to support all of these factors (Hong & Diamond, 2012). As a result, Lightbown & Spada (2013, p. 193) mentioned that "students in content-based and immersion classes develop comprehension skills, vocabulary, and general communicative competence in the new language". Hence, there is a significant emphasis on the importance of classroom instructions in science classrooms, that learners need to understand to support their understanding of what is going around them in that lesson. Thus, teachers carry a vital responsibility in facilitating their understanding within the lesson (Hong & Diamond, 2012; Hernández, 2012). Thus, more considerable efforts are to be paid by the teacher to teach science vocabulary and the functions of these words in science learning (Carrier, 2013). However, teachers must understand the fact that students' assessments in science learning require a balance in assessing the content as well as the language used, which is also based on what is covered during the taught syllabus (Shanahan & Shea, 2012; Carrier, 2013). Some studies have shown that when teachers plan and set objectives for content-based instructions lesson, they tend to struggle in planning language objectives more than the science content ones (Baecher, Farnsworth & Ediger, 2014).

Science and Teaching in the UAE

Science teachers in the UAE are prepared to cope with the recent improvement in the curriculum, in order to be expertise in the subject and help students get the scientific knowledge (Tairab, 2010). Hence, preparing Emirati teachers to be proficient in science teaching in government schools was one of ADEC's goals. Therefore, a lot of hiring processes for teachers were giving the priority to teachers who are trained and studied teaching science in the second language (Kadbey & Dickson, 2014). Teachers must be confident when applying new teaching strategies and methods such as Inquiry-Based learning, which might end up impacting the learning outcomes of the science lesson (Kadbey & Dickson, 2013). Accordingly, textbooks must be designed to promote and provide teachers with the essential aspects of inquiry to facilitate students learning. In addition, these books must not add additional efforts to teachers since the main focus is teaching science (Kadbey & Dickson, 2014; Al Naqbi, 2010). On the other hand, a lot of investigations were examining the effectiveness of textbooks in the UAE to judge the content efficiency in embarking scientific concepts (Al Naqbi, 2010). Regarding the primary generalist teachers in the UAE, they are expected to be prepared to teach more than one subject in the elementary section. This will include teaching science with another subject such as English so that the teachers are expected to be developing the language and content proficiency simultaneously. Hence, a balance must be developed in order to not let one aspect exceed the teaching of another (Buldu & Shaban, 2010; Kadbey & Dickson, 2013; Kadbey & Dickson, 2014).

Teaching Science in English via Interactive Teaching Methods

Learners of science tend to grasp and understand the concept faster if it was applied through interactive learning techniques, which is facilitated through structured process (Holmes, 2006; Harmer, 2012). Many people think that this can be achieved only in English language teaching classes. At the same time, a science lesson can include games or interactive activities to support acquiring the vocabulary and concepts. Despite the fact that many teachers think that science is only a subject that is based on experiments and observations (Hong & Diamond, 2012; Isaila, 2016). Teaching vocabulary through interaction and integrating fun into learning helps to ease the learning process, which also can develop learning through building intrinsic motivation in the subject (Cameron, 2010). Additionally, Inquiry-Based learning is considered as a new critical method in teaching science, where learners are becoming active participants in learning (Dibiase & McDonald, 2015). The application of IBL in teaching science using foreign language helps to develop the students' cognitive and metacognitive abilities because it encourages them to investigate and experience using the language at the same time. The teacher's job in this class is to scaffold and feedback on students' performance in a particular activity (Lee, 2014). It also helps to grow the interest in the students' learning the science subject, and at the same time it is considered as a useful tool to assess students' learning. Moreover, the gradual movement between the stages help to enhance learning and add further knowledge and concept in the students' minds (Grover & Stovall, 2015).

RESEARCH METHODOLOGY

A qualitative ethnographical method is the research method that is used to collect the data needed for the stated topic. Therefore, as Creswell (2014, p.205) states "in qualitative research, we identify our participants and sites on purposeful sampling, based on places and people that can best help us understand our central phenomenon". Hence, the tool used for this research is considered as artifacts of textbooks used in MAG systems to provide a closer insight into how English was used to teach science in government schools. I believe that finding the textbooks supports the ideas of how the scientific concepts were taught and integrated with English language teaching. Considerably, several points will be discussed as part of the analysis of these textbooks, also will provide related pedagogical implications that will be based on the outcomes resulted from the research question. A textbook was selected for the analysis and was used by teachers in MAG schools, which is from the "Science Fusion" collection, which is divided into the teachers' and students' editions. These books consist of different units that deal with varying topics of science. The use of these books differed from one MAG school to another depending on the unit plan and each school. The main aim of the selection of the textbook is to help me identify variables to support the evidence resulted from the data analysis.

Data Analysis and Findings

This section will discuss the science curriculum followed previously by MAG systems in UAE government schools. Math and science were intended to be taught to students in grades 1-6 in the English language, whereas it changed into Arabic in the recent two years. The science textbooks supplied by the Ministry of Education in the UAE to schools were from the collection of "Science Fusion" (Dispezio el al., 2012). I will also try to consider the other ideas provided in the teachers' edition, which are provided in detail with the methods and strategies to be followed. I also intended to examine the science textbook following five different themes which are: content, structure and objectives, English skills, methods and strategies. These themes will help me answer the stated research question:

> *How can the use of the English language facilitate teaching and learning science in government schools?*

Content, Structure and Objectives

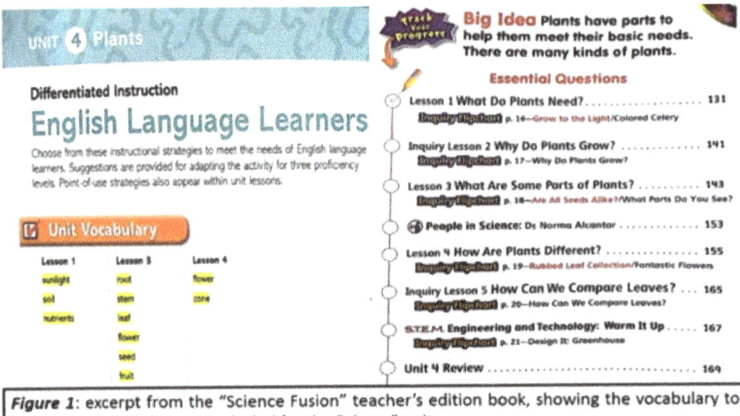

Figure 1: excerpt from the "Science Fusion" teacher's edition book, showing the vocabulary to be taught and the lessons included for the "plants" unit.

Regarding the general content of the "Science Fusion" collection of the MAG system, it was all presented through the English language as the language suites the level of EFL learners. As shown in the examined version that the language was suitable to the target level as well as using visual aids to support learning the scientific concept. It is also notable that the content was divided into themes such as; work like a scientist, technology and our world, all about animals, all about plants, environment for living things, earth and its resources, etc. Additionally, I found that each unit includes different lessons and lessons of inquiry, and each lesson included specific items of the language to be learned and science concepts (figure 1).

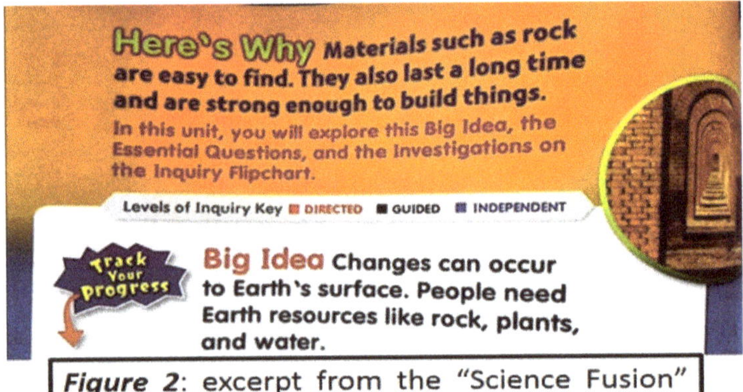

Figure 2: excerpt from the "Science Fusion" textbook showing the big idea.

Another point noticed is that each unit starts with the "big idea" which I would consider as the main aim, since it must be achieved by the end of the unit (figure 2).

The main aim shows a greater focus on the scientific concept which is considered as the target learning outcome, while the focus on the language is a supplemental element to support acquiring the concept.

As shown in the teacher's edition that each lesson in the unit included specific objectives related to the lesson stage of inquiry (figure 3).

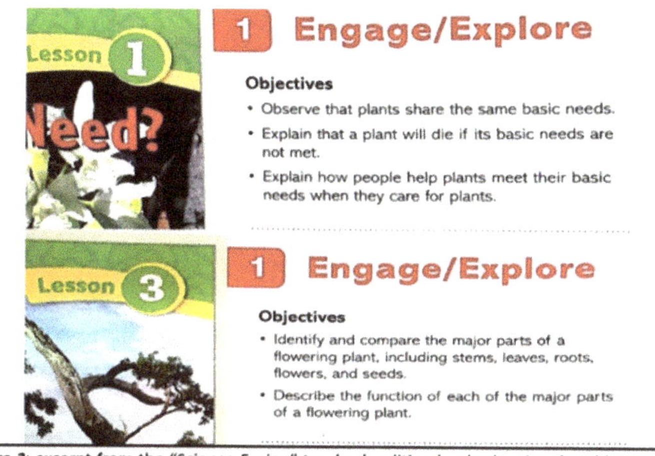

Figure 3: excerpt from the "Science Fusion" teacher's edition book, showing the objectives for different lessons depending on the type of inquiry.

Figure 4: excerpt from the "Science Fusion" teacher's edition book, showing the first objective that deals with the language vocabulary to be taught to students.

Therefore, the objectives were describing the scientific aspects to be learned, but there were minor touches on the language aspects (figure 4). Accordingly, I observed that the first objective shows that the vocabulary words mentioned are essential to understand the significant parts of flowering a plant. As a result, these vocabulary words were discussed through the activities of lesson 3 to make the students use them in a meaningful and comprehensible content (figure 5). This point is linked to the literature review that using comprehensible input through content help learners identify the information and review prior knowledge.

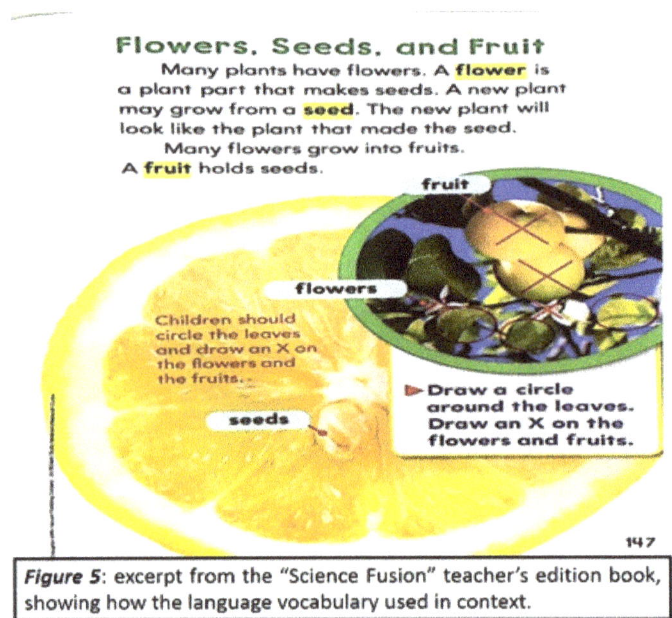

Figure 5: excerpt from the "Science Fusion" teacher's edition book, showing how the language vocabulary used in context.

English Skills

Regarding the English language skills practice for the EFL students, I found in the textbook that there was a continuous emphasis on vocabulary terms needed to understand the scientific concept (figure 1). This is done to ensure that the learners will get enough exposure to the language to proceed to learn science. Another point realized is encouraging students to read further through the "Active Reading" and to use annotations to jot down ideas. As a result, students will read to understand the new vocabulary and to comprehend the new concept simultaneously. For example, in (figure 6) the reading text provided shows the essential ideas of what plants need to grow, and also presenting 'sunlight' as a new term. Furthermore, science teachers are also provided with ideas on how to teach words with multiple meanings as part of language learning (figure 7).

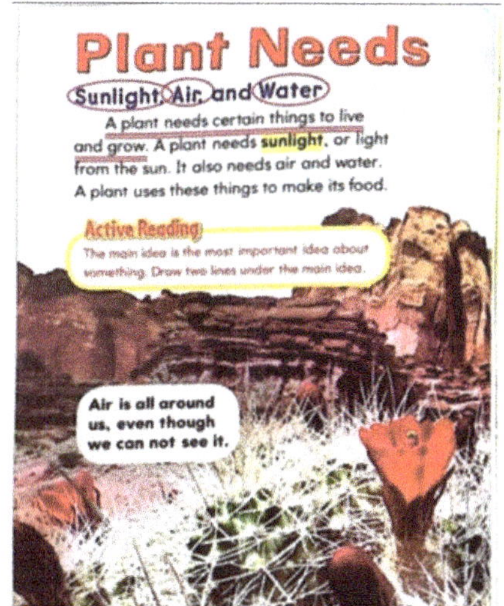

Figure 6: excerpt from the "Science Fusion" showing the active reading and how students are asked to annotate the text and identify new vocabulary.

English Language Learners

Understand Multiple-Meaning Words Children may be confused by the multiple meanings of the word *soil.* Explain that when *soil* is used as a noun, it can mean "the top layer of Earth." When it is used as a verb, it can mean "to make dirty." Then say sentences such as *That soil has many rocks in it. Did you soil your shirt?* Ask volunteers to identify the meaning of *soil* in each sentence. Then have children make up and share oral sentences using the different meanings of *soil.*

Figure 7: excerpt from the "Science Fusion" presenting how to explain science terms with multiple-meanings to students.

Through looking at the teacher's edition, part of learning science is to combine and reflect on what students learned by writing a paragraph following the correct structure of paragraphs, such as writing a topic sentence and details (figure 8). Therefore, it is considered as a revision and reflection stage of what students' learning in terms of language and scientific knowledge.

Writing Connection

Write a Paragraph Review how plants get what they need to live and grow. Ask children to use this information to help them write a paragraph with a topic sentence and at least three details describing the needs of plants. Children can illustrate their paragraphs and then share them with partners.

Figure 8: excerpt from the "Science Fusion" asking the teachers to encourage students to write a paragraph.

I could assume that speaking and listening skills were integrated through learning activities, where students discuss and share ideas (figure 9). Hence, in activities in which students generate questions and communicate with either a partner, group or teacher, they are expected to tell about what they learned and listen to what has been discussed.

English Language Learners

Make a Concept Web Help children summarize what they learned on these pages in a graphic way by making a web of the concepts. Draw a circle on the board or chart paper and write "How People Help Plants" inside it. Draw additional circles around that center circle. As children suggest ways that people can help plants, write their ideas in the surrounding circles. Talk about how a graphic organizer like the one you made together can help organize ideas and make them easier to remember.

Develop Inquiry Skills

COMMUNICATE **Point out that making a list is a form of communication because it involves using words and sometimes pictures to tell ideas. Ask partners to make two lists. The first list should name ways that people help plants. The second list should name ways that plants help people. Then have partners share and compare their lists in groups of four. Groups should then combine their lists and share them with the class.**

Figure 9: excerpt from the "Science Fusion" representing how students can share and listen to ideas during the activities.

Methods and Strategies

Examining the book, indicated that it was based on the Inquiry-Based learning method (IBL) as mentioned in the first section. Thus, the use of the target language is highly recommended to progress through each phase of the method. As shown in the appendices an example of one of the IBL lessons, which presents the use of the language in all the stages. *Figure 10* shows an outline of each IBL lesson components as well as how language is integrated through the lesson and phases.

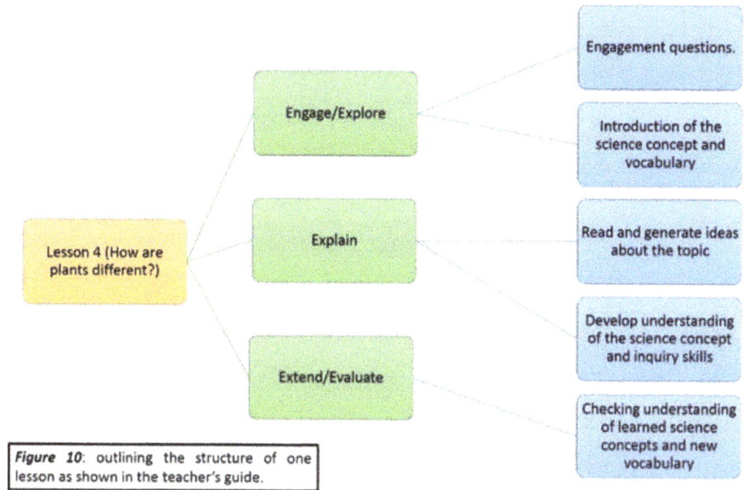

Figure 10: outlining the structure of one lesson as shown in the teacher's guide.

In addition, I discovered that the book promotes the use of different learning styles to support language and concepts learning. As shown in (figure 11) that teachers and students use visuals to enhance understanding and facilitate the use of the language. A further useful strategy found by examining the students' and teacher's books is the ability to extend students' understanding and challenge learning using the suggested ideas in the book. The book allows the teacher to use differentiated-leveled questions to engage and extend students' learning (figure 12).

Interpret Visuals

Focus children's attention on the banana leaf. **Why do you think this photo shows children on either side of the leaf?** It helps show how large the banana leaf is.

What do the many different leaves on the left page show? They show that leaves come in many shapes and sizes.

Figure 11: excerpt from the book showing an example of how to help students interpret visuals.

Differentiation—Leveled Questions

Extra Support

Why do plants need roots and stems? Plants need roots to hold them in the ground and to take in water and other things from the soil. Plants need stems to hold them up and carry water from the roots to the other plant parts.

Enrichment

Why might cutting the stems of flowers make a bouquet last longer? Cutting will keep the stems open so they can carry water up to the flowers to keep them fresh.

Figure 12: example of differentiated-leveled questions.

I observed through the book that language is taught through games for the purpose of consolidating and revising the language learned by EFL learners (figure 13). This point is linked to what is mentioned earlier in the literature review that integrating fun into learning is highly recommended in teaching both language and other subjects taught in English.

Answer Strategies

Word Play

After reading the directions with children, review the words in the word box. Remind children that they will use each word only once to label the parts of the plant.

Children can look back at the illustrations showing roots, stems, leaves, and flowers for guidance or review. Remind them that while these plant parts will not be a perfect match for what they see in the illustration, they should be able to use them to help identify the correct label for each part.

To help improve children's recall of plant parts and their functions, have pairs of children take turns using each plant part in a sentence. In their sentence, encourage them to identify how each plant part helps a plant get what it needs to live.

Figure 13: example of how to play and review words at the same time.

Discussion of the Results

One of the significant evidence resulted from examining the MAG science curriculum is the fact that English plays a vital role in teaching science (Kadbey & Dickson, 2014; Tairab, 2010). As a result, the language must match the target students' level to help them understand the content (Cameron, 2010; Lightbown & Spada, 2013). Another evidence is the use of visual aids supports students' comprehension of both the language and scientific concepts. Thus, pictures and visual resources can facilitate learning and provide learners with extra prompts about the topic (Holmes, 2006; Harmer, 2012). Additionally, the introduction of different topics within the science curriculum extends students' horizons in learning of varied themes as well as learning additional terms in English (Cameron, 2010; Dibiase & McDonald, 2015). I agree that it is crucial to inform the students and prepare them for learning through the unit and lesson aims. However, the students must be informed about the language aspects to be learned in each unit. The reason is to inform and update learners about what they are expected to learn and test their prior language knowledge around the topic (Baecher, Farnsworth, & Ediger, 2014). The findings represented that stating objectives that are clear and flexible provide a fundamental foundation for the teaching of science using the English language (Baecher, Farnsworth, & Ediger 2014; Harmer, 2012). This helps the teacher to clearly identify areas to be taught and supports the lesson sequence.

Supplying the students with the needed vocabulary is highly recommended as one of the evidence resulted from the textbook examination. Learning the content vocabulary and mainly terms required in order to understand science content is essential to facilitate and proceed with the learning experience (Cameron, 2010; Slavin, 2014). However, the teacher must be acknowledged that both teaching science and the English language must be given fair and well-balanced opportunities in the class. Therefore, the teacher can use the language effectively to teach science and make meaning accessible to learners (Cameron, 2010; Slavin, 2014; Harmer, 2012). Consequently, teaching English cannot only be as vocabulary in isolation but should focus on other skills such as reading, writing, listening and speaking (Banuelos, 2008; Tan, 2011). The use of the reading skill can help the students comprehend meanings through comprehensible input. While the use of writing can provide written evidence and reflect students' actual learning through summarizing what they learned, also make use of the vocabulary in writing meaningful sentences explaining experiments. Although some teachers may think that the listening skills are not as important in teaching science, students can employ that skill when listening to the teacher or their peers (Lee et al., 2009; Tan, 2011). Speaking also cannot be separated from listening as students need to use the terminology learned while learning or conducting experiments. Accordingly, considering the essential components of each skill and the structure of language to be taught through science content, can reinforce the correct use of language structure when learning science.

Since IBL is a popular science teaching method that most of the teachers are following, I am assuming that it is a successful strategy to teach English and science at the same time. The reason is that it consists of stages were learning the scientific concept requires a gradual understanding and experimentation of the new idea, so the language skills can be used to facilitate learning and progress into the next step (Lee, 2014; Grover & Stovall, 2015). Furthermore, a curriculum must consider facilitating learning through activities for different learning styles. This will enable learners to learn according to their preferred styles and trigger extra learning opportunities for them (Winebrenner, 2006). The final evidence is that scaffolding students in learning science through language extend learning and understanding of new concepts learned. This can prove to the teacher the amount of what the students learned, so the teacher can build upon that and support those who needed extra help (Grover & Stovall, 2015; Lee, 2014).

Pedagogical Implications for Teachers

This study has enlightened me on several pedagogical implications that I could suggest for science teachers teaching second language learners in English. The first implication is that teachers must acknowledge learners and parents about the expected learning outcomes in order to understand the strengths and weaknesses that might occur in the lesson. Another implication is to provide a meaningful context for learners to motivate learning and teaching science using the target language. In addition, teachers must model the correct use of the language and how to use it effectively through the science context. I would also suggest integrating fun into science lessons, whether in teaching the needed vocabulary or science topics and experiments. This can motivate learning and teaching science and make learning accessible through meaningful contexts. Learning will also be developed through the enhancement of the students' intrinsic motivation not only in the science subject but also in the English language.

Regarding the UAE context, the teacher can support students acquiring science and English through employing the surrounding context that is related to students' culture and background. This would increase the learning opportunities as it is similar to what students are frequently experiencing in the surrounding environment. Teachers teaching science through English must have good background knowledge of the language in order to employ that in teaching and learning. This can be done through intensive training or integrating this in the higher Education systems for future teachers' studies. As a further implication, the teacher can assess learning through the use of the different English skills to test the amount of learning either through formative or summative assessments.

In addition, following only one method in teaching science in English cannot ensure having successful lessons, so teachers must use a variety of teaching methods that would suit the level of the students and help enhance their understanding. The implementation of questions and eliciting during the lesson can benefit both the teacher and students, the teacher can regularly promote the learning and use of the language. Teachers must identify opportunities to add additional language and science aspects to scaffold students and challenge them a bit through the learning experience. Thus, teachers teaching the subjects can share their experiences and knowledge to support and build up on learning and teaching.

Conclusion and Recommendations

This study helps me to understand how the textbooks used in MAG systems are structured and constructed to suit second language learners. The examination of the qualitative data tool enabled me to answer the research question and supported my area of focus. Several evidence were identified through the data analysis and the examination of the textbooks. I agree that starting my study with the literature review helped to extend my understanding of the research and based on data analysis on previous studies implemented in the field. This also prepared me as a researcher for any obstacle that I might face while conducting the study.

On the other hand, some limitations restricted the conduction of this research and the analysis process. One of the limitations is considering one type of data collection, which was the qualitative research method, mainly examining MAG system textbooks. I do believe this type of study needs further supports from other types of data tools, such as interviewing previous MAG science teachers or considering if the assessments as additional artifacts. If I would like to extend the idea of this research, I would conduct a survey or Likert scale to measure students', parents' and teachers' satisfaction in teaching science through English. It was challenging to find teachers as the MAG system has been canceled and all subjects were shifted into Arabic.

Furthermore, there are other components for analyzing a curriculum that can be taken into consideration. A final limitation is that this research was meant to discuss further topics with other researchers, but they withdraw due to the workload they had. I also tried to access and find data to compare students' learning science in English and learning in Arabic. However, I could not reach these data as they needed a longer time to be requested and provided after getting the approval from MOE. Additionally, the time when I conducted this research did not support the conduction of the research, since all schools were on term one holiday and it was difficult to find enough science textbooks.

For additional future researches, I am intending to look closely at the use of IBL in teaching science to EFL learners, as well as focusing on the type of skills to work best with this method. Another research is to compare curriculum elements taught in Arabic and English and what differences are arising in teaching them in government schools. Accordingly, examining some issues that might occur when teaching a subject such as math or science to students using the second language. I would also focus on the types of assessments done by the MAG system to test science in English. Then, I am planning to compare the results of assessments done on a similar topic in both Arabic and English for EFL students. The comparison will lead to drawing conclusions on the effectiveness of using English in teaching science and will support the notion of whether teaching in the first language is more effective than the second language.

REFERENCES

Al-Naqbi, A. (2010). The degree to which UAE primary science workbooks promote scientific inquiry. *Research in Science & Technological Education*, 28 (3), 227-247. https://doi:10.1080/02635143.2010.506316

Arslan, U. & Akbarov, A. (2012). EFL learners' perceptions and attitudes towards English for the specific purposes. *Acta Didactica Napocensia*, 5 (4), 25-30.

Baeshcer, L., Farnsworth, T. & Ediger, A. (2014). The challenge of planning language objectives in content-based ESL instruction. *Language Teaching Research*, 18 (1), 118-136. https://doi:10.1177/1362168813505381

Banuelos, R. (2008). *Educating elementary-aged English learners in science: scientists and teachers working together*. Ph.D. Thesis. Stanford University.

Buldu, M. & Shaban, M. (2010). Visual arts teaching in kindergarten through 3rd-grade classrooms in the UAE: teacher profiles, perceptions, and practices. *Journal of Research in Childhood Education*, 24 (4), 332-350. https://doi:10.1080/02568543.2010.510073

Buri, C. (2012). Determinants in the choice of comprehensible input strategies in science classes. *Journal of International Education Research*, 8 (1), 1-18.

Cameron, L. (2010). *Teaching languages to young learners*. Cambridge University Press.

Carrier, S. (2013). Elementary preservice teachers' science vocabulary: knowledge and application. Journal of Science Teacher Education, 24 (2), 405-425. https://doi:10.1007/s10972-012-9270-7

Creswell, J. (2014). *Research design: qualitative, quantitative, and mixed methods approaches*. Sage Publications.

DiBiase, W. & McDonald, J. (2015). Science teacher attitudes toward inquiry-based teaching and learning. *The Clearing House: A Journal of Educational Strategies, Issues and Ideas*, 88 (2), 29-38. https://doi:10.1080/00098655.2014.987717

Dispezio, A., Frank, M., Heithaus, R., Ogle, D., & Houghton Mifflin Harcourt Publishing Company. (2012). *Science fusion*. Houghton Mifflin Harcourt.

Grover, K. & Stovall, S. (2015). Enhancing student experience in plant sciences through inquiry-based learning. *NACTA Journal*, 59 (1), 88-89.

Harmer, J. (2012). *Essential teacher knowledge*. Pearson Longman.

Hernández, A. (2012). Teaching science in English through cognitive strategies. *Gist: Revista Colombiana de Educación Bilingüe*, 129-146.

Holmes, M. (2006). Integrating the learning of mathematics and science using interactive teaching and learning strategies. *Journal of Science Education and Technology,* 15 (4), 247-256. https://doi:10.1007/s10956-006-9011-9

Hong, S. & Diamond, E. (2012). Two approaches to teaching young children science concepts, vocabulary, and scientific problem-solving skills. Early Childhood Research Quarterly, 27 (2), 295-305. https://doi:10.1016/j.ecresq.2011.09.006

Isăilă, N. (2016). Interactive-creative teaching and learning using educational games. *Knowledge Horizons. Economics,* 8 (1), 136-138.

Kadbey, H. & Dickson, M. (2014). Emirati pre-service teachers' experiences of teaching science during college internships. *Education, Business and Society: Contemporary Middle Eastern Issues,* vol. 7 (4), pp. 216-228. https://doi:10.1108/EBS-01-2014-0008

Kadbey, H. & Dickson, M. (2013). What kind of future science teachers might they be? pre-service primary school teachers in Abu Dhabi, United Arab Emirates amidst Educational Reform. *Journal of Turkish Science Education,* 10 (4), 136-150.

Lee, H. (2014). Inquiry-based teaching in second and foreign language pedagogy. *Journal of Language Teaching and Research,* vol. 5 (6), pp. 1236-1244.

Lee, O., Maerten-Rivera, J., Buxton, C., Penfield, R. & Secada, G. (2009). Urban Elementary Teachers' Perspectives on Teaching Science to English Language Learners. *Journal of Science Teacher Education,* 20 (3), 263-286.

Lightbown, P. & Spada, N. (2013). *How languages are learned.* Oxford University Press.

Mills, J. (2011). *Action research: a guide for the teacher researcher*. Pearson.

Moore, M. (2007). Language is science education as a gatekeeper to learning, teaching, and professional development. *Journal of Science Teacher Education,* 18 (2), 319-343. https://doi:10.1007/s10972-007-9040-0

Shanahan, T. & Shea, L. (2012). Incorporating English language teaching through science for K-2 teachers. *Journal of Science Teacher Education,* 23(4), 407-428. https://doi:10.1007/s10972-012-9276-1

Slavin, R. (2014). *Education psychology theory and practice.* Harlow: Pearson Education Limited.

Tairab, H. (2010). Assessing science teachers' content knowledge and confidence in teaching science: how confident are UAE prospective elementary science teachers? *International Journal of Applied Educational Studies,* 7 (1), 59-69.

Tan, M. (2011). Mathematics and science teachers' beliefs and practices regarding the teaching of language in content learning. *Language Teaching Research,* vol. 15 (3), 325-342. https://doi:10.1177/1362168811401153

NOTES ON CONTRIBUTORS

Prof. Sufian Forawi is a science education professor who works at the British University in Dubai. Prof. Forawi obtained an Ed.D. in Science Education from the University of Massachusetts Lowell, Massachusetts, USA. He obtained a bachelor's degree in biology from the University of Alexandria, Egypt and a master's degree in Education from Omdurman Islamic University, Sudan. His area of expertise spans through topics of STEM education, nature of science, multicultural education, school improvement and critical and creative thinking. He was awarded the distinguished US Fulbright scholar exchange at the UAE University. Prof. Forawi has been a member of several science education organizations such as the National Association of Research on Science Teaching (NARST) and the European Science Education Research Association (ESERA). He has published widely in peer-reviewed journals and participated in book publications.

Dr. Rehaf Madani earned a PhD in Science Education from the British University in Dubai, master's degree in Clinical Nutrition and a bachelor's degree in Biochemistry from King Abdul-Aziz University- Jeddah. She is a lecturer of Health Sciences and Nutrition courses and a certified coach and TOT at several institutes in Saudi Arabia and United Arab Emirates. She also has several publications in the field of Science Education.

Dr. Lara Abdallah is a school Executive Principal at Dubai Modern Education School. She earned her PhD in Education from the British University in Dubai. She obtained a bachelor's degree in applied mathematics (Statistics) from the Lebanese University, Beirut and a master's degree in Education from British University in Dubai. Her area of expertise is in transforming schools, mathematics, curriculum development and education training and coaching. She is an active member in National Council of Teachers of Mathematics (NCTM), National Council of Science Teachers (NSTA), and National Council of English Teachers (NCTE). She has authored a series of mathematics books for K-12 based on the common core state standards. She attended and presented at many international and local conferences. She won best paper award at BDRC 2016 Dubai, UAE.

Dr. Elaine Al Quraan earned a doctorate of Education in Leadership and Management and a master's of education both from the British University in Dubai. She is currently working in the Ministry of Education in the UAE as a change manager and leadership professional specialist. Dr. Al Quraan is a dedicated and compassionate educator, a school inspector, and a professional development specialist in the field of leadership, strategic planning, policy design, smart learning and innovation management. She was able to demonstrate exceptional strategic planning, organizational and fiscal management skills, and to achieve high levels of performance using research as an evidence-based decision making. Dr. Al Quraan has published in peer-reviewed journals and participated in publications in STEM Education and policy books.

Dr. Areej ElSayary is an Assistant Professor in the Department of Education at Zayed University, UAE. She obtained a PhD from the British University in Dubai in Education Leadership and Management. She has wise experience in teaching and research in areas of STEAM education, curriculum design and development, teaching and learning, assessment, and school accreditation. She is an approved accreditation visitor from New England Association of School and Colleges (NEASC) and Council of International School (CIS). Dr. Areej is an accomplished researcher, having published on cognitive development, Interdisciplinary STE(A)M curriculum, and assessment and presenting in several regional conferences.

Noura Assaf is a PhD candidate at the British University in Dubai, Department of Science Education, United Arab Emirates. She is also the head of the K-12 Science department in one of the reputable private schools in Abu Dhabi. She holds both a Bachelor and master's degree in molecular chemistry form the Lebanese University, Beirut, Lebanon.

Lames Abdul-Hadi earned a bachelor's degree in Pure Mathematics and Computer Science and a master's degree in Science Education from the British University in Dubai. Currently, she is pursuing a doctorate in educational leadership and international policy. Her research area of interest is in metacognition and assessment. key role as director of assessment and strategic plan is to lead the assessment department and develop the school's key performance indicators. I am responsible for assessment auditing and following the UAE National Agenda Parameters. She has been involved with CIS as an evaluator for CIS-ACE accreditation preparation, school inspection, and self-evaluation.

Fatima Ahmed Abazar is a Sudanese physics instructor at Abu Dhabi University, United Arab Emirates. She earned a B.Sc. from the United Arab Emirates University and an M.Sc. from Cairo University, Egypt. Now she pursues a doctorate in science education at the British University in Dubai. Ms. Abazar worked for two years as a research assistant in the Electrophysiology laboratory at the Faculty of Medicine at the UAE University. Then she filled a position as a Medical Physicist at the National Institute of Cancer at El Gezira University in Sudan and recently, at Fatima College for Health Sciences, UAE. Fatima is interested in human reasoning and physics teaching pedagogy.

Hadya Abboud Abdul-Fattah has a master's degree in Leadership Education from Abu Dhabi University. She is a PhD candidate at the British University in Dubai

(UAE), International Board of Certified Lactation Consultants Course (IBCLC). B.Sc. She also has a degree in Nursing from the University of Jordan (Amman), the UAE Board of Registered Nursing Exam (MOH), the Saudi Board of Nursing Council Exam, and the Jordan Board of Nurses and Midwives Council. Registered in the Jordanian Nurses and Midwives Council. Registered Nurse (RN) in the Jordanian MOH. Currently a Senior Lecturer at FCHS, Hadya has broad interests in health promotion and a specific research interest in UAE cultural beliefs and values, and contextualizing curricula.

Nimmy Thomas is a skilled Science/Chemistry educator who has been sharing her love of science with young minds for over 14 years now, in reputed schools in India, Oman and the United Arab Emirates. She has demonstrated history of working in Indian, British and American curriculum schools, facilitating learning for secondary level students from diverse backgrounds over these years. She holds a research-based M.Ed. in Educational Leadership and Management from the British University in Dubai, a B. Ed. in science education from Kannur university and a BS in Chemistry from Mahatma Gandhi University in India.

Hind Kassir is a doctoral student at the final stage of thesis submission at the University of Liverpool, UK. Ms. Kassir is the CEO and founder of SEEDS with the vision to develop human capital through professional consultation services seeking innovative change with the spirit of well-being for youth our core focus. She has experience of 22 years in the education sector as educational advisor and curriculum design expert. She worked heavily with the UNGSII and known as a Key speaker for youth empowerment in Espoo (Finland) and in several Universities working to implement the sustainable development goals, keynote speaker to support youth readiness and preparedness through promoting digital literacy and go back to family roots and social cultural values.

Dr. Sura Sabri earned a PhD degree in science education, a Master of Science Education, both from the British University in Dubai, and a BSc in Biology. She has been in the teaching field for over 20 years. Dr. Sabri worked as teacher for physical sciences and biology, then a curriculum developer in the Institute of Applied Technology. Currently, she has been the Vice Principal of Academics in the Applied Technology High Schools in UAE. She was involved in planning and implementing several teacher training and professional development and actively publishing articles in science education.

Dr. Marwa Eltanahy is a Curriculum Developer who has an extensive experience in the educational fields with specific expertise in teaching, coordination, training new and in-service teachers, and school accreditation. In addition to her experience as a researcher and a peer reviewer in academia, she recently earned a Ph.D. in education from the British University in Dubai. Her research interest includes STEM education,

Science education, curriculum and instruction. She worked in different projects concerning inquiry-based learning which resulted in published papers in peer reviewed journals and conference presentations. She developed an E-STEM model to incorporates entrepreneurial practices into STEM education to enhance students' entrepreneurial competencies.

Mona Mohamed has participated in different professional development sessions: Stanford Introduction to Food and Health, Improvement of Science Education- University of Michigan, Constructivism and Mathematics, Science, and Technology Education- University of Illinois and Introduction to Family Engagement in Education- Harvard University.

Dr. Mina Radhwan has a doctorate degree in the area of Education Leadership and Management at the British University in Dubai, UAE. She had a MEd degree in Science Education from same university and previously a BS in Biology from Al-Mustansiriya University in Iraq. Currently, Ms. Radwan works as an Educational Evaluator at the Ministry of Education - Evaluation and Quality Directorate in the UAE. She has participated in internal conferences at the MoE and external conferences at the British University in Dubai and additionally delivers lectures as a guest speaker for Master classes at the same university. She has several publications in areas of science education, cooperative learning, and formative assessment.

Rania Samir Alayli is an educator of Nursing and Health Sciences. She has completed her bachelor's degree in Nursing Sciences from Makassed University in Beirut. Her postgraduate certificate and Master of Education are from University of Southern Queeensland and she is currently pursuing her doctoral studies in Health Education at the British University in Dubai, UAE.

Nada Albarabri is an eighteen-year veteran mathematics teacher, strongly believes in the importance of student-centered education and continually seeking to improve learning by integrating technology into class instruction. She holds a master's degree in Science and Mathematics Education from the British University in Dubai. She taught mathematics for high school in several schools in Jordan. Currently she teaches mathematics at AL Ghurair University in UAE. She has been a member of her school improvement team and math curriculum developer and served as mentor for new teachers.

Dr. Nagib Balfakih currently works at the Estrella Mountain Community College in Arizona, USA. Dr. Balfakih obtained a bachelor's degree in chemistry from the United Arab Emirates University, a master's degree in Education from University of Idaho, USA, and an Ed.D. in Science Education from the same university in 1996. His area of expertise is in the teaching science methods and the integration of technology in science teaching. Dr. Balfakih has many years of science teaching and coordination experience in higher education. He publishes widely and has been a member of several science education organizations.

Nesrin Tantawy is a professional educator who has been in the educational field for more than 15 years teaching EFL/ESL learners. In 2016, she obtained a master's degree in TESOL, with distinction, from the British University in Dubai. She is currently an EdD student at the University of Exeter. She also has several publications in a number of peer reviewed journals.

Fatema Al Awadi is a member of the Education Faculty at the Higher Colleges of Technology in UAE and a former Bachelor graduate teaching within the Higher Colleges of Technology system. A TESOL master's degree holder from the British University in Dubai, and at the process of applying for a PhD in one of the universities in UK. She attempted to write several research papers related to the field of education during her B.A. in Education and MA studies. She is looking forward to raising the teaching and learning standards in her country, the UAE.

www.ingramcontent.com/pod-product-compliance
Lightning Source LLC
Chambersburg PA
CBHW042312210326
41598CB00042B/7368

* 9 7 8 1 8 6 3 3 5 2 2 4 6 *